SHAPE CASTING:

3rd International Symposium

2009

SHAPE CASTING: 3RD INTERNATIONAL SYMPOSIUM

Related titles include:

- *Shape Casting of Metals I (2005) Shape Casting: The John Campbell Symposium (2005)*
- *Shape Casting of Metals II (2007) Shape Casting: The John Campbell Symposium (2007)*
- *Simulation of Aluminum Shape Casting Processing: From Alloy Design to Mechanical Properties (2006)*

HOW TO ORDER PUBLICATIONS

For a complete listing of TMS publications, contact TMS at (724) 776-9000 or (800) 759-4TMS or visit the TMS Knowledge Resource Center at http://knowledge.tms.org:

- Purchase publications conveniently online and download electronic publications instantly.
- View complete descriptions and tables of contents.
- Find award-winning landmark papers and webcasts.

MEMBER DISCOUNTS

TMS members receive a 30 percent discount on TMS publications. In addition, members receive a free subscription to the monthly technical journal *JOM* (both in print and online), free downloads from the Materials Technology@TMS digital resource center (www.materialstechnology.org), discounts on meeting registrations, and additional online resources to name a few of the benefits. To begin saving immediately on TMS publications, complete a membership application when placing your order or contact TMS:

Telephone: (724) 776-9000 / (800) 759-4TMS

E-mail: membership@tms.org or publications@tms.org

Web: www.tms.org

SHAPE CASTING:
3rd International Symposium
2009

Proceedings of a symposium sponsored by
the Aluminum Committee of
the Light Metals Division (LMD) and
the Solidification Committee of
the Materials Processing & Manufacturing Division (MPMD) of
TMS (The Minerals, Metals & Materials Society)

Held during TMS 2009 Annual Meeting & Exhibition
San Francisco, California, USA
February 15-19, 2009

Edited by

John Campbell

Paul N. Crepeau

Murat Tiryakioğlu

A Publication of

TMS

A Publication of **The Minerals, Metals & Materials Society (TMS)**
184 Thorn Hill Road
Warrendale, Pennsylvania 15086-7528
(724) 776-9000

Visit the TMS Web site at
http://www.tms.org

ISBN Number 978-0-87339-734-6

If you are interested in purchasing a copy of this book, or if you would like to receive the latest TMS Knowledge Resource Center catalog, please telephone (724) 776-9000, ext. 270, or (800) 759-4TMS.

TABLE OF CONTENTS
Shape Casting: Third International Symposium

Shape Casting: Third International Symposium

Properties

Processes

Characterization

Novel Methods and Applications

Modeling

Foreword

The science of solidification and the technology of casting production, as twin waifs and strays, have over the years suffered erratic growth. This growth, perhaps slower on average than many might wish, has been the outcome of devoted researchers and foundry people over the past decades. Although often separated, when brought together, the coalescence of knowledge contributed by each twin promises the burgeoning of a fast growing and powerful, if not monumental, alliance.

Those adopting and nurturing our twins into maturity have the foundation of knowledge provided by books (although I personally believe I can legitimately claim to be the author of the world's most unread foundry text books), but in addition, require regular updating by reading many of the world's excellent metallurgical journals. In particular, our twin waifs are served by two research journals: one from the UK, The International Journal of Cast Metals Research, and the other from the American Foundry Society, The International Journal of Metalcasting. The correspondence columns of both journals are an education in the most recent thinking and are recommended to all. The current debates have the potential to create a sea-change in our approach to solidification and casting.

From my personal perspective, the wind of change blows especially strongly from the direction of those externally introduced defects in liquid metals that appear to control many of the aspects of microstructure, and the mechanical and corrosion behavior of castings. Many of our researchers in the field of solidification continue to overlook the possibility that (i) entrained bifilms exist in their materials and (ii) may actually be in control. The author is unrepentant for introducing this turmoil to our previously settled and contented lives. He maintains that our current science alone cannot explain either the structures or the property response of metallic materials, and that only extrinsic factors, particularly the endemic populations of bifilms, can develop our twins into a cogent science and an effective technology.

Time will tell. This symposium is yet another in our *Shape Casting* series in which these concepts, unloved by some, but welcomed by others, can be debated. Passion is permitted. But Truth is sought, and, it is to be hoped, will prevail.

On this occasion the symposium has been unusually problematic to organise as the result of a number of technical issues, and it is therefore appropriate to thank all our session chairs for the non-trivial task of reviewing submissions to their sessions. Above all, particular thanks go to Paul Crepeau and Murat Tiryakioglu, without whose monumental efforts this symposium would never have happened. Thank you Paul and Murat.

JC
22 October 2008

Symposium Organizers/Editors

John Campbell has retired from his post at The University of Birmingham and his editorship of the International Journal of Cast Metals Research. He keeps in touch, retaining an Emeritus Status at the University.

Since these moves he has mainly occupied himself with practical work in foundries around the world, applying, testing, and extending the latest technology to upgrade quality and reduce costs. Despite some casting successes in which the author is grateful and proud, he has also accumulated a few scrapped castings that confirm the technology of filling castings is still not fully developed. Good, targeted research into the most valuable means to reduce scrap, increase quality and improve profit margins remains systematically neglected for lack of vision and funds. In China he has recently met foundries with the courage and entrepreneurial spirit to tackle huge castings using revolutionary technology. In Mexico the application of correct filling systems promises to revolutionise profitability. These foundries are manoeuvring themselves into world leading positions. There are salutary lessons here.

Thus the globetrotting activity constitutes valuable education. However, longer-term updating of existing books and the writing of further books has been indefinitely delayed. In the meantime, *Castings, 2ed* (2003) and *Castings Practice* (2004), in addition to the proceedings the first two Shape Casting Symposia, all remain bargains!

Paul N. Crepeau is a GM Technical Fellow in Advanced Materials Engineering at General Motors Powertrain Group in Pontiac, MI USA. He supports aluminum intensive engine programs and leads a multidisciplinary team merging CAE and Materials Engineers to advance structural FEA of engine components. Dr. Crepeau received his B.S. in metallurgical Engineering at the University of Alabama (1978) and, after a 5-year respite at an iron foundry, both M.S. in Metallurgy (1985) and Ph.D. in Metallurgical Engineering (1989) from the Georgia Institute of Technology. He has published in the areas of fracture mechanics, molten metal processing, quantitative metallography and image analysis, aluminum heat treatment, Monte Carlo simulation of fatigue test methods, and material property database strategy. Dr. Crepeau is a registered professional engineer and former chairman of both the AFS Aluminum Division and the TMS Aluminum Committee. Dr. Crepeau was editor of *Light Metals 2003* and currently chairs the TMS Aluminum Processing Committee.

Murat Tiryakioğlu is Professor in the Department of Engineering at Robert Morris University. He received his B.Sc. in Mechanical Engineering from Boğaziçi University (1990), M.S. (1991) and Ph.D. (1993) in Engineering Management from the University of Missouri-Rolla, and another Ph.D in Metallurgy and Materials (2002) from the University of Birmingham, England.

Dr. Tiryakioğlu grew up in his family's foundry, which continues to thrive in Istanbul, Turkey. This has led to research interests in process design for high quality castings, aluminum heat treatment modeling and optimization, process-structure-property relationships in metals, statistical modeling and quality and reliability improvement, on which he has written a chapter, published over 80 technical papers, and edited 5 books.

Dr. Tiryakioğlu has received a *Certificate of Outstanding Performance* from The Boeing Company and a *Certificate of Appreciation* from the American Society for Quality (ASQ). He was selected a TMS *Young Leader* in Light Metals and was awarded the SME *Eugene Merchant Outstanding Young Manufacturing Engineer*. He is a member of the ASM International, TMS, and a senior member of ASQ, and is an ASQ *Certified Quality Engineer*.

Session Chairs and Lead Reviewers

Properties Session

Glenn Byczynski is Manager of Research and Development for Nemak's European Business Unit and is currently based in Germany. He received his Ph.D. in Metallurgy and Materials Science from the University of Birmingham in U.K. in 2002. His Masters (Materials Science in 1997) and Bachelor's (Mechanical Engineering in 1994) were conducted at the University of Windsor. In his 18 year career in metal casting he has held several R&D and Engineering positions within Nemak and Ford Motor Company including Manager of the Nemak Engineering Centre (Windsor) and Engineering Manager at Nemak's Windsor Aluminum Plant. He was Chairman of the Detroit-Windsor Chapter of the American Foundry Society (AFS) in 2006-2007, was a director and regional chairman of the Foundry Educational Foundation and is a registered Professional Engineer in the Province of Ontario. He is also an Adjunct Professor at the University of Windsor. He enjoys spending time with his wife and two sons.

Process Session

Alan Druschitz is Research Professor in the Department of Materials Science and Engineering at the University of Alabama at Birmingham. He received his PhD in Metallurgical Engineering in 1982 from the Illinois Institute of Technology, Chicago, IL. He was previously a Staff Research Engineer General Motors Research Laboratories and the Corporate Director of Materials R&D Intermet Corporation. He is a co-founder of BAC of VA, LLC, a small company that provides design support and castings for the military and specialty vehicle market. He is an SAE Fellow, the Vice-Chairman Alabama Section of SAE, a past president of the Ductile Iron Society, former Vice-Chairman of the Governors Board of Transportation Safety for the Commonwealth of Virginia and a current member of AFS, ASM International, ASTM and AIST.

Characterization Session

Sumanth Shankar is founding director of the Light Metal Casting Research Centre at McMaster University, Hamilton, ON, Canada. His expertise encompasses solidification processing, aluminum alloy development and microstructure-property-perfor¬mance charac-terization. Current projects include determination of structure of liquid alloys, rheology of liquid aluminum and magnesium alloys, heat treatment of aluminum alloys, rheocasting via CDS process, microstructure characterization of aluminum alloys and solidification simulation of binary alloys. Significant contributions have been in the molten aluminum-ferrous diffusion reactions, evolution and modification of eutectic phases in Al-Si alloys and development of novel Al-Si-Mg alloys for structural shaped casting applications. He has co-authored over 40 journal publications and presented at several technical symposia.

Srinath Viswanathan is Associate and FEF Key Professor of Metallurgical and Materials Engineering at The University of Alabama, responsible for Foundry and Metal Casting Research and the Ray Farabee Metalcasting Laboratory. He focuses on modeling and characterization of casting and solidification behavior, prediction of defects and microstructure, and development of advanced materials and processes. He was formerly Principal Member of the Technical Staff, Sandia National Laboratories, and Senior Research Staff Member, Oak Ridge National Laboratory. Dr. Viswanathan is a Fellow of ASM International, and member of AFS, NADCA, and TMS. He has over 80 publications on casting, casting modeling and processing, has edited two books and has six patents. He was awarded an R&D 100 award for the Metal Compression Forming Process in 1997, and has awards for technical accomplishment from Oak Ridge National Laboratory, the AFS, and the IMS.

Novel Methods & Applications

Mahi Sahoo, FASM, received his B.Sc. (Eng.) degree from the Indian Institute of Science, Bangalore in 1967 and a Ph.D. in Physical Metallurgy from The University of British Columbia, Vancouver in 1971. After six years at Queen's University, Kingston, he joined the Physical Metallurgy Research Laboratories of CANMET as a research scientist in 1977. At CANMET he later became Assistant Director, Research Program Office, Head of the Foundry Section in the Metals Technology Laboratories, and Program Manager, Sustainable Casting at the Materials Technology Laboratory (MTL). He is now a senior research scientist at MTL. His research interests are in casting, solidification, alloy and process development. He is a recipient of the John A. Penton Gold Medal from the American Foundry Society and the Dofasco Award from the Metallurgical Society of CIM for his outstanding contributions to non-ferrous casting technology. He was the President of the Metallurgical Society of CIM in 2002.

Modeling Session

Christof Heisser received the equivalent of a Masters Degree in Foundry Technology at the Technical University of Clausthal in Clausthal/Germany. After his first employment as Leader of Research & Development at Thyssen Feinguss, an Aluminum Investment Casting Foundry in Soest/Germany, he joined MAGMA GmbH in Aachen/Germany in a Marketing & Support position. Christof moved to the Chicago office of MAGMA Foundry Technologies, Inc. in 1995 as Foundry Application Engineer. He now is the President of MAGMA Foundry Technologies, Inc. Christof is a member of AFS, where he participates in several AFS committees. He also is a member of DIS and SAE and has authored several technical papers.

Novel Methods & Applications

[illegible]

Modeling Session

[illegible]

SHAPE CASTING:
3rd International Symposium
2009

Properties

Session Chair:
Glenn Byczynski

Shape Casting: The 3rd International Symposium
Edited by: John Campbell, Paul N. Crepeau, and Murat Tiryakioğlu
TMS (The Minerals, Metals & Materials Society), 2009

INTRINSIC AND EXTRINSIC METALLURGY

John Campbell[1]

[1] University of Birmingham, Dept. of Materials and Metallurgy, Edgbaston B15 2TT, UK

Keywords: Bifilms, Extrinsic Defects, Entrained Defects, Flake Graphite, Spheroidal Graphite.

Abstract

The intrinsic metallurgy of materials is well understood, forming the basis of physical metallurgy, and leading to the classical concepts of nucleation and growth of phases in a matrix. This highly successful physics has unfortunately generally overlooked the contribution of process metallurgy, in which extrinsic factors in the form of entrainment defects can affect the structure and properties of materials. The limitations of intrinsic metallurgy are seen mainly in the presence of pores and cracks that lead to various kinds of failure in materials, but which cannot be explained by intrinsic mechanisms. For instance solidification as a simple phase change is unable to nucleate a pore or a Griffiths crack (by either homogeneous or heterogeneous nucleation as will be proven) because of the extremely high interatomic forces. Only extrinsically entrained defects can explain these features, and therefore provide understanding of the fundamental causes of failures in tensile, creep, fatigue modes, and probably some if not all, corrosion pitting failures. Furthermore, the assumption of the presence of extrinsic defects in the form of bifilms predicts type 'A' flake, undercooled and spheroidal graphite forms of cast irons, and in Al-Si alloys, unmodified and modified eutectic, and primary silicon forms. An accurate metallurgical understanding of cast and wrought alloys requires both intrinsic and extrinsic contributions.

Introduction

The most superficial consideration of metals and solidification quickly illustrates that metals should never fail. This perhaps surprising conclusion follows from the well-known impossibility of homogeneous nucleation of defects in liquids based on the mechanical equilibrium $P = 2\gamma/r$ where P is the internal (positive) pressure inside a bubble (or the external negative pressure, or hydrostatic tensile stress, outside the bubble) γ is the surface tension, being a measure of the interatomic forces, and r is the radius of the bubble. The surface tension γ for liquid Fe is in the region of 2 N m^{-1}. Assuming an embryonic nucleus of radius approximately one atomic diameter, giving a bubble volume of approximately 8 atomic volumes, P is of the order of 40 GPa. This is the 'fracture pressure' or 'tensile strength' of liquid iron.

Similarly for solid Fe, assuming a surface energy of around 2 J m^{-2}, if 1 m^2 of atoms were separated by 0.2 nm to form a uniform crack of tip radius approximately 0.1 nm, one atomic radius, the stress to achieve this would be 40 GPa. Thus the strengths of both liquid and solid phases are similar. This is to be expected because the interatomic distances in both are about the same. These stresses are known to be reasonably accurate from measurements of the strengths of whiskers that have sufficiently small dimensions as to have a good statistical chance of containing no defect.

These extremely high stresses for the 'homogeneous' nucleation of pores or cracks might, of course, be reduced in the presence of a poorly-wetted substrate that would allow 'heterogeneous' nucleation. (It is probably worth pointing out that the solid/liquid interface is of course well wetted, being in perfect atomic contact with both liquid and solid phases, and so not a favoured substrate for the creation of volume defects). However, for conditions of the worst possible wetting, assuming the highest contact angles ever normally recorded, in the region of 160 degrees, the nucleation stress is predicted to be reduced by a factor of nearly 20 [1]. Thus the fracture stresses for liquids and solids reduce to approximately 2 GPa. (Corresponding stress levels in aluminium are predicted to be about half of these values.)

Significantly, the stresses remain over 20,000 times higher than can be met during solidification, since, as every foundry person knows, a poorly fed casting can collapse under only atmospheric pressure, 0.1 MPa, which indicates the limit at which internal tensile stress can be supported. Thus the tensile stresses to create volume defects in castings cannot be generated, simply because inter-atomic forces are too high to allow this.

Even in solids at room temperature there has been direct evidence for over 40 years that cracks and pores cannot form. Transmission electron microscope observations of the condensation of vacancy clusters in face-centered-cubic metals such as Al never found pores or cracks. The clusters always collapsed under the high inter-atomic forces, producing dislocation rings or stacking fault tetrahedra [2]; the atoms could never be prized apart; elastic and/or plastic collapse of the lattice always intervened to prevent this.

Thus classical physical (intrinsic) metallurgy would predict that an Al-Si eutectic alloy undergoing a tensile test would not exhibit a failure of a single Si particle. If the Si particles were sufficiently strong they would cluster together as the Al matrix plastically flowed until the Si particles impinged. Alternatively, the Si particles themselves might at some stage start to plastically flow until the whole specimen finally parted by necking down to zero.

Clearly, we can conclude that solidification cannot form volume defects. This has the interesting consequence that there will be no features such as Griffiths cracks to initiate failures by cracking in castings [3]. Similarly there will be no porosity or cracking or decoherence of phases to initiate ductile failure [4]. The absence of failure initiation mechanisms will necessarily result in extensive plastic flow in tensile tests, necking down to 100 % reduction in area in tensile tests.

Practical Failure Mechanisms

The theoretical prediction of 100 % ductility of cast metals raises the fascinating question, 'how do failures by fracture occur in practice?'

Once again, the answer is clear and logical: a metal can only initiate failure from interfaces that are unbonded (since atomic bonds are too strong to be broken). Since unbonded surfaces cannot be formed by solidification, such interfaces have to be introduced from *outside* the metal. These are necessarily *extrinsic* features. The one such feature that always necessarily forms by a folding action, ensuring an extensive unbonded interface, is the bifilm. This is an introduced section of the surface film on the liquid metal. The bifilm is, of course, usually an oxide, but can on occasions be a film of carbon, or nitride etc. Its unbonded inner faces, and perfectly wetted exterior faces have all the features required to explain their mechanical and metallurgical roles in the casting.

In classical intrinsic metallurgy the initiation of tensile failure or the formation of a shrinkage pore in a liquid involves nucleation at an atomic level; the high stresses leading to explosive growth as a result of the sudden release of the elastic energy of the liquid [1]. The nucleation of a gas pore is expected to be rather more gentle because the initial growth phase, although still relatively rapid, will be controlled by the rate of diffusion of gas into the pore.

The scenario in which extrinsic unbonded defects are present is quite different. No nucleation is involved since the relatively macroscopic sized (compared to atomic dimensions) unbonded crack effectively already exists. Only a gentle growth occurs by the unbonded surfaces gradually separating under the action of stress or influx of gas.

The continued presence of the unbonded bifilm defects frozen into the solid also appears to explain many features of the tensile failure of metals. The bifilms tend to survive considerable plastic working. Al alloys retain their unbonded regions even after the severe extrusion required to produce window panes, as is evident from the filiform corrosion that follows the unbonded tunnels that happen to intersect the surface. The survival of the bifilms is probably the result of the reservoirs of air that remain trapped in the rucks and folds between the oxide surfaces, so that extension of the area of the interfaces by working is accompanied by simultaneous oxidation and nitridation of the freshly-created surfaces, preventing bonding until all the air is consumed. The unbonded surfaces in wrought alloys probably constitute the Griffiths cracks necessary for brittle failure [3], and the population of 'pores' or 'decohered' surfaces from intermetallics and seconds phases (that seem almost invariably to precipitate on oxide bifilms) that are required to initiate the dimples in the fracture surface of a ductile failure [4].

Other strength-controlling features such as grain refinement [5] and various brittle fracture processes [1] can be similarly re-interpreted in terms of bifilm theory, but space does not allow us to present these important phenomena here.

Extrinsic Control of Solidification Microstructure

Flake or Nodular Cast Irons

De Sy has shown that liquid cast iron generally contains oxygen in solution in excess of its solubility [6]. He concluded, on the basis of careful and rigorous experiments, that the undissolved fraction of oxygen was present as SiO_2 particles. We can go further to conclude that the particles almost certainly would not be compact spheres or cubes etc, but would most likely be in the form of films. Only films would have a sufficiently low

Stokes velocity to remain in suspension for long periods of time. Additionally, of course, the folding-in of the surface film of SiO_2 during melting in the cupola as droplets of iron rained down, or during pouring etc. would ensure a natural population of SiO_2 bifilms.

In view of precipitates of many kinds finding bifilms as preferred substrates for precipitation, we can propose that the suspension of SiO_2 bifilms in grey irons leads to the morphology of flake grey irons. The flake form type 'A' forms ahead of the solidification front because of the favoured formation of graphite on the silica substrates in suspension (Figure 1). The growth morphology of graphite, extending the length of its basal plane, would favour the straightening of the bifilm, and the freedom of growth in the liquid means that the flakes can grow straight [7]. Some of these flakes can be seen incorporated into the solid. However, much of the graphite precipitates as a eutectic, forming a regular macroscopically planar front, and with its spacing now controlled by the diffusion of carbon in the liquid, and forming at the somewhat lower temperature to drive the growth of the eutectic. A facetted growth phase forced to grow alongside neighbors, and so perpetually changing its growth direction, will be a 'coral' type. Thus the morphology of 'A' type graphite and the various types of 'undercooled' or 'coral' graphites may start to be explicable for the first time. Most irons will be expected to contain these two populations of graphite forms in different proportions. Figure 2 shows an extreme example of extrinsically nucleated, and a eutectic (i.e. intrinsic) form of graphite.

On the addition of Mg to the melt, the SiO_2 films are immediately reduced to Si metal (that is taken into solution in the melt) and MgO particles. The magnesium oxide together with magnesium sulphides form compact inclusions which are well-known and widely observed to be favourable substrates for graphite nucleation. The absence of planar substrates and the new arrival of compact substrates dictate that the graphite will now form as spheroids. The growth form simply follows directly from the form of the favoured substrate (Figures 3 and 4). The relatively poor mechanical properties of grey iron may be more to do with the presence of bifilm cracks down the centers of graphite flakes rather than any intrinsic weakness of the graphite itself. A crack down the center of a flake is seen in Figure 5 [8].

Modification of Al-Si Alloys

The author has proposed that Si will form on oxide bifilms in Al-Si alloys, the growth of the silicon being fairly planar, and so straightening the bifilm crack [9], and so reducing properties. Modification by Sr appears to deactivate the bifilms as favoured substrates, forcing the formation of Si as a eutectic, now growing at a lower temperature on a substantially planar (or cellular) front. The elimination of bifilms in other ways, using very clean alloy, will also promote modification, as will strong chilling in a high temperature gradient, that freezes the liquid as a eutectic at its low growth temperature while the liquid in the center of the casting remains at too high a temperature for precipitation on bifilms.

Thus 'modified' silicon occurs simply as a result of eutectic growth, and 'unmodified' silicon occurs akin to a primary phase precipitating on a favoured substrate that it can straighten as it grows, thus forming apparently brittle (because of its center bifilm crack) flake morphology.

The unconstrained freedom of growth of Si in the liquid results in a morphology with almost no growth defects, whereas the modified Si phase, now growing as a continuous 'coral' form (the natural form of a facetted phase growing in a eutectic) has to grow surrounded by other fibres of silicon, all continuously changing direction (with difficulty because of their facetted growth directions) as a result of the constraints of their neighbours, thus naturally developing a high density of defects such as growth twins [10].

The porosity often suffered after modification with Sr appears most likely to be the result of the release of bifilms (no longer entombed in coarse Si particles) that act to block interdendritic flow, and provide the easily opened interfaces to initiate porosity [11].

For hypereutectic alloys, the formation of silicon on AlP (or possibly AlP_3) particles is also well understood and documented [12]. Because these particles are compact, the silicon forms around them, and also grows as a compact particle. Once again, the morphology of the primary phase is dictated by the form of the favoured substrate.

The Mechanical Properties of Al-Cu Alloys

In Al-Cu alloys the freezing of the $CuAl_2$ phase appears to happen at grain boundaries (Figure 6). It seems likely, however, in common with other intermetallics that the $CuAl_2$ phase is forming on bifilms that happen to have been pushed to the grain boundaries during the growth of dendrites. A central fine line can be seen associated with the $CuAl_2$ particles that may be the bifilm substrate (the jagged form seen in Figure 7 might be the result of a limited short-range straightening effect by the growing particles). Careful examination reveals interdendritic regions that clearly have high levels of segregated copper (not easily discerned in Figure 6), but no $CuAl_2$. By chance, this appears to be the result that no bifilm substrate exists along this interdendritic region. Thus it seems possible that, in general, an apparent grain boundary may actually be (i) a grain boundary, or (ii) a grain boundary containing one or more bifilms, or (iii) simply a bifilm. Without the bifilm grain boundaries would not be expected to be common sites for the precipitation of second phases, nor sites for failure by hot tearing in the partially solidified state, nor by creep or cavitation in the solid state, since they would be expected to be naturally strong.

It is significant that the $CuAl_2$ phase has a cubic structure, so that its high crystal symmetry allows it to grow in almost any direction; no direction is highly favoured. Thus in this alloy the bifilms are not subjected to a strong straightening mechanism. It seems probably that this is the fundamental reason for the Al-Cu alloys to enjoy a high combination of ductility and fatigue resistance [13]. In contrast, the Al-Si alloys have a bifilm-straightening mechanism built-in to the alloy in the form of the precipitating Si crystal. Its highly idiomorphic growth morphology is effective in straightening bifilms and thus enlarging cracks, and so reducing properties. Thus Al-Si will only ever compete with Al-Cu alloys if the natural bifilm content is reduced. There are signs that the industry is making progress to achieve this by better attention to metal quality and the improved design of filling systems.

Alpha-Fe and Beta-Fe Phases in Al Alloys

The mechanism by which Fe impurity reduces the mechanical properties of Al alloys appears to be the extremely effective straightening of the bifilm substrate (Figure 9) when forming the monoclinic βFe precipitates [14]. When converting the βFe to αFe (by for example the addition of Mn) the αFe still forms favourably on the bifilm, but now, as a result of its cubic structure that can easily grow in any direction, the αFe simply wraps up the bifilm in its convoluted state, encasing it in a compact precipitate. The complex internal cracks often seen in αFe crystals (Figure 8) are the 'skeleton' of its originating bifilm substrate [15].

Conclusions

1. Extrinsic phenomena, so far largely neglected, appear to exert a major control over the morphologies of cast metals. Flake and ductile iron microstructure, and the structures of Al-Si alloys may be understood for the first time as adopting forms of the second phase based on the form of their substrate, whether film or particle. Many microstructures will contain the two populations of second phases (intrinsic and extrinsic initiated) side by side.
2. Grain boundaries may be constituted by, or may contain, bifilms, explaining much second phase precipitation and occasional so-called grain boundary failures.
3. Properties may be controlled by bifilms. If cleaner metals were cast, the properties of relatively inexpensive alloys would increase to levels as yet not achieved. The benefits would accrue to both the cast and wrought metal industries.

References

1. J Campbell "Castings" 2003 Elsevier. p180.
2. D Kuhlman-Wilsdorf; in "Lattice Defects in Quenched Metals" 1965 Academic Press, New York.
3. A H Cottrell; "The Griffith Centenary Meeting – The Energetics of Fracture" 1993 The Institute of Materials.
4. P F Thomason, "Ductile Fracture of Metals" Pergamon Press 1990 pp 19-29.
5. J Campbell; Materials Sci & Technol 2008 Letter 7 pp in press
6. A de Sy; TAFS 1967 75 161-172
7. Y X Li, B C Liu and C R Loper; TAFS 1990 98 483-488
8. S I Karsay; Ductile Iron II; Engineering Design Properties Applications. 1971 Quebec Iron & Titanium (QIT) Corporation.
9. J. Campbell; AFS Internat J Metalcasting 2008 letter in press.
10. M Shamsuzzoha, L M Hogan and J T Berry; TAFS 1993 101 999-1005.
11. J Campbell and M Tiryakioglu; Materials Sci & Technol submitted April 2008
12. J Campbell; AFS Journal of Metalcasting 2008 discussion
13. T Din and J Campbell; Materials Science and Technology 1996 12 644-650.
14. X Cao and J Campbell; Met & Mat Trans A 2003 34A 1409-1420
15. X Cao and J Campbell; Internat J Cast Metals Research 2004 17 (1) 1-11.
16. M Hillert, S Rao; 'The Solidification of Metals' Brighton Conf. 1967 ISI publication 110 pp 204-212.

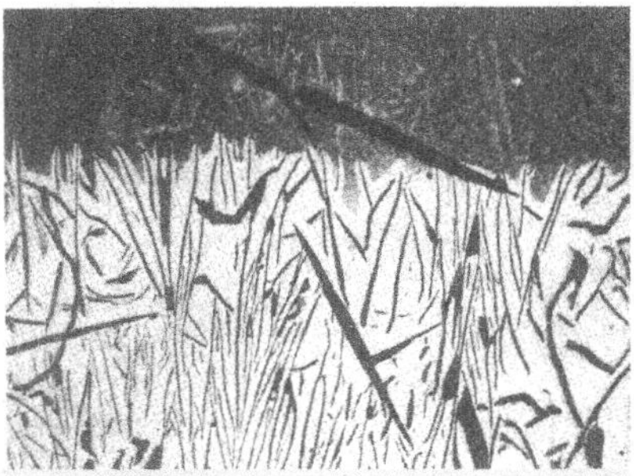

Fig 1. Extrinsic initiation of straight graphite flakes in the liquid, ahead of cooperative growth of eutectic [7].

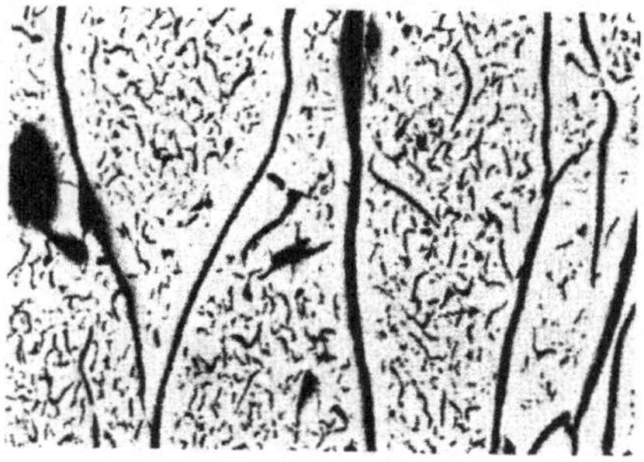

Fig 2. An extreme example of extrinsic flakes and intrinsic coral (undercooled) eutectic [16].

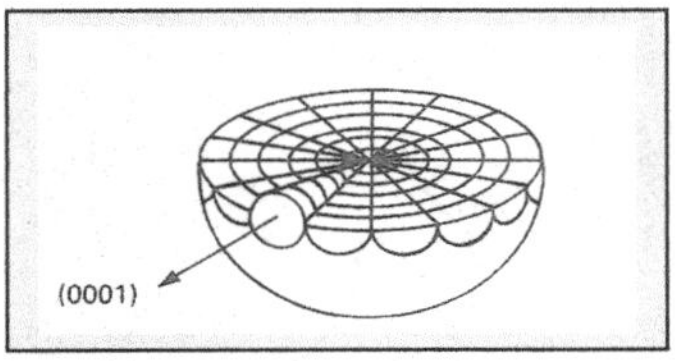

Fig 3. The probable crystal structure of a graphite nodule.

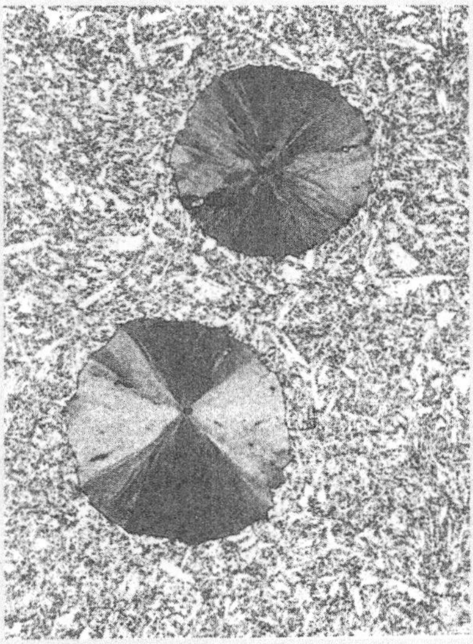

Fig 4. Graphite nodules in an austempered iron indicating nucleation on a small central inclusion

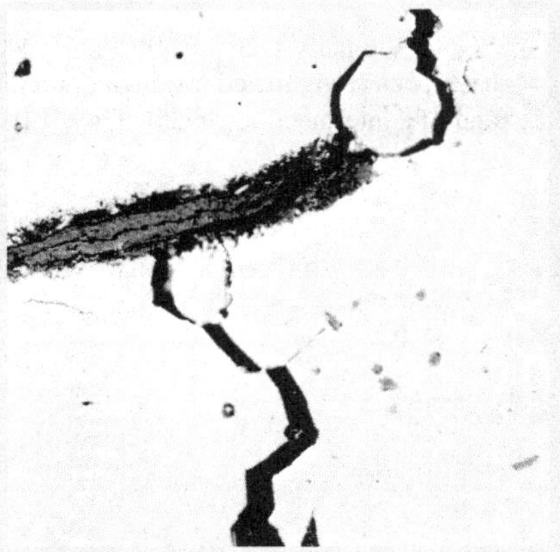

Fig 5. Graphite flake exhibiting a central crack (the solid state precipitation of surrounding temper graphite is also fractured off) [8].

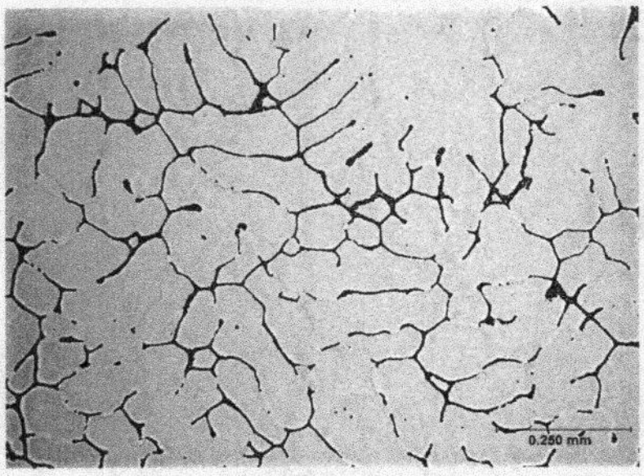

Fig 6. Interdendritic regions all showing Cu segregation but precipitation of $CuAl_2$ in only some regions (Figure courtesy Alotech).

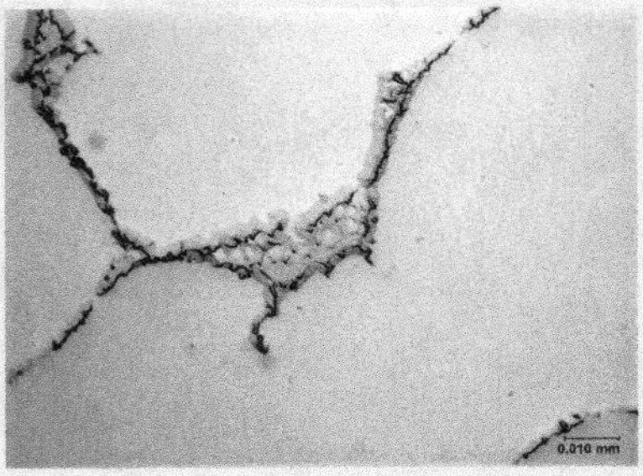

Fig 7. So-called grain boundary precipitates of $CuAl_2$ in 206 alloy showing central black line (Figure courtesy Alotech).

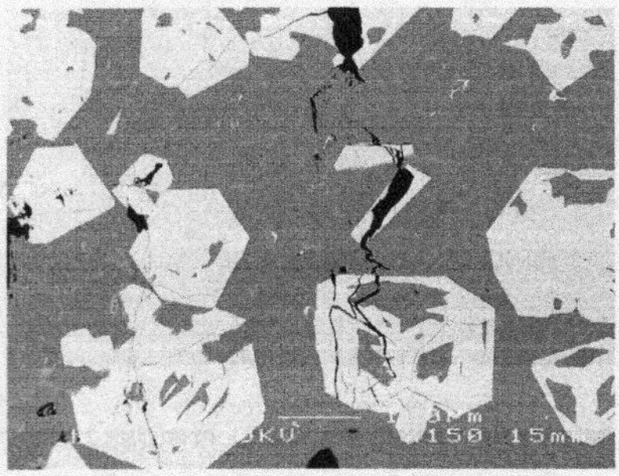

Fig 8. Crystals of αFe have grown around and enclosed originally compact bifilms as a labyrinth of cracks [15].

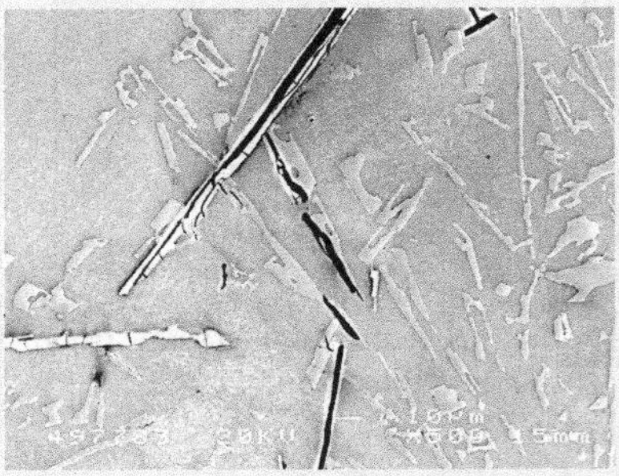

Fig 9. Originally crumpled bifilm cracks have been straightened inside Si particles and βFe intermetallics in 356 alloy [14].

Shape Casting: The 3rd International Symposium
Edited by: John Campbell, Paul N. Crepeau, and Murat Tiryakioğlu
TMS (The Minerals, Metals & Materials Society), 2009

QUALITY INDICES FOR CAST ALUMINUM ALLOYS

Murat Tiryakioğlu[1], John Campbell[2]

[1] Robert Morris University, Dept. of Engineering, Moon Township, PA 15108, USA
[2] University of Birmingham, Dept. of Materials and Metallurgy, Edgbaston B15 2TT, UK

Keywords: Porosity, Bifilms, Work Hardening, Ductility

Abstract

Several indices are available in the literature to assess the structural quality of cast Al alloys, especially Al-7%Si-Mg alloys, based on tensile test results. Some of these indices, most notably the one developed by Drouzy *et alia* provide a number that do not necessarily have a physical meaning, while the others are a measure of what fraction of the expected tensile property is achieved. These indices are discussed and the concept of maximum potential ductility is introduced. A new quality index that uses this maximum ductility potential concept is introduced for Al-7%Si-Mg as well as two Al-Cu alloys.

Introduction

Aluminum castings have been rarely used in critical applications due to concerns about the variability in properties, especially in elongation and fatigue life. This high level of variability is a consequence of structural defects in castings, i.e., pores and oxide bifilms, which degrade mechanical properties; they cause premature fracture in tension [1] and fatigue [2], resulting in low ductility, tensile strength and fatigue life [3]. In addition, the presence of major structural defects results in increased variability in properties [4,5]. Hence minimization and even elimination of the structural defects is vital for wider use of aluminum castings in structural applications in aerospace and automotive industry.

Foundry engineers striving to resolve quality concerns in aluminum castings, such as low ductility, often try to change the heat treatment procedure, assuming that ductility can be increased mainly by trading off strength. These efforts are usually ineffective unless the root cause of low ductility, i.e., structural defects, is addressed. Hence when oxide bifilms and porosity are present in castings, it is at best inefficient and at worst fruitless to approach the problem from a ductility-strength compromise point of view.

In the quest for improved properties it is helpful for the foundry engineer to have a metric to measure the degree of improvement that they make. The so-called quality indices developed over the years are intended to serve this need. This paper reviews the quality indices developed in the past and provides a new quality index based the concept of ductility potential of cast aluminum alloys.

Background

Drouzy et al. [6] introduced an empirical equation that defines the relationship between yield strength (σ_Y), tensile strength (S_T) and elongation (e_F) of cast Al-7%Si-Mg alloy castings;

$$\sigma_Y = S_T - 60\log_{10}(e_F) - 13 \qquad (1)$$

Equation 1 is valid for e_F>1%. The authors also introduced a quality factor, Q_{DJR}, for underaged and peak-aged alloys;

$$Q_{DJR} = S_T + 150\log_{10}(e_F) \qquad (2)$$

Din et al. [7] investigated the change in the tensile properties of A356 and A357 castings with artificial aging time. The authors found linear relationship between elongation and yield strength. Based on this result, the authors introduced a quality index, Q_{DC}, which is the yield strength extrapolated to zero elongation, based on the linear relationship between e_F and σ_Y that they observed:

$$Q_{DC} = \sigma_Y + 50e_F \qquad (3)$$

Alexopoulos and Pantelakis [8] developed a quality index, Q_{AP}, that takes the compromise in yield strength and fracture toughness into account. To represent fracture toughness, the authors used the plastic strain energy density, the area under the true stress-strain curve, which is alternatively referred to as toughness [9], ψ;

$$Q_{AP} = \left(\frac{\sigma_Y}{\sigma_{Y(max)}} + \frac{\psi}{\psi_{(max)}}\right)(\sigma_Y + 10\Psi) \qquad (4)$$

where the subscript (max) represents the maximum property achievable. None of the quality indices mentioned so far compares the performance of the casting with that expected from a defect-free casting.

Cáceres [10] developed a quality index after explaining the physical basis of Q_D. Cáceres' quality factor, Q_C, is a ratio of e_F to elongation expected of the specimen if it were free from structural defects, $e_{F(e)}$:

$$Q_C = \frac{e_F}{e_{F(e)}} \qquad (5)$$

Cáceres assumed that cast aluminum alloys follow the Ludwik-Hollomon equation:

$$\sigma = C\varepsilon_p^n \qquad (6)$$

where σ and ε_p are true stress and true plastic strain, respectively, C is the strength coefficient and n is the strain hardening exponent. When the Considere criterion is met, true uniform strain is equal to *n*, if the material deforms following Equation 6 [9]. In Al-7%Si-Mg aluminum aerospace castings, McLellan [11] observed that fracture takes place without almost any necking and, hence, fracture occurs at the nominal uniform elongation value, being approximated as the uniform engineering strain $e_{F(e)} \approx n$.

In the development of a new quality index, Q_{TSC} based on energy absorbed by a specimen prior to fracture, Tiryakioğlu et al. [12] estimated $e_{F(e)}$ and the expected toughness , $\psi_{(e)}$, by using work hardening characteristics, namely the Stage III Kocks-Mecking work hardening model [13,14] and the Voce equation [15];

$$Q_{TSC} = \frac{\psi}{\psi_{(e)}} \quad (7)$$

It was shown [16,17] that e_F and ψ have a strong linear relationship in cast Al alloys. Therefore the Q_C and Q_{TSC} are built on the same concept that the energy absorbed by the specimen before fracture is an indication of the structural quality of the casting. They both supply a measure in terms of the ratio of the current to achievable performance. Both indices determine the achievable (or target) quality from the work hardening characteristics of the specimen. The main difference between the two indices is how the defect-free performance is estimated. The approach taken in both indices, however, underestimates the true ductility potential of the alloy [18] because the late stages of work hardening, where the Considere criterion is met, cannot be estimated accurately from early stages [16]. Hence, if a specimen fractures prematurely due to the presence of structural defects, such as porosity and/or oxide bifilms, the extrapolation of work hardening characteristics to higher strains underestimates elongation. This is partially because structural defects reduce the observed work hardening rates significantly [1,16]. Therefore a more fundamental approach is needed to estimate the true ductility potential of cast aluminum alloys. Such an approach is reported in this study.

Nyahumwa et al. [19,20] introduced the concept of fatigue life potential and applied it to an Al-7%Si-Mg alloy. The authors argued that although structural defects would almost always dictate the fatigue life in accordance with the findings in the literature [9,21], occasionally a specimen would be obtained without a defect, i.e., without the weakest link. Such data can be obtained when molten metal is treated carefully and the filling system for the casting is designed correctly. Such outliers, then, can be used as a measure of the fatigue life potential or intrinsic fatigue life of the alloy. Nyahumwa et al. showed that the fatigue life of castings that are free from structural defects is several orders of magnitude higher than those with structural defects.

Mechanical property data from mostly premium quality (aerospace) castings are analyzed to find trends in *maximum* values. Those data are then analyzed to (i) estimate the ductility potential of cast Al-7%Si-Mg and Al-Cu alloys, and (ii) introduce a new quality index.

Data Analysis and Discussion

Eighteen datasets for yield strength-elongation to fracture were analyzed in this study. Details of the datasets are provided in References 22,23,24. The analysis included 323, 438 and 431 tensile data for Al-7%Si-Mg (A356, 357), 201 and 206 alloys, respectively.

The σ_Y-e_F data from the sources shown in Table 1 are presented in Figure 1. Note that yield strength is plotted in the x-axis because of its relative insensitivity to the presence of structural defects. In Figure 1, the scatter is mostly vertical due to the varying structural quality of the specimens. Note in Figure 1 that the highest points follow a linear trend. The line drawn in the figure follows the same form found previously by Din et al.:

$$e_{F(int)}(\%) = \beta_0 + \beta_1\sigma_Y \qquad (8)$$

The values of the coefficients β_0 and β_1 are given in Table 1 for the cast aluminum alloys investigated in this study.

Table 1. The coefficients of Equation 8 for Al-7%Si-Mg, 201 and 206 alloys.

	β_0	β_1 (MPa^{-1})
A356-357	36.0	0.064
201	34.5	0.047
206	47.8	0.085

It is significant that maxima of data taken from different sources indicate such a consistent trend with yield strength for the alloys investigated. To the authors' knowledge, it is the first time that the ductility potential of cast Al alloys is reported, especially with such a large number of data. Equation 8 represents the true strength-ductility compromise in cast aluminum alloys. This ductility potential estimated by Equation 8 can now be used in lieu of $e_{F(e)}$ in Equation 5, such that;

$$Q_T = \frac{e_F}{\beta_0 + \beta_1\sigma_Y} \qquad (9)$$

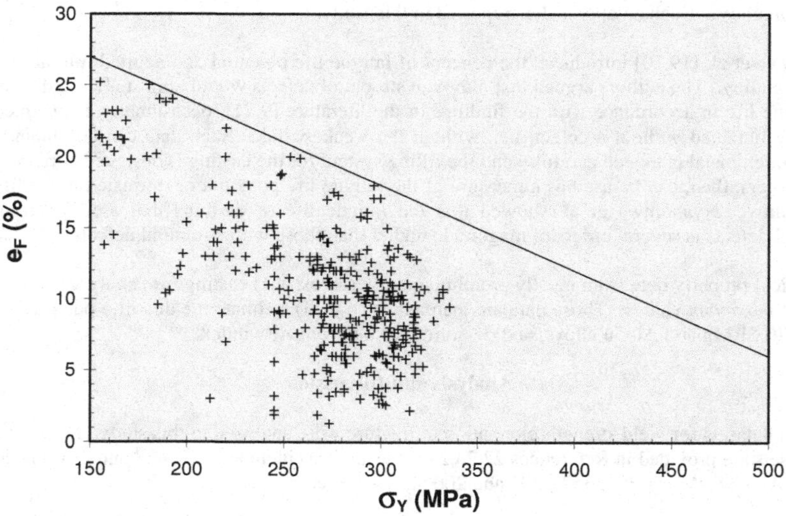

(a)

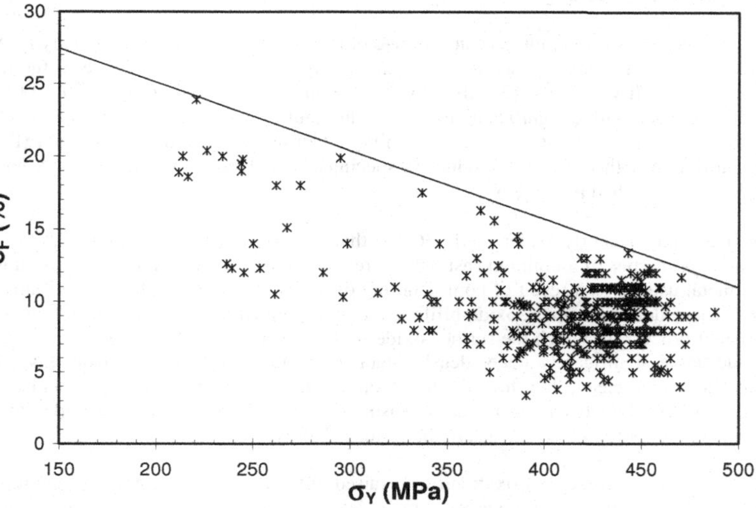

(b)

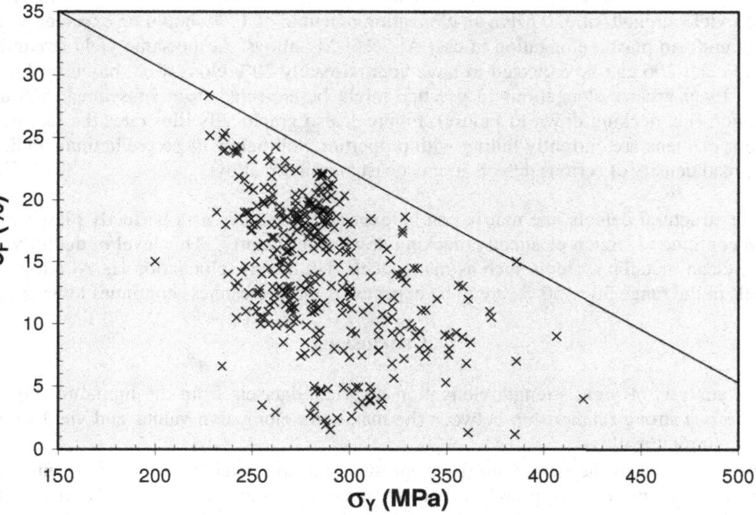

(c)

Figure 1. Yield strength-elongation relationship in (a) Al-7%Si-Mg, (b) 201, and (c) 206 alloy castings.

Extrapolations of the Results

It should be noted that Equation 8 is an estimate of the uniform elongation, especially for the Al-7%Si-Mg alloys. Therefore it should be taken as only a lower bound estimate of the intrinsic total elongation. It was found [25] that powder metallurgy Al-Si-Mg alloys with 10 to 20%Si and 0.5%Mg, even with a significant amount of aluminum oxide in them, necked and deformed non-uniformly past the point of tensile instability. Consequently, reduction in area (RA) was significantly higher than e_F. For instance, a specimen with 15% Si with σ_Y = 358 MPa, a e_u = 2.3% and e_F = 5.5% had a 14% RA.

It has been only recently understood [26,27] that the degradation of and variability in the mechanical properties of aluminum castings are related to the defects that are introduced into the molten metal usually as a result of poor handling of the molten metal and/or poor filling system design. These defects, namely oxide bifilms, are incorporated into the bulk of the liquid by an entrainment process, in which the surface oxide folds over itself. Unlike steel castings in which the oxide has a significantly lower density than the metal thus floating out quickly and thus leaving the metal clean, the folded aluminum oxide in aluminum has practically neutral buoyancy, so that defects tend to remain in suspension. The layer of air in the folded oxide can grow into a pore, or remain as a crack in the solidified alloy.

As a result of the awareness of oxide bifilms, raised in the last 15 years, castings continue to be produced to higher quality, with increased melt cleanness and careful filling system design. The elongation-to-fracture, once limited to only 1 or 2 %, is now steadily increasing, and now commonly achieves at least 10% in combination with good strength [28]. Figure 1.a illustrates that at a yield strength of 320 MPa, an elongation potential of 17% should be expected as a result of only uniform plastic elongation in cast Al-7%Si-Mg alloys. At the same yield strength level, both 201 and 206 can be expected to have approximately 20% elongation, based on Figure 1.b and c. Even greater elongations in practice might be expected from subsequent non-uniform elongation (i.e. necking down to failure). Figure 1 also graphically illustrates the fact that most tensile specimens are currently failing with properties well below these predictions, indicating a widespread density of serious defects in most cast aluminum alloys.

Without structural defects, the matrix can be expected to behave as a perfectly plastic material, should continue to stretch plastically, necking down significantly. This level of ductility is often seen in clean metallic systems such as many steels. Ultimately, elongations in Al alloys, typical of steels in the range 30 to 50 %, are to be expected if metal cleanness continues to be enhanced.

Conclusions

- The analysis of yield strength-elongation bivariate datasets from the literature showed that there is a strong relationship between the maximum elongation values and yield strength in cast aluminum alloys.
- The line fitted to the maximum data represents the true strength-ductility potential and the inverse relation (the 'compromise') between strength and ductility in cast aluminum alloys.
- A new quality index is introduced based on those proposed by Cáceres and Din et al. and the observed ductility potential of cast aluminum alloys. The new quality factor is calculated as;

$$Q_T = \frac{e_F}{\beta_0 - \beta_1 \sigma_Y}$$

- The data of the vast majority of the quality of cast aluminum specimens indicate that mechanical performance falls well below these predictions, implying evidence of a population of defects.
- At a yield strength of 320 MPa, an upper bound of 17% and 20% elongation is predicted for Al-7%Si-Mg and Al-Cu alloys, respectively. Even higher elongations, allowing a further regime of extension in which necking to failure may occur, seem likely to be achievable with adequate cleanness.

References

---

1. M. Tiryakioğlu, J. Campbell, J.T. Staley: *Scripta Mater.*, 2003, vol. 49, pp. 873–878.
2. C. Nyahumwa, N.R. Green, J.Campbell: *Metall. Mater. Trans. A.*, 2001, vol. 32A, pp. 349-358.
3. J.T. Staley, Jr., M. Tiryakioğlu, J. Campbell: *Mater. Sci. Eng. A*, 2007, vol. A465, pp. 136–145.
4. N.R. Green, J. Campbell: *Mater. Sci. Eng. A*, 1993, vol. A173, pp. 261-266.
5. X. Dai, X. Yang, J. Campbell, J. Wood: *Mater. Sci. Eng. A*, 2003, vol. A354, pp. 315-325.
6. M. Drouzy, S. Jacob and M. Richard: *AFS Intl. Cast Metals. J.* 1980, vol. 5, pp. 43-50.
7. T. Din, A.K.M.B. Rashid, J Campbell: *Mater. Sci. Technol.*, 1996, vol. 12, pp. 269-273.
8. N.D. Alexopoulos, S.G. Pantelakis: Metall. Mater. Trans. A, 2004, vol. 35A, pp. 301-308.
9. G. Dieter: *Mechanical Metallurgy*, Mc-Graw Hill, 1988.
10. C.H. Cáceres: *Intl. J. Cast Metals Res.*, 1998, vol. 10, pp. 293-299.
11. D L McLellan: *J. Testing Eval.*, 1980, vol. 8, pp. 170-176.
12. M. Tiryakioğlu, J.T. Staley, J. Campbell: *Mater. Sci. Eng. A*, 2004, vol. A368, pp. 231–238.
13. U.F. Kocks: J. Eng. Mater. Tech., 1976, vol. 98, pp. 76-85.
14. H. Mecking, U.F. Kocks: *Acta Metall.*, 1981, vol. 29, pp. 1865-1875.
15. E. Voce: *Metallurgia,* 1955, vol. 51, pp. 219-226.
16. M. Tiryakioğlu, J.T. Staley, Jr., J. Campbell: *Mater. Sci. Eng. A*, 2008, vol. A487, pp. 383-387.
17. M. Tiryakioğlu, J. Campbell, Mechanisms and Mechanics of Fracture: Symposium in Honor of Prof. J.F. Knott, TMS, 2002, pp. 111–115.
18. M. Tiryakioğlu, J. Campbell, J.T. Staley: *Mater. Sci. Eng. A*, 2004, vol. A368, pp. 205–211.
19. C. Nyahumwa, N.R. Green, J.Campbell: In 'Advances in Aluminum Casting Technology', (ed. M. Tiryakioğlu, J. Campbell), 225-233; 1998, ASM International.
20. C. Nyahumwa, N.R. Green, J.Campbell: *J. Mech. Behavior Mater.*, 1998, vol. 9, pp. 227-235.
21. H. Dowling: *Mechanical Behavior of Materials*, Prentice-Hall, 1993.
22. M. Tiryakioğlu, J. Campbell, N.D. Alexopoulos: submitted to *Metall. Mater. Trans. A.*
23. M. Tiryakioğlu, J. Campbell: *Mater. Sci. Technol.*, in press.
24. M. Tiryakioğlu, J. Campbell, N.D. Alexopoulos: submitted to *Mater. Sci. Eng. A.*
25. W.H. Hunt, Jr., *Microstructural Damage Processes in an Aluminum Matrix-Silicon Particle Composite Model System*, Ph.D. Dissertation, Carnegie Mellon University, 1992.
26. J. Campbell: '*Castings'*, 2nd Edition, Elsevier.
27. J Campbell: *Mater. Sci. Tech.*, 2006, vol. 22, pp. 127-145 and 999-1008.
28. J. Campbell: "The Bifilm Concept: Prospects of defect-free castings", World Foundry Congress, Chennai, India 2008.

Shape Casting: The 3rd International Symposium
Edited by: John Campbell, Paul N. Crepeau, and Murat Tiryakioğlu
TMS (The Minerals, Metals & Materials Society), 2009

USE OF 'STANDARD' MOLDS TO EVALUATE METAL QUALITY AND ALLOY PROPERTIES

Geoffrey K. Sigworth[1] and Tim A. Kuhn[2]

[1] Alcoa Primary Metals, P.O. Box 472, Rockdale, Texas 76567 USA
[2] Alcoa Primary Metals, 5601 Manor Woods Rd., Frederick, Maryland 21703 USA

Keywords: Casting quality, Melt Treatment, Aluminum, Standard Molds

Abstract

Several mold designs have been proposed as standards for aluminum castings. The two most commonly used in North America are the ASTM B108 test bar, and a 'step' casting proposed by the Aluminum Association (AA). The history of these molds is reviewed and mechanical properties are presented for A356-T6 alloy castings. The B108 test bar is prone to shrinkage. Measures that help to minimize shrinkage are discussed. The AA mold is also prone to shrinkage, but a judicious selection of sample locations avoids much of the problem. In spite of their limitations, the two molds can be used to evaluate melt treatment procedures and metal quality in the foundry. The introduction of better degassing practices has resulted in a significant improvement in casting quality during the last thirty years.

Introduction

When one talks with buyers of castings and foundries, it is common to hear the following story. A cast component is obtained from foundry 'A' with good results. However, after some time the purchase agent puts out the part for bid and foundry 'X' comes in with a lower price. The order is awarded to 'X', but the mechanical properties are significantly lower. What has caused this change? We are using the same alloy. We might even be using the same tooling. Why do the material properties change? This story illustrates why net-shaped castings sometimes have a poor reputation compared to wrought products. It also illustrates why most handbook values for mechanical properties should be taken with a 'grain of salt'. The engineering data on aluminum castings seldom has a proper 'pedigree'. Little is said about the casting process or melt treatment, other than it was a permanent mold casting or a sand casting.

Considering this problem, how can we develop better engineering information for foundries and casting users? And what is the best way to develop test data when we are trying to design or evaluate a new alloy composition? A common approach is to employ a test casting. This may be a production casting with a great deal of past history. It may also be a 'standardized' casting using a specified mold design. This document evaluates two standardized castings, and the mechanical properties obtained with these molds.

Aluminum Association Mold

The Aluminum Association mold was introduced about twenty years ago.[1] The mold and the resulting casting are shown in Figure 1. This mold was used by the Aluminum Association to develop property data in six different alloys: 319.0-F, 333.0-F, 355.0-T6, C355.0-T6, 356.0-T6 and A356.0-T62. Tensile test samples were taken from five areas, each having a different solidification rate and section thickness. Section 4 has the fastest solidification rate and produces a dendritic structure (DAS) which is roughly equivalent to a separately cast ASTM B-108 (1/2" diameter) test bar. The available mechanical property data may be found in a report published by the Aluminum Association[2]. The data tabulated for A356-T6 alloy castings is given below in Table 1, together with properties for separately cast test bars. The dendrite arm spacing (DAS) observed at each of the five locations in 356 alloy is also given. This DAS has been used to estimate the local solidification times. It is readily seen that the elongation and UTS decrease as the solidification time increases.

Figure 1. Aluminum Association permanent mold test casting (Location of test bar samples is shown at the right)

Table 1. Average tensile properties of A356-T62 alloy in AA mold[2]

Location (Thickness)	UTS (MPa)	Yield (MPa)	Elongation (%)	Dendrite Arm Spacing inches	Dendrite Arm Spacing microns	Solidification Time[3] (sec)
1 (1-3/8")	246	216	2.7	0.0020	51	100
2 (2")	252	221	3.0	0.0018	46	63
3 (7/8")	274	230	4.6	0.0015	38	36
5 (3/8")	274	229	4.9	0.0013	33	25
4 (1/2")	288	233	6.5	0.0010	25	11
test bars	294	242	4.8	0.0012	30	20

It should be emphasized that these are ***average*** values obtained when the same mold was sent to a number of different foundries. It is useful to consider the range of alloy compositions used and tensile properties obtained. Table 2 shows the maximum and minimum values recorded.

Table 2. Minimum/Maximum values for A356-T62 alloy in AA study[2]

% Si	% Fe	% Mg	UTS (MPa)	YS (MPa)	Elongation	Location
6.8/7.45	0.12/0.18	0.28/0.40	235/276	166/242	1.8/4	1
			231/283	166/35typo?	1.5/4.5	2
			252/297	173/162	3/7.7	3
			248/293	173/162	3/7.5	5
			259/314	166/269	3.5/9.5	4

Some foundries evidently had much better practices than the others. There was a significant difference in tensile properties, especially elongation. This is a good example of the problem one faces when mechanical property data lacks the proper 'pedigree' regarding melting and casting practices. There is no information about the degassing process used, if any. Likewise, other important treatments are unknown, such as filtration and grain refinement or modification practices. These have an important influence on mechanical properties. The above information is interesting, but what does it really mean? Is it possible to do better? Absolutely. It will be useful to compare the above results with castings having a better pedigree.

We first consider castings produced by Ken Whaler at Stahl Specialty company.[3] A heat of A356 alloy with 0.07 % Fe, 0.36 % Mg and 0.08 % Ti was melted in a double chamber dry hearth furnace. The metal in the dip out well was degassed by porous plugs and maintained at a low gas level, as determined by reduced pressure samples taken every thirty minutes. A filter crucible was placed in the furnace well and all metal was ladled from this crucible. (In other words, all metal was filtered to remove oxides and inclusions.) The metal temperature inside the crucible was held between 1350 and 1380 F (730 to 750 C). The metal was modified with Sr and grain refined with small additions of 5Ti-1B rod. Thermal analysis samples were taken every 30 minutes to ensure that the grain refinement and modification were not lost.

The resulting castings were solution heat treated for six hours at 1000 F (538 C) and water quenched. The solution furnace had a door on the bottom, so castings could be dropped rapidly into a rail-movable quench tank. This insured the castings were quenched from the furnace solution temperature within six seconds. After a 24 hour age at room temperature the castings were aged in a gas-fired furnace for 6 hours at 310 F (160 C). The resulting mechanical property data are given in Table 3. The tensile properties were considerably improved over those found in the Aluminum Association Study.

Table 3. Tensile properties of A356-T6 alloy (Stahl Specialty)[3]

Location (Thickness)	**UTS (MPa)**	**Yield (MPa)**	**Elongation (%)**	**DAS (microns)**	**Quality Index (MPa)**
1 (1-3/8")	270	193	6.2	51	389
2 (2")	269	197	6.2	46	388
3 (7/8")	292	197	12.3	38	455
5 (3/8")	290	198	10.1	33	440
4 (1/2")	308	210	14.3	25	481

Values of the quality index were calculated and are shown in Table 3. This index is the best way to evaluate casting quality and to compare different sets of data on castings. The quality index was first proposed by French foundrymen,[4] and is defined by the formula:

$$Q = UTS + 150 \log E \qquad (1)$$

where Q and UTS are given in MPa and the elongation to fracture, E, is in percent. It has been calculated[5] that the maximum quality index value obtainable in A356-T6 alloy castings is about 490 MPa. The properties of most rapidly solidified sections of the AA casting (in Table 3) come very close to this value. The quality index drops by about 20 % (100 MPa) at longer solidification times.

It should be noted that the properties of the AA castings given before (in Table 1) had a significantly lower overall level of quality. The best value of Q was 380 MPa; the worst was 293 MPa.

Although useful, there are problems with the AA 'step' mold. The thick section (No. 2) is fed through a thinner section (No. 1). The result is shrinkage in the top portion of this casting. This is obvious from the surface appearance. A 'frosted' surface is found at the top, center portion of the casting. (See middle of Figure 1.) This 'frosted' surface is produced by poor feeding and a negative pressure inside the casting. The negative pressure 'sucks' liquid metal away from the mold surface, leaving the 'frosted' or 'dry' appearance at the surface; as opposed to the smooth or 'wet' appearance at other locations of the casting. Figure 2 shows the result of a liquid penetrant inspection along the cut (centerline) section. It is readily seen that the center portion of location 1 contains a good deal of shrinkage. A small amount also appears in location 2.

Figure 2. Shrinkage in AA casting

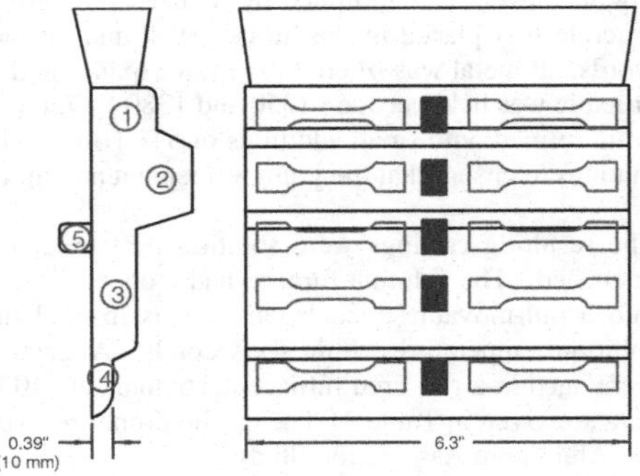

Figure 3. Sample locations

To a certain extent the shrinkage problem can be avoided by judicious selection of sample locations for the tensile test bars. We have cut samples from the locations indicated in Figure 3. Small tensile samples are cut from each side of the casting. A sample between each tensile bar (indicated in blue in the figure) is reserved for metallurgical evaluation of DAS and grain size. Shrinkage can also be minimized by using a hotter mold temperature; and by pouring metal into the top of the riser just before it solidifies. In our casting trials the AA mold temperature was 630-700 F (330-370 C).

Recently Alcoa conducted casting trials at Littlestown Hardware and Foundry. The chemistry of the base alloy used is shown in Table 4.

Table 4. Chemistry of base A356 alloy

Si	Fe	Mn	Cu	Mg	Ti	Sr	V
7.13	0.07	0.001	0.009	0.34	0.01	0.012	0.007

The dissolved Ti level was varied between 0.01 and 0.15 wt. % by adding Al-6% Ti rod and waiting a half hour for the Ti to dissolve. Further grain refiner additions were either not made, or made as Al-3Ti-1B rod at an addition level of 20 ppm B. The melt was degassed 30 minutes with a rotary impeller degasser. Gas samples were taken at the beginning and end of each cast with a copper Ransley mold and analyzed for hydrogen by a LECO analyzer. The measured gas content of the melt samples were between 0.08 and 0.14 cc/100 grams. Castings were solution heat treated for six hours at 540 C, quenched into 60 C water, held 8-9 hours at room temperature, and aged six hours at 155 C.

An analysis of the resulting tensile data showed that the dissolved Ti content and the boron addition had no significant effect on casting quality. The average values (56 samples for each location) found for tensile strength in the AA 'wedge' casting are given in Table 5.

Table 5. A356-T6 alloy tensile properties in Aluminum Association mold

Location	Yield (MPa)	UTS (MPa)	Elongation (%)	Quality Index (MPa)
1	236.3 ± 4.2	304.3 ± 10.9	7.9 ± 2.2	433.3 ± 31.8
2	235.4 ± 3.1	312.1 ± 5.9	10.7 ± 1.9	463.8 ± 18.3
3	236.3 ± 3.7	318.8 ± 4.0	14.3 ± 1.7	491.1 ± 10.8
5	238.5 ± 3.6	321.7 ± 3.4	14.3 ± 1.5	494.2 ± 8.8
4	240.2 ± 3.6	325.2 ± 3.4	15.5 ± 2.0	502.9 ± 10.9

All three sets of tensile property data have been plotted in Figure 4. The elongation is shown on a logarithmic scale, so that constant values of quality appear as straight lines. The iso-quality lines are shown in red and the values of quality are given in MPa. The average results from the Aluminum Association report[2] are labeled 'AA'. Ken Whaler's data is labeled 'KW' and our results from castings poured at Littlestown Hardware and Foundry are labeled 'LHF'.

The importance of melt treatment in casting quality is obvious from this plot. The 'AA' data is for castings produced 30 years ago, when degassing was a haphazard affair and the need for melt treatment was not widely known. At this time most foundrymen melted ingot and poured the metal into the mold without any treatment. Many shops did not even do a chemical analysis. The castings produced at Stahl Specialty ('KW') were filtered and degassed, and showed a significant improvement in properties. However, Stahl used a short degassing treatment with a porous plug. The most recent experiment ('LHF') used a 30 minute treatment with a rotary impeller degasser, a more efficient degassing method.[6] Very low gas contents were found in this metal, and the best mechanical properties were obtained.

ASTM B108 Test Bar

Another standard mold commonly used in North America is specified in ASTM B108. This is a gravity-fed permanent mold casting. (Figure 5.) This mold was designed more than thirty years ago by Ken Whaler at Stahl Specialty Company. He machined Plexiglas models of the mold and filled them with colored water to study the flow patterns. The final design uses a ribbon sprue, 3.6 mm thick at the base.

The narrow sprue controls metal flow into the mold and eliminates turbulence. However, it makes the mold difficult to pour. The mold only fills properly at temperatures above 700 F (370 C), and the standard mold operating temperature is 830-860 F (440-460 C). Various modifications of the original design have been proposed, but the new designs do not appear to deliver improved castings.[7, 8]

An example will show how the mold may be used to evaluate metal quality. Experimental data was kindly supplied by Ken Whaler.[9] An A356 alloy containing 7.0% Si, 0.03 % Fe, 0.36% Mg, 0.02% Zn, 0.08% Ti and 0.0002% P was melted in a reverbatory furnace, degassed, filtered and cast into ASTM B108 permanent mold test bars. The Cu and Mn in this alloy were below the limits of detection. The alloy was modified with 0.012% strontium and grain refined with a 5Ti-1B master alloy. Duplicate heats were made by adding small amounts of Fe to the base alloy.

All castings were given a T4 solution heat treatment (8 hours at 1000F), water quenched, pre-aged 24 hours and aged for times between 2 and 18 hours at 310 F (155 C). The tensile properties obtained are plotted in Figure 6. The iron content of the alloy and the aging time used

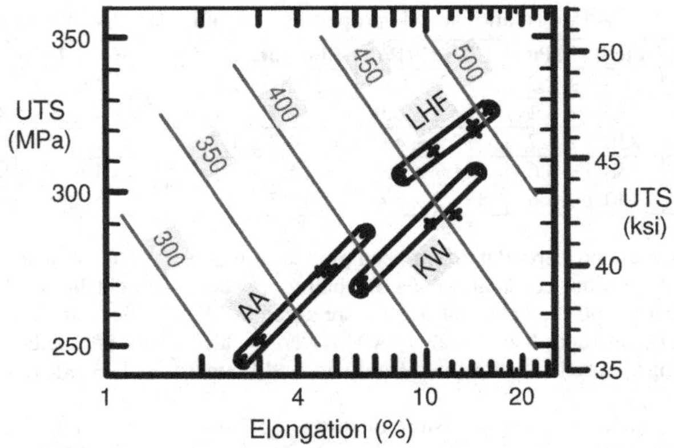

Figure 4. A356-T6 Tensile properties in AA test casting (this figure would benefit from a legend – as the references to are on another page)

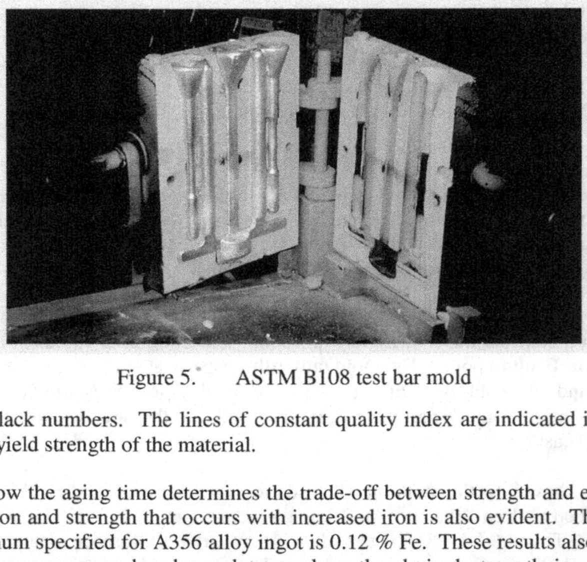

Figure 5. ASTM B108 test bar mold

are shown in black numbers. The lines of constant quality index are indicated in red, and blue lines show the yield strength of the material.

Fig. 6 shows how the aging time determines the trade-off between strength and elongation. The loss of elongation and strength that occurs with increased iron is also evident. This is the reason why the maximum specified for A356 alloy ingot is 0.12 % Fe. These results also show how the heat treatment process may be changed to produce the desired strength in a casting alloy. Generally, longer aging times produce higher tensile strength, with a corresponding loss of elongation.

In the casting trials, whose results for the AA mold were reported in Table 5, we also poured B108 test bar castings. The mechanical properties in the test bars were lower than in the more rapidly solidified sections of the AA 'wedge' casting. The average tensile properties of the test bars are given in Table 6.

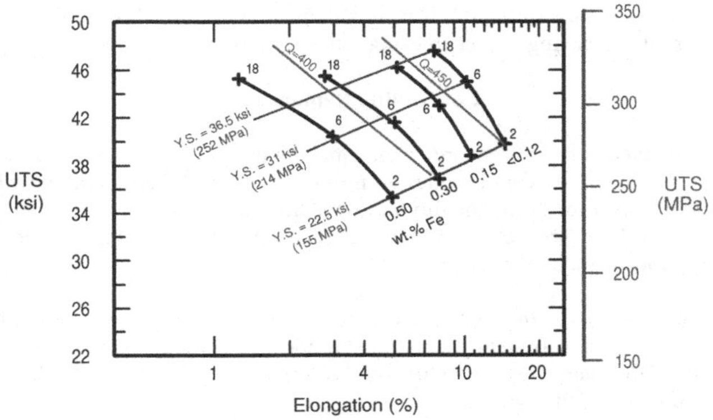

Figure 6. Mechanical properties of 356 alloy containing various iron levels and aged 2, 6 and 18 hours at 310 F (155 C). [Data from Ken Whaler [9]]

Table 6. A356-T6 alloy tensile properties of ASTM B108 test bars

20 ppm boron added				No boron added			
UTS (ksi)	YS (ksi)	Elongation	Quality Index	UTS (ksi)	YS (ksi)	Elongation	Quality Index
319	236	8.9	461.7±14.1	318	234	9.5	463.8±16.5

In the test bar castings grain refinement had no effect on the casting quality. It is also evident that the quality of the test bars is less than in the AA 'wedge' casting. This result puzzled us, until we cut a thin strip out of the test bar and x-rayed it. (Figure 7.) Small shrinkage pores were observed in the gage section of the bar. Discussions with other foundrymen confirm that this mold is prone to shrinkage.

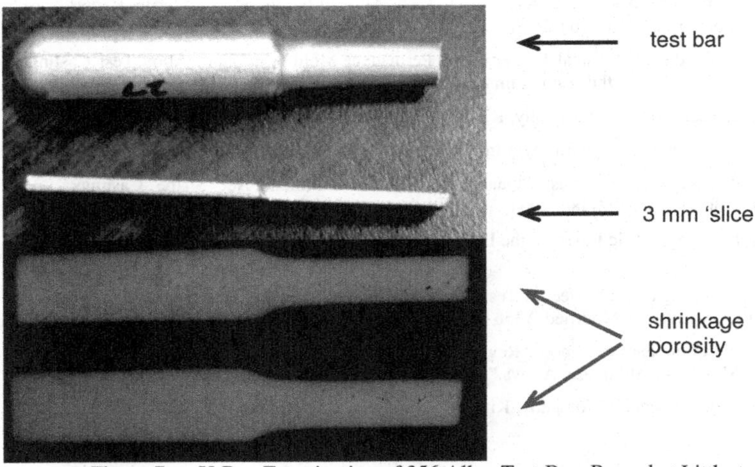

Figure 7. X-Ray Examination of 356 Alloy Test Bars Poured at Littlestown

It should be noted that this shrinkage was found in A356 – an alloy which has excellent castability. Other alloys, having longer freezing ranges, would be even more prone to shrinkage in this mold.

Concluding Remarks

In spite of their limitations both 'standard' molds may be used to determine the quality of metal in the foundry. Twenty five years ago degassing practices were inefficient and filtration was almost nonexistent. As a result, the quality index of 'average' castings was 100-150 MPa less than today's current best practice. Much better results can be obtained now, with the widespread use of rotary impeller degassing.

Both of the molds produce castings which are susceptible to shrinkage formation. The porosity in the B108 test bar is often sufficiently small, that it does not show up on x-rays when the entire bar is examined. However, the sensitivity of the radiographic examination is improved by cutting a thin section out of the bar.

The shrinkage porosity in the test bars can be minimized by proper mold coating practices and by pouring with hot metal. Neither is a completely satisfactory solution to the problem, however. It would be better to improve the design of the mold, so results are less sensitive to operator-controlled casting practices.

Acknowledgements

The authors gratefully acknowledge Littlestown Hardware and Foundry Company, who made one of their furnaces available and assisted us in our casting trials. We also acknowledge the assistance of Jen Lin, Xinyan Yan and Marlene Reisinger of Alcoa, who helped in the evaluation of our castings.

References

[1] E.W. Miguelucci: "The Aluminum Association Cast Alloy Test Program: Interim Report," AFS Transactions, Vol. 93, pp. 913-916 (1985).

[2] "Special Report on the Mechanical Properties of Permanent Mold Aluminum Alloy Test Castings," Jobbing Foundry Division of the Aluminum Association, Washington, D.C., © November, 1990.

[3] Ken R. Whaler, Stahl Specialty Company, Kingsville, Missouri, private communication (1995).

[4] M. Drouzy, S. Jacob and M. Richard: AFS International Cast Metals Journal, Vol. 5, pp 43-50 (1980).

[5] G.K. Sigworth and C.H. Caceres: "Quality Issues in Aluminum Net Shape Castings," AFS Transactions, Vol. 112, pp. 373-386, 2004.

[6] G. K. Sigworth: "A Scientific Basis for the Degassing of Aluminum," AFS Transactions, vol. 95, pp. 73-78 (1987).

[7] D. Emadi, L.V. Whiting *et al.*: "Effect of Test Bar Mould Design and Heat Treatment Parameters on the Mechanical Properties of Sr-Modified A356.2 Alloy," AFS Transactions, Vol. 109, pp. 1-12 (2001).

[8] D. Emadi, L.V. Whiting and M. Sahoo: "Revisiting the ASTM B108 Test Bar Mold for Quality Control of Permanent Mold Cast Aluminum Alloys," AFS Transactions, Vol. 112, pp. 225-236 (2001).

[9] Ken R. Whaler, Stahl Specialty Company, Kingsville, Missouri, private communication, 2003.

Shape Casting: The 3rd International Symposium
Edited by: John Campbell, Paul N. Crepeau, and Murat Tiryakioğlu
TMS (The Minerals, Metals & Materials Society), 2009

PROPERTIES OF B356-T6 ALUMINUM CAST VIA PERMANENT MOLD AND ADVANCED SQUEEZE CAST (ASC) PROCESSES

Gerald Gegel[1], David Weiss[2], William Edney[3]

[1]Material & Process Consultancy, 740 Columbus Avenue, Morton, IL
[2]Eck Industries, Inc., 1602 North 8th Street, Manitowoc, WI
[3]Prototype Cast Manufacturing, Inc., 51500 Schoenher Drive, Shelby Twp., MI

Keywords: aluminum, squeeze casting, mechanical properties

Abstract

The mechanical properties of cast products are a function of alloy composition, solidification rate and porosity content. The tensile and fatigue properties of AA B356 cast using low pressure permanent mold and advanced squeeze casting processes are compared to illustrate the mechanical property advantages accrued as a result of solidification under pressure. The ASC method used low pressure to till the die and then applies squeeze pressure directly to the entire volume of the component. As this technology is new, we will describe the design and operation of this production-viable too-ton machine. The design of the machine permits the use of the same tooling to produce both LPPM and ASC castings. This DOE sponsored research and development project has provided a production-viable machine and process technology that will improve the strength and reliability of cast components.

Introduction and Background

Squeeze casting occurs when a metal solidifies in a closed die under pressure. The resultant casting has improved properties and a more uniform microstructure as compared to those produced by traditional molten metal fabrication techniques. The higher properties are achieved through controlled entry of the metal into the die through large gates, which reduces turbulence, and high solidification rates with resultant refinement of microstructural features. The two types of squeeze casting, direct and indirect, are distinguished by the method of pressure application. Direct squeeze casting applies pressure directly on the casting. This contrasts with indirect squeeze casting where pressure is applied via the gating system. The indirect method is prone to partial solidification of the gating system prior to complete casting solidification and may result in casting porosity. The direct method has traditionally been accomplished by pouring a known weight of liquid metal alloy into a cavity and applying a pressure of 65 MPa or higher via a top punch assembly. This is sometimes called liquid metal forging. Both methods suffer from poor metal handling from the furnace into the mold or gating system, resulting in additional defects caused by oxides.

Early attempts [1, 2] to develop a Metal Compression Forming (direct squeeze casting) process and a casting machine with higher production rate capability demonstrated that gravity pouring molten metal into a gating system produced a lot of turbulence and generated oxide films in the resultant castings. To attain quiescent fill, low-pressure bottom filling was incorporated into the next version of a machine. Mating a low-pressure furnace with a horizontal die cast machine solved the metal delivery issue but did not result in a production capable casting machine.

Substantial mechanical property improvements were demonstrated using the direct squeeze approach. Specimens excised from 357-T6 castings exhibited properties generally higher than those achieved with conventional low pressure casting processes; an average yield strength of 276 MPa, an average tensile strength of 356 MPa, and total elongation of 12.3% [3]. Use of Metal Compression Forming to cast metal matrix composite alloys was not as successful. The properties achieved were much less than those attainable via conventional sand or permanent mold casting [4]. SEM micrographs of fractures revealed large numbers of oxide films, probably due to metal handling issues.

Clearly the direct squeeze method was both theoretically and practically an approach to lower porosity and higher mechanical properties in castings. Problems related to machine design such as difficulty with the gate shut off mechanism, and structural rigidity of the die and casting machine prevented the concept from being volume production capable. The goal of the current program was to design and build a production capable direct squeeze cast machine having improved metal filling capability and produce pore-free cast components with properties comparable to forged parts, at considerably lower cost.

Advanced Squeeze Cast (ASC) Machine Design

To obtain controlled fill of the die and pressurized solidification of near-net-shape castings the ASC (Advanced Squeeze Cast) machine utilizes the metal delivery system of a low-pressure casting machine and the structural strength and rigidity of a high-pressure die casting machine. In this case, the machine is vertically oriented. The 600 ton capacity production machine is designed to provide squeeze pressures of up to 103 MPa.

Furnace and Metal Delivery System

The metal delivery system used is a typical low-pressure furnace. Gas pressure (up to 1 bar) in an enclosed furnace is used to move liquid metal through a riser tube into a die or mold. The metal is drawn from under the melt surface minimizes the entrapment of oxides in the casting. The equipment is usually pressurized using dry air but is designed to allow the use of carbon dioxide, nitrogen, argon or other gases when casting metals that must be held under a protective atmosphere. When used in the squeeze cast mode, the ASC machine's low-pressure furnace is used only to fill the die.

Machine Design Consideration

The machine's structure is designed to maintain accuracy and withstand the forces used to pressurize the die cavities (up to 600 tons). This necessitates a structure similar to a die casting machine with some modifications. The machine has four platens – a lower platen to which the drag of the mold is attached, a die platen to which the cope is attached, a moving platen that establishes the starting or fill position of the die, and an upper platen – and an additional hydraulic cylinder to apply the squeeze pressure (see Figure 3). Under normal conditions an attempt to squeeze metal in the cavities would result in the die opening with excess metal escaping to the parting lines. In this equipment a hydraulically controlled system mechanically locks the moving platen in a position relative to the fixed platen while the squeeze pressure is applied.

Figure 1. ASC Machine Components

A computerized data acquisition system is interfaced with the displacement, pressure and temperature sensors located on the ASC press and casting tooling. Data like those shown in Fig. 4 are used to determine the interrelationship of the process parameters. For the example shown, the time at which maximum die temperature occurs (approximately 18 seconds) indicates when the die cavity was filled. The graph also indicates that the squeeze pressure was applied when the metal temperature was approximately 493°C. These types of data were used to fine tune the timers that control operation of the ASC press functions.

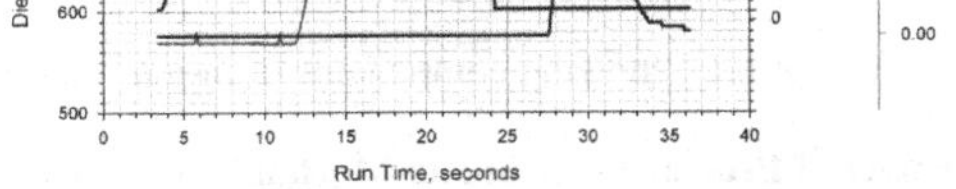

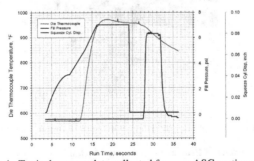

Figure 4. Typical process data collected from an ASC casting run

Die Design

The die must accomplish two functions – contain a controlled volume of metal provided by the low-pressure metal delivery system, and allow pressurization of the solidifying metal. Since the goal is to pressure feed liquid metal to offset solidification shrinkage the initial metal volume in the die cavities is equivalent to part volume plus solidification shrinkage volume. Thus the cavity volumes must be variable. The drag portion of the mold is pretty conventional. However, the cope is very unconventional as it must "expand" to provide the additional liquid metal volume needed to offset solidification shrinkage volume. The cope cavities are constituted of die block inserts and floating retainers (see Figure 5). The floating retainers are spring mounted on retainer bolts. When the die is positioned in the "fill position" the spring pressure holds the retainers against the drag and provides a liquid metal seal between the stationary (drag) and moving (cope) halves of the die. When the squeeze cylinder is activated the retainers retract and the die is fully closed. If the in-gates are not closed when the squeeze cylinder is activated the metal will be squeezed out of the die cavities via the in-gates. Therefore, the in-gates are closed- off before pressure is applied with a centralized in-gate shut-off system (see Figure 7). Thermocouples are imbedded into the die cavities in various locations to sense metal presence and measure metal temperatures in the die cavities.

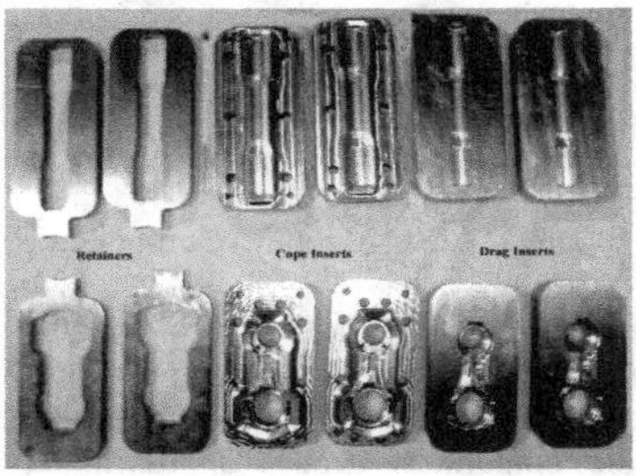

Figure 5. Floating retainers and inserts for ASC dies

Figure 6. View of drag showing components of in-gate shut-off system

Effect of Pressure Casting on Mechanical Properties

Production of Test Castings

To determine the effect of direct squeeze casting on static and dynamic properties, the ASC machine and the permanent mold tooling were used to produce B356 alloy castings via both LPPM (low pressure permanent mold) and the ASC (advanced squeeze casting) process. With the exception of the application of squeeze pressure, the process parameters used to cast both sets of specimens were the same. The B356 alloy was modified by the addition of 0.015% Sr to the degassed melt. Melt temperature was controlled to 760 ± 6°C. The process profile used to make the squeeze castings is shown in Figure 8. Each machine cycle produced four parts – two connecting rods (Figure 9) and two tensile bars having a 19.1 mm diameter test section. This initial production run achieved a production rate of 120 castings (30 cycles) per hour. This is equivalent to the rates attainable during conventional low-pressure casting of the same parts.

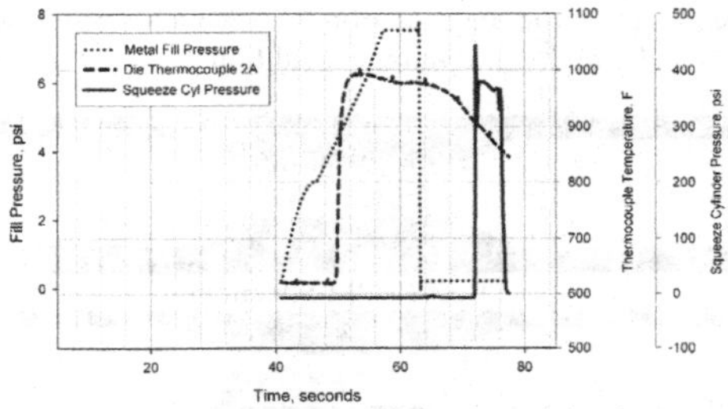

Figure 8. Typical operating conditions used during production capability run.

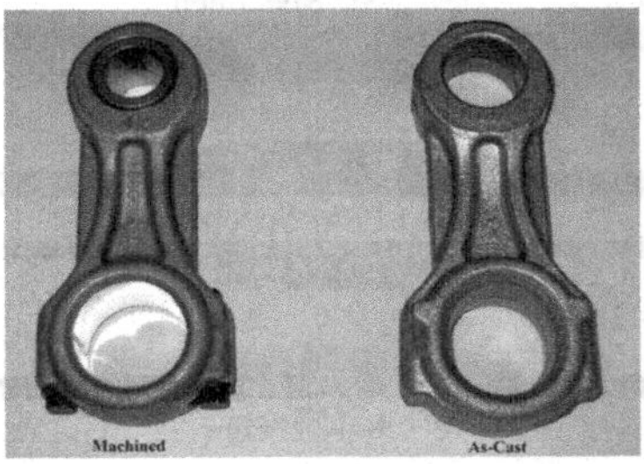

Figure 9. Machined and as-cast (in-gate removed) B356 connecting rods.

Evaluation of Mechanical Properties

The as-cast tensile bars were heat treated to a T6 condition and radiographed to determine their quality. Only those specimens that had a Grade B or better quality level in accordance with ASTM E505 were machined for determination of room temperature tensile and fatigue properties. Tensile and fatigue test specimens were machined from the 19.1 mm diameter section of the cast tensile bars. The radiographic images in Figure 10 show that the quality of the ASC specimens is better than that of the LPPM specimens.

Room temperature tensile tests were accomplished on a universal load frame at a crosshead speed of 0.013 mm/second. Stress controlled fully reversed loading (R = -1) at a loading rate of 10 Hz was used to determine fatigue lives at several stress levels.

Discussion of Results

The results of room temperature tensile tests are shown in Table 1. As expected, the casting process used had little effect on the room temperature tensile properties of the

B356-T6 test castings. The results of the fatigue tests (see Figure 11) show the high cycle (10^7 cycles) fatigue strength of the ASC cast samples to be 31 percent higher (62.3 MPa vs. 47.5 MPa) than that of the LPPM cast samples. Metallographic examination of both the LPPM and ASC fatigue specimens revealed the presence of porosity (Figure 13). The morphologies are, however, different. The pores in the LPPM specimens have the expected irregular shape typical of shrinkage porosity. The pores in the ASC specimens are smaller and more rounded and would exhibit lower stress concentrations than those in the LPPM samples. We believe this to be the major reason why the ASC processed castings demonstrated higher fatigue strength under high cycle fatigue conditions.

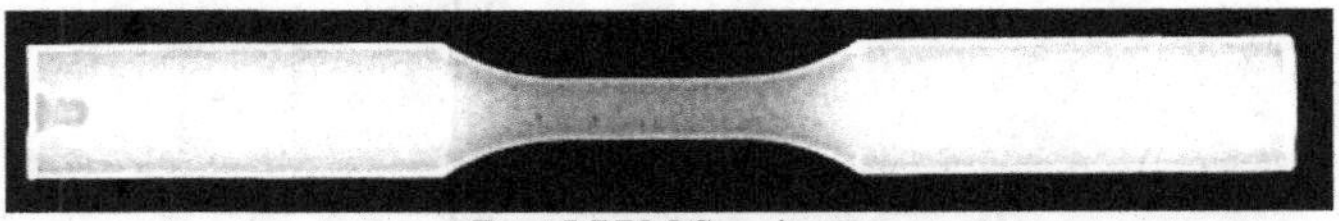

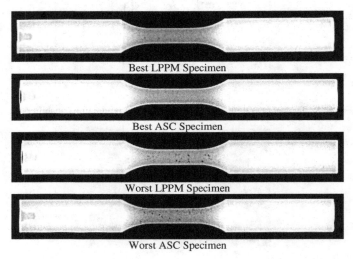

Best LPPM Specimen

Best ASC Specimen

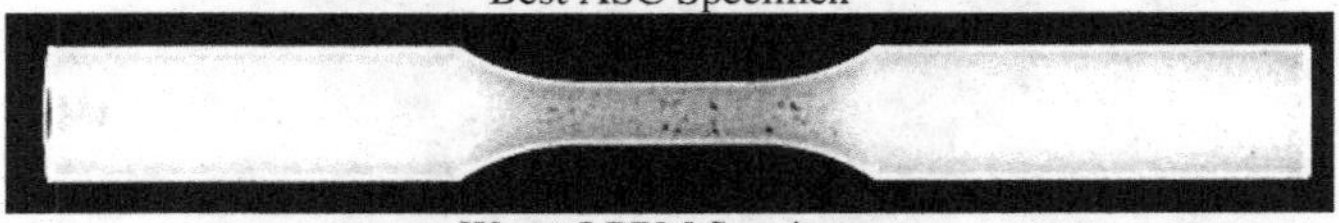

Worst LPPM Specimen

Worst ASC Specimen

Figure 10. Range of radiographic quality of LPPM and ASC Specimens

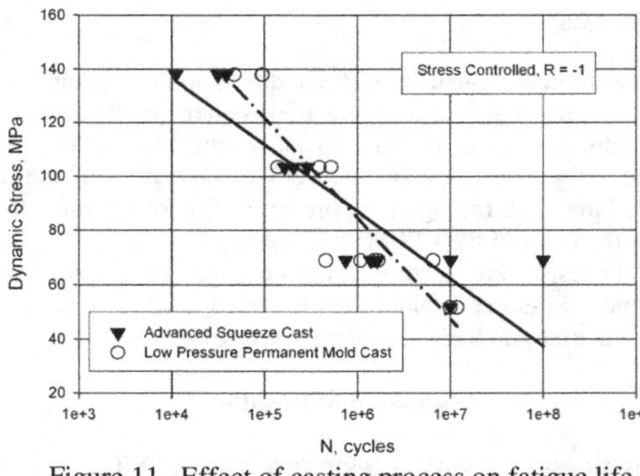

Figure 11. Effect of casting process on fatigue life.

Table I. Room temperature tensile properties

Casting Process	Specimen ID	Ultimate Strength, MPa	0.2% Yield Strength, MPa	Elongation %
ASC	9-2	267.8	245.3	1.1
ASC	22-2	271.1	252.9	1.0
ASC	23-2	301.0	256.6	2.3
ASC	31-4	295.4	261.4	1.8
ASC	33-2	296.6	261.6	1.7
ASC	36-2	289.6	232.4	4.3
	Average	**286.9**	**251.7**	**2.0**
LPPM	440-11-4	268.4	245.8	1.2
LPPM	450--7-2	278.2	238.2	2.0
LPPM	450-11-4	267.8	229.9	1.9
LPPM	450-12-2	255.7	213.7	3.5
LPPM	450-13-2	275.4	241.2	1.7
LPPM	450-15-2	274.7	247.9	1.3
	Average	**270.0**	**236.1**	**1.9**

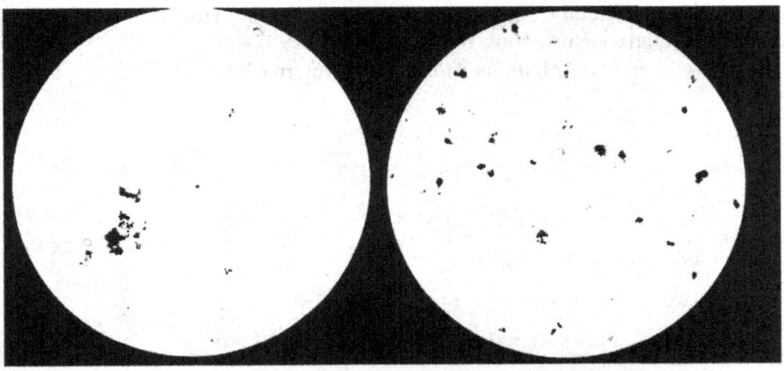

Figure 12. Typical morphology of porosity in LPPM (left) and ASC samples

A squeeze cast part would expectedly contain little or no porosity. How did the structure shown in Figure 12 for the ASC castings form? In the ASC process the die is "over filled" by an amount equivalent to the expected solidification shrinkage volume. Die closure when the metal is above liquidus temperature (610°C) generates high pressure that facilitates mass feeding until the metal fully solidifies at about 530°C. However, the process is not isothermal; dies are at a temperature of about 325°C, so metal will start to solidify as the temperature at the die-metal interface reaches 610°C. Thus, optimum pressure application would be at a metal temperature greater than the metal's liquidus. The process data (Figure 8) shows that squeeze pressure was applied at a temperature of about 493°C; the metal was mostly solid. Pressurization essentially forged the nearly solid casting and the "rounded" shrinkage pores formed when the remaining liquid solidified.

Conclusions

The requirement for higher mechanical properties in cast components drives the need for unique casting methods. Direct squeeze casting is one method to achieve those needs for certain types of components. Further optimization of the process parameters and machine operational sequence will expectedly result in even higher fatigue strength values than those already achieved.

Acknowledgements

The authors wish to acknowledge the DOE and Oak Ridge National Lab for financial and testing support for this program under Contract 4000022893. Others supporting the program included Empire Casting and Bendix.

References

1. Viswanathan, S., Porter, W. D., Ren, W., Brinkman, C. R., Sabau, A. S., Purgert, R. M., Metal Compression Forming of Aluminum Alloys and Metal Matrix Composites, CRADA Final Report, No. ORNL95-0363, Oak Ridge National Laboratory (1995).
2. Viswanathan, S., Porter, W. D., Ren, W., Purgert, R. M., "Application of the Metal Compression Forming Process for the Production of an Aluminum Alloy Component", Ed. S. K. Das, TMS-AIME, Warrendale, PA, p 87-94 (1997).
3. Viswanathan, S., Porter, W. D., Brinkman, D. R., Purgert, R. M., "Production of A357 Motor Mount Bracket by the Metal Compression Forming Process", Transactions North American Die Casting Association, Rosemont, IL, pp. 353-358 (1997).
4. Davis, J.R., *Aluminum and Aluminum Alloys*, ASM International, (1993).

Shape Casting: The 3rd International Symposium
Edited by: John Campbell, Paul N. Crepeau, and Murat Tiryakioğlu
TMS (The Minerals, Metals & Materials Society), 2009

THE RELATIONSHIP BETWEEN DEFECT SIZE AND FATIGUE LIFE DISTRIBUTIONS IN AL-7%SI-MG ALLOY CASTINGS

Murat Tiryakioğlu

Robert Morris University, Dept. of Engineering, Moon Township, PA 15108, USA

Keywords: Porosity, Bifilms, Gumbel, Paris-Erdoğan Law, Crack Propagation

Abstract

A new statistical distribution function was developed by combining the size distribution for the fatigue-initiating defects and a fatigue life model based on the Paris-Erdoğan law for crack propagation. Two datasets for the fatigue-initiating defects in Al-7%Si-Mg alloy castings, reported previously in the literature, were used to demonstrate that (i) the size of fatigue-initiating defects follow the Gumbel distribution, (ii) the crack propagation model developed previously provides respectable fits to experimental data, and (iii) the statistical distribution function developed in the present study provides excellent fits to the datasets.

Introduction

Structural defects, such as porosity and oxide inclusions, affect fatigue life of aluminum alloy castings by causing premature failure [1,2] and increasing the variability in properties. It has been of interest to engineers to model variability in fatigue life, so that parts can be designed accordingly. In a majority of the cases found in the literature, the variability in the fatigue life of aluminum alloy castings has been modeled by either the lognormal [3] or 2-parameter Weibull [2,4,5], the cumulative probability (P) functions of which are written as;

$$P(N_f) = \int_0^{N_f} \frac{1}{N_f \omega \sqrt{2\pi}} \exp\left[\frac{-(\ln(N_f) - \theta)^2}{2\omega^2}\right] dN_f \qquad (1)$$

$$P(N_f) = 1 - \exp\left(-\left(\frac{N_f}{N_0}\right)^q\right) \qquad (2)$$

respectively. In Equation 1, N_f is number of fatigue cycles until final fracture (fatigue life), ω and θ are the standard deviation and average of $\ln(N_f)$, respectively. In Equation 2, N_0 is the scale parameter and q is the shape parameter, alternatively referred to as the Weibull modulus. Although in most cases, these two distributions provide acceptable fits to the data, they disregard the root cause of the scatter observed in fatigue life: the variability in the defect sizes in the specimens.

There have been several attempts to link the distribution of defect sizes with the fatigue life of aluminum castings. Casellas et al. [4] followed the approach taken by Jayatilaka and Trustrum [6] to fit a power equation to the upper tail of the defect size distribution, which was originally developed for ceramics [7]. Casellas et al. showed that the Weibull modulus for the distribution of fatigue life is a function of the power of the equation fitted to the upper tail of the defect size distribution. Although this approach can be taken as a good approximation, size distribution of largest defects should theoretically follow one of the extreme value distributions, and the power fit by Jayatilaka and Trustrum does not constitute an extreme value distribution.

In another study, Yi et al. [8,9] assumed that pore size in A356 castings follows the lognormal distribution, which is consistent with their histograms as well as statistical analysis of pore sizes for Mg alloy castings [10]. The authors, using fatigue crack propagation models and maximum pore size data, attempted to estimate the distribution of fatigue life of A356 castings. Although the results were promising, cumulative probability plots indicated systematic lack of fit.

Theoretical Background

To establish the relationship between fatigue life and defect size distributions, one needs to (i) determine the size distribution of failure-initiating defects, and (ii) a crack propagation model based on fracture mechanics to estimate fatigue life. These two aspects are discussed first.

Defect Size Distribution

All fatigue models based on the microstructure need the size distribution of defects as an input. A survey of the literature has shown that various statistical distributions have been used for the size of the fatigue-initiating defects in metals. If the largest defects (upper tail of the defect size distribution) are responsible for initiating cracks as suggested by Murakami [11], then their size has to follow an extreme value distribution, based on mathematical statistics [12,13]. Tiryakioğlu [14] analyzed the equivalent defect diameter, d_{eq}, of fracture-initiating defects found on the fracture surfaces of four different Al and three Mg casting from seventeen datasets in the literature. By using the General Extreme Value distribution, the author found that sixteen of the datasets followed the Gumbel distribution, for which the cumulative probability, P, can be written as;

$$P = \exp\left(-\exp\left(\frac{d_{eq} - \lambda}{\delta}\right)\right) \quad (3)$$

where λ and δ are location and scale parameters, respectively. Tiryakioğlu pointed out that the results are consistent with the size of inclusion defects measured on the fracture surfaces of steel castings, which have been shown to follow the Gumbel distribution [15,16,17,18].

Crack Propagation

For a part that contains a crack-like defect of initial length a_i, the crack length increases to a value, *a*, at any given number of stress cycles, N. The fatigue crack growth rate in the power-law or steady-state stage, as expressed by the Paris-Erdoğan law [19], can be written as,

$$\frac{da}{dN} = C(\Delta K_{eff})^m \quad (4)$$

where C and m are Paris-Erdoğan constants and ΔK_{eff} is the effective stress intensity factor range, which can be written as;

$$\Delta K_{eff} = 2UY\sigma_a\sqrt{\pi a} \quad (5)$$

where U is the crack closure factor, Y is the compliance calibration factor and σ_a is alternating stress amplitude. Taking Y as independent of defect size with propagating crack introduces only minor error to the model [20]. U can also be assumed to be only a function of the stress ratio, R [5]. Inserting Equation 5 into Equation 4, the Paris-Erdoğan equation can be integrated as:

$$\int_{a_i}^{a_f} a^{-\frac{m}{2}} da = 2^m C U^m Y^m \sigma_a^m \int_{N_i}^{N_f} dN \tag{6}$$

where N_i is the number of cycles to initiate a fatigue crack, and a_f is the final crack length just before final fracture at N_f cycles. After integrating, we obtain:

$$\frac{2}{2-m} a^{\frac{2-m}{2}} \Big|_{a_i}^{a_f} = 2^m C U^m Y^m \sigma_a^m N \Big|_{N_i}^{N_f} \tag{7}$$

$$\frac{2}{2-m}\left(a_f^{\frac{2-m}{2}} - a_i^{\frac{2-m}{2}} \right) = 2^m C U^m Y^m \sigma_a^m (N_f - N_i) \tag{8}$$

Because $a_f >> a_i$, Equation 8 can be simplified as

$$a_i^{\frac{2-m}{2}} = (m-2) 2^{m-1} C U^m Y^m \sigma_a^m (N_f - N_i) \tag{9}$$

Let us now assume that the projected area of the defect, A_i, can be written as βa_i^2 where β is a coefficient determined by the geometry of the defect. Consequently Equation 9 can now be written as;

$$A_i^{\frac{2-m}{4}} = (m-2) 2^{m-1} \beta^{\frac{2-m}{4}} C U^m Y^m \sigma_a^m (N_f - N_i) \tag{10}$$

After rearranging and further simplification, we obtain

$$N_f = N_i + B \sigma_a^{-m} A_i^{\frac{2-m}{4}} \tag{11}$$

where

$$B = \frac{2^{1-m}}{m-2} \beta^{\frac{m-2}{4}} C^{-1} U^{-m} Y^{-m} \tag{12}$$

In several studies, cracks were observed to grow from structural defects at or shortly after the first stress cycle [21,22,23]. Therefore N_i can be taken as zero in aluminum castings.

Equation 12 is a simplistic but useful tool to estimate fatigue life based on stress levels and defect size distributions. The validity of Equation 12 (with N_i=0) was verified by Davidson et al. [24] and Wang et al. [25]. In these studies, the authors provided A_i vs. N_f plots and found that the slope of log(A_i) vs. log(N_f) is approximately -0.5, indicating that (i) N_f is related to $\sqrt{A_i}$ in cast Al-Si alloys and (ii) m ≈ 4. Similarly, Murakami and Endo [26,27] used $\sqrt{A_i}$ as a parameter to model the fatigue limit of surface and internal defects in steels. Moreover they showed that the fatigue strength is governed by one critical inclusion, which usually has the largest size, not by the presence of many inclusions.

Estimating the Fatigue Life Distribution from the Defect Size Distribution

Equation 3 can be rearranged to obtain the inverse function of the Gumbel distribution for equivalent defect diameter as;

$$d_{eq} = \lambda + \delta(-\ln(-\ln(P))) \tag{13}$$

Hence A_i can be calculated as;

$$A_i = \frac{\pi d_{eq}^2}{4} = \frac{\pi(\lambda + \delta(-\ln(-\ln(P))))^2}{4} \tag{14}$$

Inserting into Equation 11 and solving for P, we obtain

$$P = \exp\left(-\exp\left(\frac{\lambda}{\delta} - \frac{2}{\delta\sqrt{\pi}}\left(\frac{N_f - N_i}{B\sigma_a^{-m}}\right)^{\frac{2}{2-m}}\right)\right) \tag{15}$$

In Equation 15, P represents the exceedance probability because a low defect size implies long fatigue life. Hence the cumulative statistical distribution of fatigue life, $P(N_f)$, is found by subtracting Equation 15 from unity;

$$P(N_f) = 1 - \exp\left(-\exp\left(\frac{\lambda}{\delta} - \frac{2}{\delta\sqrt{\pi}}\left(\frac{N_f - N_i}{B\sigma_a^{-m}}\right)^{\frac{2}{2-m}}\right)\right) \tag{16}$$

Data Analysis, Results and Discussion

Two datasets were used to examine the relationship between defect size and fatigue life distributions:

- A356 aluminum alloy castings with high (0.20-0.23 ppm) hydrogen levels by Yi *et al.* [8,9,28,29]. The specimens were tested at a stress ratio (R) of 0.1 and two maximum stress levels: 120 and 150 MPa (σ_a being 54 and 68 MPa, respectively). A total of 56 specimens were tested.
- Cast Al-7wt.%Si-0.6wt.%Mg-0.11wt.%Fe castings by Davidson *et al.* [24]. The specimens were tested at R=0 and three maximum stress levels: 170, 190 and 210 MPa (σ_a being 85, 95 and 105 MPa, respectively). A total of 64 specimens were tested.

Defect Size Distribution

In both studies, fatigue-initiating defect sizes were measured. Gumbel probability plots for the two datasets are shown in Figure 1, where cumulative probability, P, was assigned to each data point using the modified Kaplan-Meier probability estimator:

$$P = \frac{i - 0.5}{n} \tag{17}$$

where *i* is the rank of the data point in ascending order and n is the sample size.

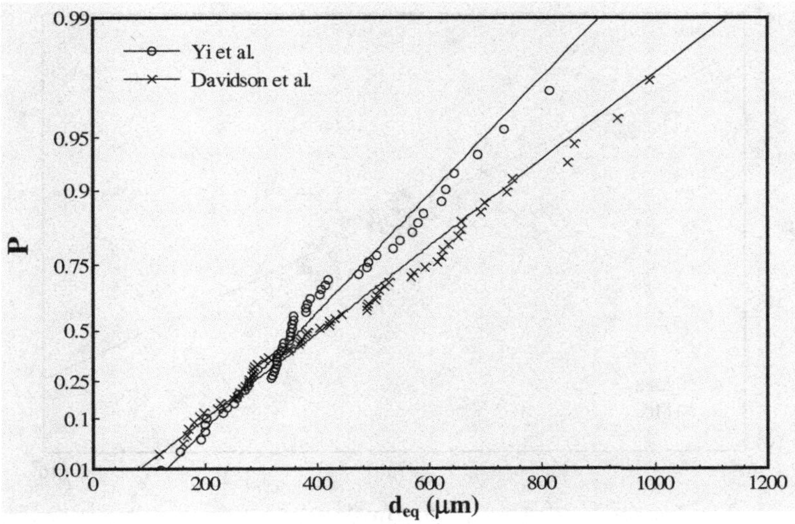

Figure 1. Gumbel probability plot for the size data of fatigue initiating defects.

The estimated Gumbel parameters are presented in Table 1, which also shows the results of goodness-of-fit hypothesis test using the Anderson-Darling [30] test statistic, A^2. Because p-values for both cases are in excess of 0.05, the hypotheses that the datasets follow the Gumbel distribution could not be rejected.

Table 1. Gumbel fits to d_{eq} data from the two datasets

	λ (μm)	δ (μm)	A^2	p-value
Yi *et al.*	323.2	125.5	0.502	0.212
Davidson *et al.*	343.7	170.4	0.379	>0.250

Effect of Defect Size on Fatigue Life

Equation 11 was fitted to the data from both studies, assuming that N_i=0. The data collected by Yi et al. at a maximum stress of 150 MPa were adjusted for equivalent fatigue lives at a maximum stress of 120 MPa. The values of B and m that yield the smallest error were determined by using the Newton-Raphson method. The same approach was followed with the data of Davidson et al., where equivalent fatigue lives at a maximum stress of 170 MPa were determined. The results are presented in Figure 2, which shows reasonable agreement between the data and Equation 11, with B and m values listed in Table 2.

That fatigue often starting from the first cycle or very early cycle can be taken as evidence that the material is pre-cracked, so that no initiation is required. The obvious pre-crack is a pore or a bifilm, both of which originate from external, entrainment events [31].

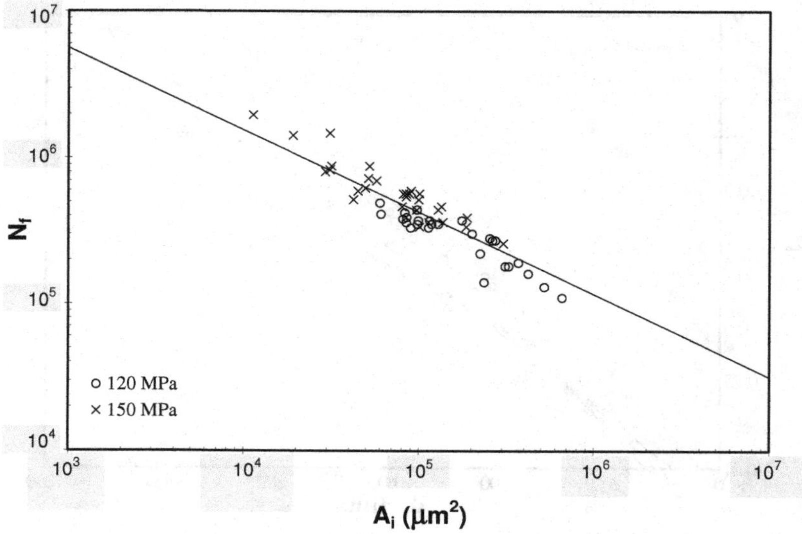

(a)

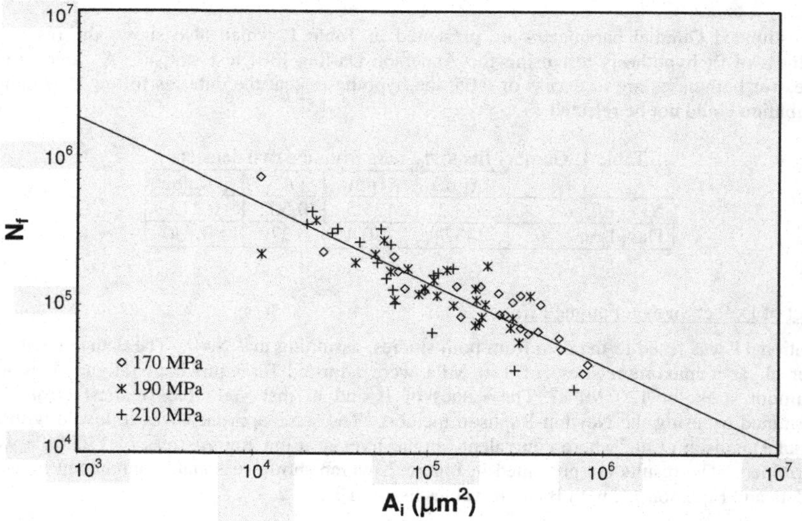

(b)

Figure 2. Fatigue life data of (a) Yi et al. at a maximum stress of 120 MPa, and (b) Davidson et al. at 170 MPa, plotted as a function of the area of fatigue-initiating defect.

Table 2. The values of m and B in Equation 11 estimated for both datasets.

	m	B
Yi *et al.*	4.25	6.16×10^{15}
Davidson *et al.*	4.21	1.14×10^{16}

Fits of Lognormal and Weibull Distributions to N_f Data

The cumulative probability plots for the two datasets and the fit by Equation 16 with parameters listed in Tables 1 and 2 are shown in Figure 3. The data were assigned probabilities using Equation 17. Note that there is excellent agreement between the data and the fatigue life distribution developed in this study. The goodness of fit of Equation 16 was evaluated using A^2 statistics. The results are shown in Table 3, with p-values for both cases being in excess of 0.250, based on the values listed by Stephens [32].

Table 3. Goodness of fit results for Equation 16 to the two datasets.

	A^2	p-value
Yi *et al.*	1.091	>0.250
Davidson *et al.*	0.678	>0.250

For comparison purposes, lognormal and 2-parameter Weibull distributions were also fitted to the fatigue life data. The fits are presented in Figure 4 and the estimated parameters as well as the results of goodness of fit tests are shown in Table 4. For both datasets, the hypothesis that the data follow the lognormal distribution could not be rejected. The 2-parameter Weibull distribution, however, did not provide acceptable fits to the data.

A comparison of Tables 3 and 4 shows that fits provided by Equation 16 are much better than those by the 2-parameter Weibull distribution and almost as good as those by the lognormal distribution. Because Equation 16 incorporates the fracture mechanics of fatigue failure, it is recommended that Equation 16 be used to characterize the distribution of fatigue life.

Table 4. Lognormal and Weibull fits to N_f data from the two datasets.

	Lognormal				2-parameter Weibull			
	θ	ω	A^2	p-value	N_0	q	A^2	p-value
Yi *et al.*	12.93	0.572	0.606	0.110	5.49×10^5	1.653	1.862	<0.010
Davidson *et al.*	11.77	0.641	0.148	0.963	1.77×10^5	1.550	1.050	<0.010

Conclusions

- The size distribution for fatigue-initiating defects in Al-7%Si-Mg alloys is Gumbel.
- The fatigue life model based on the Paris-Erdoğan equation with no crack initiation stage provides respectable fits to the two datasets from the literature.
- A new distribution function combining the size distribution of fatigue-initiating defects and fatigue life model provides fits to the two datasets that could not be rejected by goodness of fit tests.
- Goodness of fit hypothesis tests showed that the 2-parameter Weibull did not provide an acceptable fit to the fatigue life data, whereas the lognormal distribution did.
- Because the new distribution function incorporates the defect size distribution and fracture mechanics, it is recommended that it be used to characterize the statistical distribution of fatigue life of cast Al alloys.

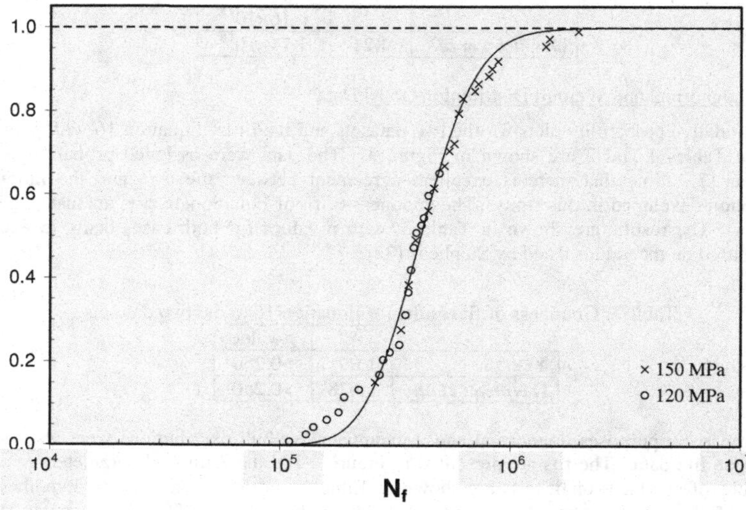

(a)

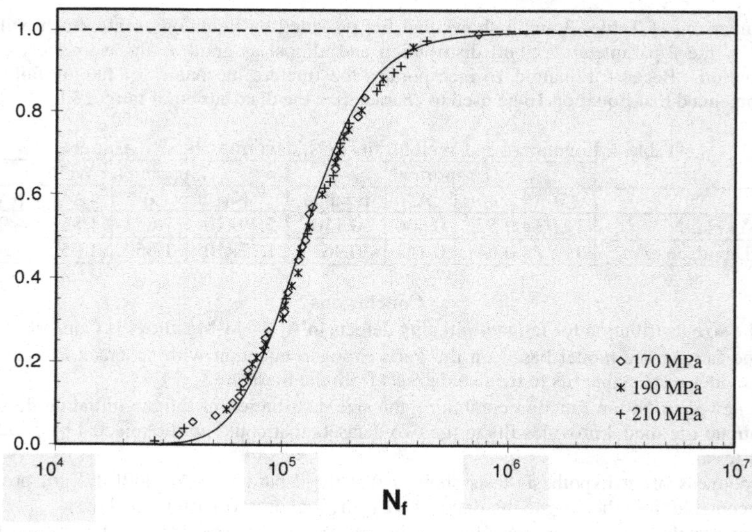

(b)

Figure 3. Cumulative probability plot for fatigue life data of (a) Yi et al., and (b) Davidson et al. The fits obtained by using Equation 16 are indicated by the solid curves.

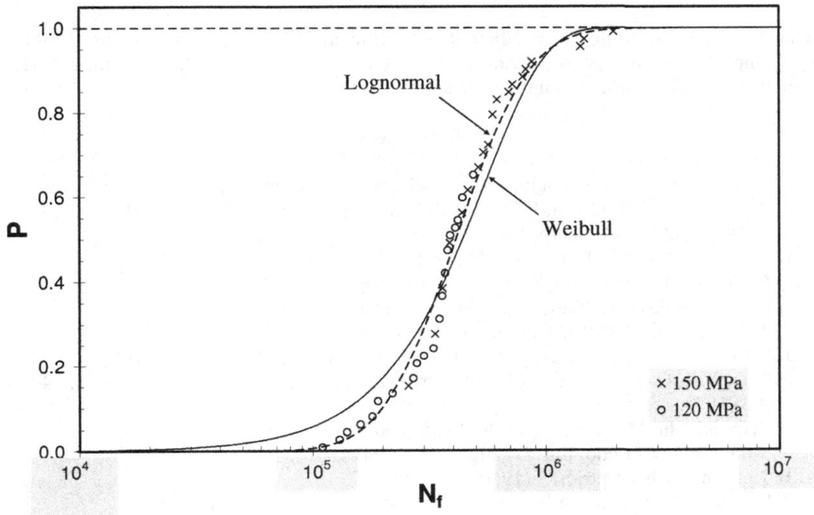

(a)

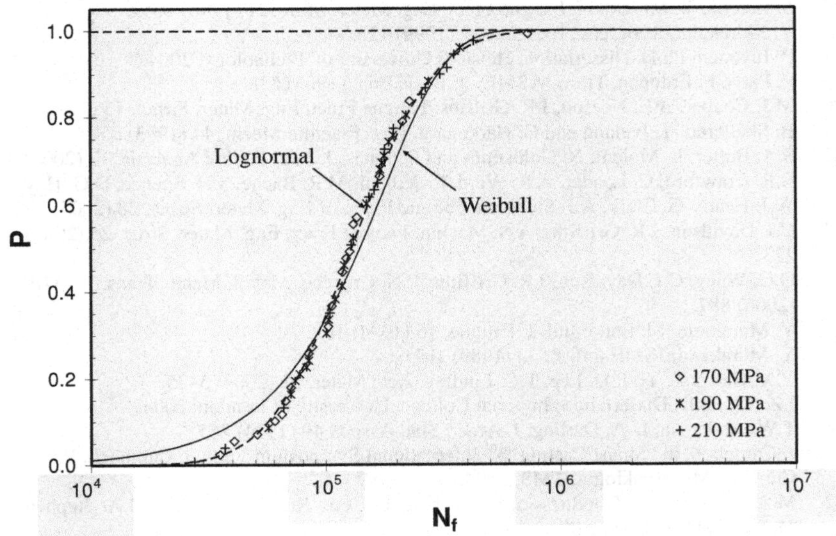

(b)

Figure 4. Cumulative probability plots for the Lognormal and Weibull fits to data of (a) Yi et al., and (b) Davidson et al.

Acknowledgements

The author would like to thank Prof. Peter D. Lee (Imperial College) for sharing his data, Profs. John Campbell (University of Birmingham, UK) and Anthony Rollett (Carnegie Mellon University) for their comments on the manuscript.

References

1. J.T. Staley, Jr., M. Tiryakioğlu, J. Campbell, Mater. Sci. Eng. A, 465 (2007) 136.
2. C. Nyahumwa, N.R. Green, J. Campbell, Metall. Mater. Trans. A, 32 (2001) 349.
3. G.R. Wakefield, R.M. Sharp. Mater. Sci. Tech. 12 (1996) 518-522.
4. D. Casellas, R. Pérez, J.M. Prado, Mater. Sci. Eng. A, A398 (2005) 171.
5. Q.G. Wang, D. Apelian, D.A. Lados, J. Light Metals, 1 (2001) 73.
6. A.De S. Jayatilaka, K. Trustrum, J. Mater. Sci., 12 (1977) 1426.
7. J.D. Poloniecki,, T.R. Wilshaw, Nature Phys. Sci., 229 (1971) 226.
8. J.Z. Yi, P.D. Lee, T.C. Lindley, T. Fukui, Mater. Sci. Eng. A, A432 (2006) 59.
9. J.Z. Yi, Y.X. Gao, P.D. Lee, H.M. Flower, T.C. Lindley, Metall. Mater. Trans. A. 34A (2003) 1879.
10. M. Tiryakioğlu, Mater. Sci. Eng. A, A465 (2007) 287.
11. Y. Murakami, M. Endo, Intl. J. Fatigue 16 (1994) 163.
12. B. Epstein, Technometrics, 2 (1960) 27.
13. E.J. Gumbel, "Statistics of Extremes", Columbia University Press, 1958.
14. M. Tiryakioğlu, Mater. Sci. Eng. A 476 (2008) 174.
15. Y. Murakami, T. Toriyama, E.M. Coudert, J. Testing Eval. 22 (1994) 318.
16. S. Beretta, Y. Murakami, Fatigue Fract. Eng. Mater. Struc. 21 (1998) 1049.
17. Y. Murakami, S. Beretta, Extremes, 2 (1999) 123.
18. P. Juvonen, Ph.D. Dissertation, Helsinki University of Technology, 2004.
19. P. Paris, F. Erdoğan, Trans. ASME, J. Basic Eng. (1963) 528.
20. M.J. Couper, A.E. Neeson, J.R. Griffths: Fatigue Fract. Eng. Mater. Struct. 13 (1990) 213.
21. B. Skallerud, T. Iveland and G. Härkegård, Eng. Fracture Mech., 44 (1993) 857.
22. S.A. Barter, L. Molent, N. Goldsmith and R. Jones: J. Eng. Failure Analysis, 12 (2005) 99.
23. B.R. Crawford, C. Loader, A.R. Ward, C. Urbani, M.R. Bache, S.H. Spence, D.G. Hay, W.J. Evans, G. Clark, A.J. Stonham, Fatigue Fract. of Eng. Mater. Struc., 28 (2005) 795.
24. C.J. Davidson, J.R. Griffiths, A.S. Machin, Fatigue Fract. Eng. Mater. Struc. 25 (2002) 223.
25. Q.G. Wang, C.J. Davidson, J.R. Griffiths, P.N. Crepeau, Metall. Mater. Trans. B., 37B (2006) 887.
26. Y. Murakami, M. Endo, Intl. J. Fatigue, 16 (1994) 163.
27. Y. Murakami, JSME Intl. J., 32 (1989) 167.
28. Y.X. Gao, J.Z. Yi, P.D. Lee, T.C. Lindley, Acta Mater. 52 (2004) 5435.
29. J.Z. Yi, Ph.D. Dissertation, Imperial College, University of London, 2004.
30. T.W. Anderson, D.A. Darling, J. Amer. Stat. Assoc., 49 (1954) 765.
31. J. Campbell, in "Shape Casting: 3rd International Symposium", eds. J. Campbell, P.N. Crepeau, M. Tiryakioğlu, TMS, 2009.
32. M.A. Stephens, in "Goodness of Fit Techniques", eds, R.B. D'Agostino, M.A. Stephens, Marcel Dekker, 1986, p.97.

Shape Casting: The 3rd International Symposium
Edited by: John Campbell, Paul N. Crepeau, and Murat Tiryakioğlu
TMS (The Minerals, Metals & Materials Society), 2009

IMPROVEMENT OF AN EXISTING MODEL TO ESTIMATE THE PORE DISTRIBUTION FOR A FATIGUE PROOF DESIGN OF ALUMINIUM HIGH-PRESSURE DIE CASTING COMPONENTS

Christian Oberwinkler, Heinz Leitner, Wilfried Eichlseder

University of Leoben, Institute of Mechanical Engineering
Franz-Josef Str. 18, 8700 Leoben, Austria, Europe

Keywords: High pressure die casting, aluminium, porosity, fatigue

Abstract

The estimation of the fatigue life time of aluminium high-pressure die casting components requires the knowledge of the pore distribution. A basic model was derived from a hpdc plate using Self-Organizing Maps and statistical tools to compute a statistical distribution of the porosity within a well defined area.
A new component (especially designed for this project) has been used to test the applicability of the existing model.
An example will be presented to visualize how the estimated porosity distribution can be included into the computation of the safety against cyclic failure.

Introduction

The common procedure to include the pores into the computation of the safety against dynamic failure is to define two material datasets – one for the pore free surface layer and one for the pore afflicted base material [1]. A typical cross section through a hpdc component does not provide a homogeneous pore distribution. The porosity increases towards the centre of the component. The difficulty is now which regions should be used in order to measure the S-N curve for the pore afflicted base material. Therefore it is important to understand the pore distribution in the component for an improved computation of the safety against cyclic failure. The sensitivity of the safety against cyclic failure on the S-N curve of the pore afflicted material is shown in Fig. 1.

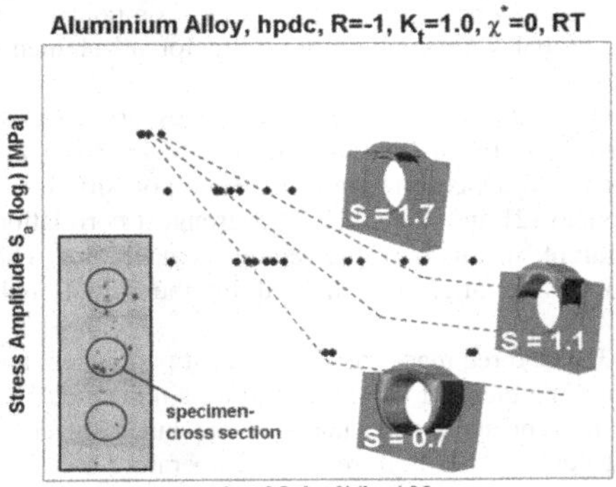

Fig. 1: Sensitivity of the safety against cyclic failure on the pore afflicted base material.

Klein and Plattenhardt [2] investigated the effect of different HPDC parameters on the density of the final component and found that the wall thickness and solidification pressure were the most important parameters. For lower solidification pressures, the temperature of the form plays also an important role in the development of porosity: the lower the temperature, the higher is the density. For thinner plates and high solidification pressures, the other parameters are of negligible effect.

Atwood and Lee [3] developed a model for the simulation of a three-dimensional gas pore growing under varying conditions (e.g., thermal history, pressure, hydrogen content). The model also takes into account solidified material and the restriction this poses on the growth of gas pores.
In the present paper an existing porosity model to compute the pore distribution throughout an aluminium hpdc component is validated using a complex geometry.

Porosity Model

The porosity model has been developed to compute the pore distribution throughout a real component for a fatigue proof design. A simple plate with the dimensions of 115x140x20 mm has been used to keep the number of influencing factors low and derive a porosity model. Three qualities have been casted with varying piston velocity and solidification pressure resulting in varying pore contents in the plates.
In order to obtain the information on the pore distribution for the empirical model one plate from each quality was cut 12 times. Each of the resulting surfaces were ground, polished and analysed for the pores in the surface. This provided a dataset with the location and size of the pores in each surface. This pore data cannot be compared directly with the results from a casting simulation. Therefore a grid of the same size as in the casting simulation was put on the surfaces and the pores within each element led to a porosity in each element as shown in Fig. 2.

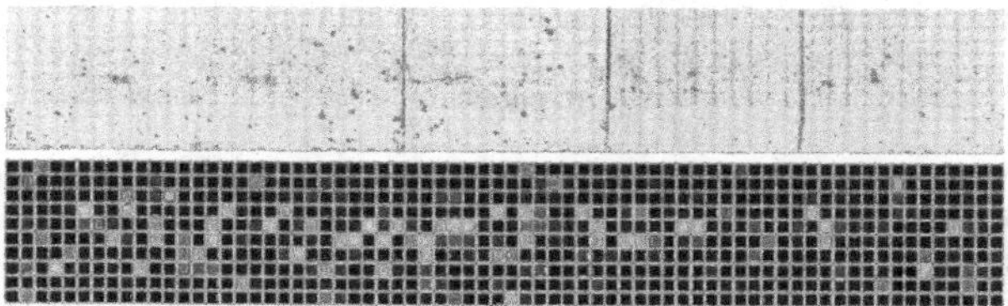

Fig. 2: Pore distribution (upper) and calculated porosity distribution (lower) for quality 2.

In addition a casting simulation was performed to compute the local conditions during the filling and the solidification using MAGMASOFT® v4.4. Finally each element has the information from the casting simulation together with the real porosity from the casted component. In addition the process parameters are known as global values which are valid for all elements within a component of a particular batch.
This data has been analysed using self-organizing maps. Self-organizing maps are used to visualize and group high dimensional data sets. The results are that the parameter from the simulation of the filling process did not show an obvious correlation to the final porosity in the hpdc plate. This finding is also supported by Klein [2] and Viets [4]. The strongest correlation could be observed to the solidification temperature, in this case the relative temperature. The relative temperature is the solidification temperature after 1 s divided by the initial melt temperature.
The final self-organizing map is shown in Fig. 3. The three maps for the different parameters are built of nodes (hexagons). Each node represents one element from the casting simulation. The node at the same location in the different maps represents the same element but a different parameter. Cluster 1 for instance shows the elements with a high relative temperature and as a consequence a low cooling rate. The resulting porosity is the highest compared to the cluster 2-8. In the opposite cluster 9 has a low relative temperature because of the high cooling rate. The resulting porosity is zero for most of the elements.

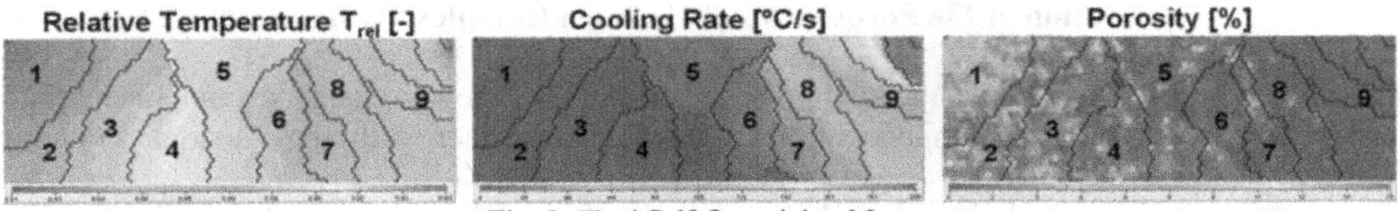

Fig. 3: Final Self-Organizing Map.

It is clear that the relative temperature and the cooling rate show an inverse trend - the higher the cooling rate the lower the relative temperature. Therefore the cooling rate has been disregarded for the further development of the porosity model.
A Weibull distribution was used to describe the porosity distribution within each particular cluster as shown in Fig. 4 (left). The presented data are from one casting quality. The lines are the regression lines to fit the real data. This has been done for all three qualities.
The final model to compute the porosity in dependence to the relative temperature (from the casting simulation) and the solidification pressure (from the casting process) is shown below.

$$\begin{aligned}
&for \quad T_{rel} \leq 0.927 \\
&\quad k_{Weibull} = \left(k_1 p_{after} + d_1\right) T_{rel} + \left(k_2 p_{after} + d_2\right) \\
&\quad d_{Weibull} = \left(k_3 p_{after} + d_3\right) T_{rel} + \left(k_4 p_{after} + d_4\right) \\
&for \quad T_{rel} > 0.927 \\
&\quad k_{Weibull} = const \\
&\quad d_{Weibull} = 0.927\left(k_{Weibull}^{T \leq 0.927} - k_{Weibull}^{T > 0.927}\right) + d_{Weibull}^{T \leq 0.927}
\end{aligned} \qquad \text{Equ. 3}$$

$k_{Weibull}$ and $d_{Weibull}$ are the slope and the y-intercept of the regression line in Fig. 4 (right) respectively. p_{after} and T_{rel} are the solidification pressure and the relative temperature. k_1 to k_4 and d_1 to d_4 are the fitting parameters for the aluminium alloy used (AlSi9Cu3).
The porosity in dependence on the solidification pressure and the relative temperature for a probability P of 90% is given in Fig. 4 (right). This means 90% of the porosity data are located below and 10% are above the computed surface. The real data are indicated by the points.

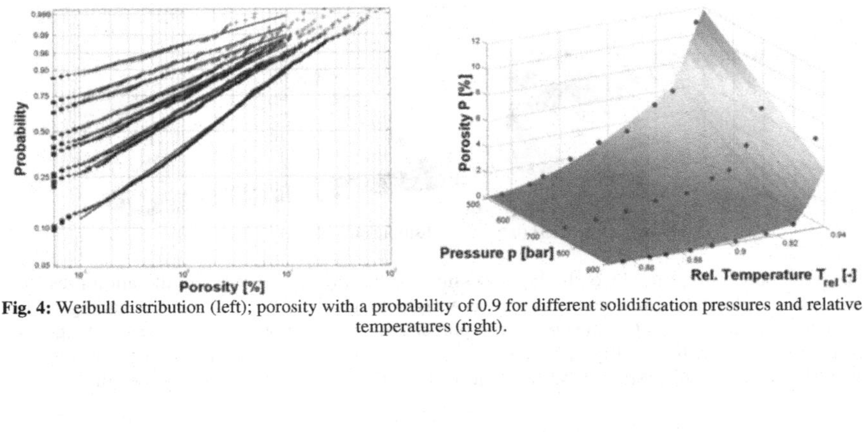

Fig. 4: Weibull distribution (left); porosity with a probability of 0.9 for different solidification pressures and relative temperatures (right).

Validation of the Porosity Model Using a Complex Component

The validation of the porosity model has been done with a newly designed component. The design process of the new component has been done together with the Austrian Foundry Research Institute (OEGI).
The porosity model was derived based on a simple plate to keep the number of influencing factors as low as possible and to derive a first basic understanding of the basic relationships. For the validation of the model a complex geometry was required. The newly designed component provides wall thicknesses from 15 to 6mm, different feeding times and entrapped air in a defined region. A comparison of simulation and reality is shown in Fig. 5 for the area with the entrapped air.

Fig. 5: Entrapped air – simulation (left), real component (right)

The investigated areas of the new component are presented in Fig. 6. Area 1 represents the volume with the biggest wall thickness (15mm). Area 2 is 12mm and area 3 8mm thick. This will allow investigating the influence of the wall thickness on the pore development.

Fig. 6: Investigated areas of the new component

The upper part of the new hpdc part is designed to perform component tests. The hole serves as a cavity for an axis to load the component. The component test and the stress distribution of the upper part are shown in Fig. 7.

Fig. 7: Component test (left); stress distribution in the upper part (right).

The components were produced by high pressure die casting at different solidification pressures and piston velocities. Each component was weight to get first "global" understanding of the influence of the process parameters on the porosity in the component. The average weight for each batch is shown in Fig. 8 (the case with a low piston velocity and a low solidification pressure has not been casted). It shows that the solidification pressure has a major influence of the component weight - the higher the pressure the higher the weight of the component. This means that if a higher pressure is applied more melt is squeezed into the solidifying volumes and the number of pores can be reduced. The velocity of the piston during the filling process seems to have only a minor influence compared to the solidification pressure.

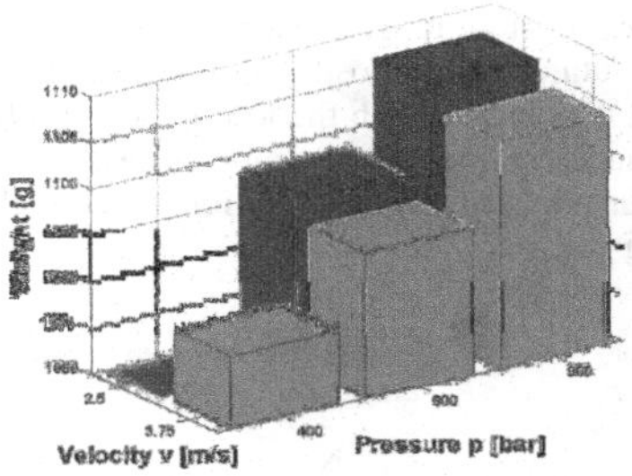

Fig. 8: Component weight in relation to the process parameters.

Therefore it was decided to take one component from selected batches (shown in Table 1) for the further investigation.

Quality	Piston velocity	Pressure
1	2.5 m/s	900 bar
2	2.5 m/s	600 bar
3	3.75 m/s	400 bar

Table 1: Process parameters of the investigated components.

The three components are referred to as quality 1 to 3 in the sections below. From each component three areas (Fig. 6) with different wall thickness were ground, polished and analysed for the pores in the cutting plane. The resulting pore distribution in the three components for area 1 is shown Fig. 9. It can be easily observed that the number and size of pores increases from quality 1 to 3. This correlates well with the applied solidification pressure.
In addition a simulation model of the casting process was created using MAGMASOFT® v4.4 to obtain the required values for the estimation of the pore distribution in the component.
As for the derivation of the porosity model the data from the casting simulation and the real porosity is merged into one dataset.

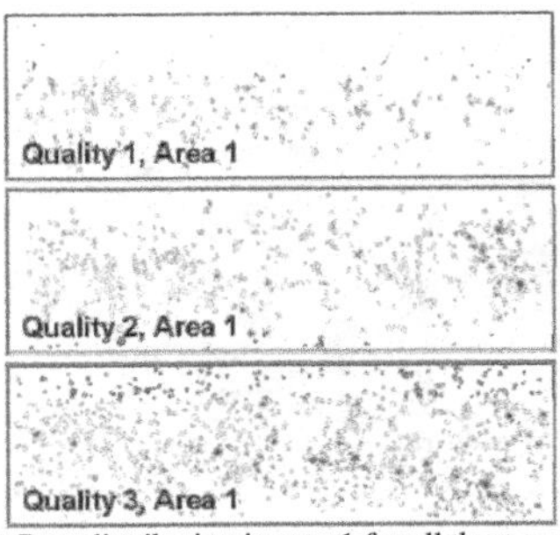

Fig. 9: Pore distribution in area 1 for all three qualities.

This dataset is used to validate the existing porosity model. The expected porosity distribution in a defined temperature range can be computed using the equations summarized in Equ. 3. In addition the real porosity data from the complex component is plotted to see if they fall into the same range as the prediction of the model. The results are shown in Fig. 10 for quality 3 for all three investigated areas. The data for area 1 and 2 shows that they are well within the predicted porosity distribution and follow the expected trends. In area 3, because of the wall thickness of only 8 mm, exhibits only the temperature range of two clusters. The inner regions with the high relative temperature and the resulting high porosity do not exist. Therefore only two lines exist from the real data representing the clusters with the lowest porosity from the SOM in Fig. 3. In this case the presented model underestimates the existing porosity in the area with the small wall thickness in the region of lower porosity. The distribution of the real porosity converges to the computed trend with increasing porosity.

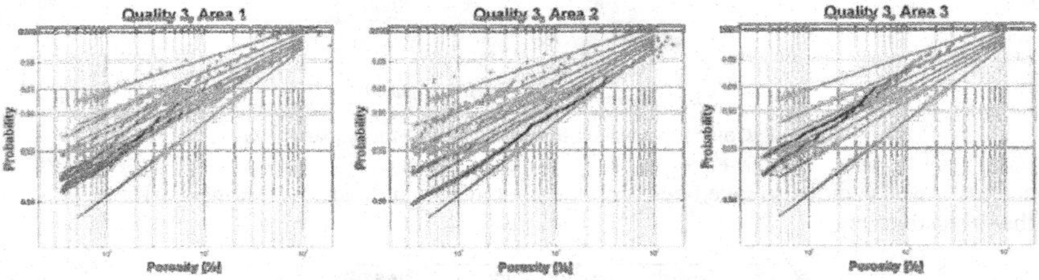

Fig. 10: Weibull distribution for quality 3 and all three investigated areas.

A 2-dimensional pore distribution has been computed for quality 1 in area 1 in order to compare it to the real pore distribution. The pore distribution can be computed using the porosity distributions from the model. Each relative temperature range has one defined porosity distribution. The number of elements within a temperature range is known. The number of elements allows to randomly distributing pores to the elements according to the porosity distribution defined by the porosity model. A good agreement has been achieved according to the results presented in Fig. 11.

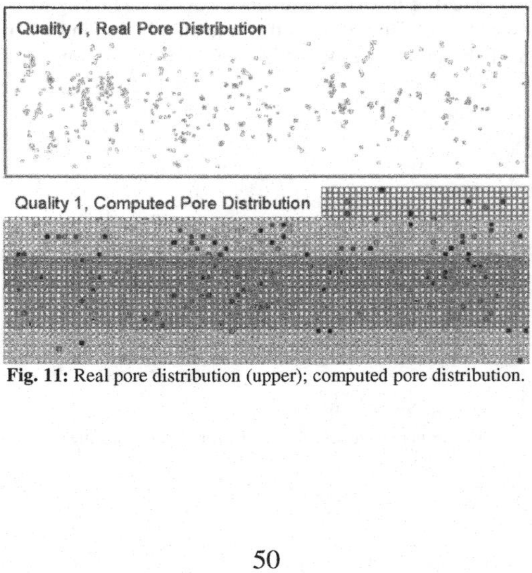

Fig. 11: Real pore distribution (upper); computed pore distribution.

Safety Against Cyclic Failure

The safety against cyclic failure is defined as the ratio of the bearable load to the real load defined for a number of load cycles. In the case of high pressure die casted components the inherent defects drastically influence the local fatigue strength.

From the porosity model the porosity distribution is known within defined temperature regions for a particular solidification pressure. As the probability of a certain porosity range is known for a defined temperature range, the porosity can be assigned on the elements in the component accordingly. This process is random, as it is not possible to predict the location of a pore in a component. The constant factor is the porosity distributions for each temperature range.

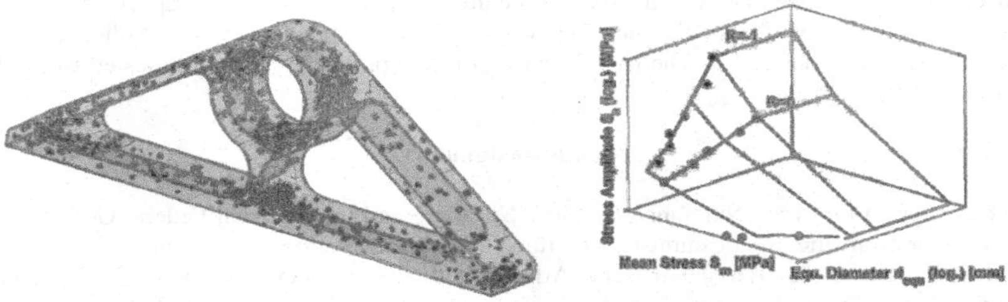

Fig. 12: Pore distribution in the component (left); Kitagawa-Haigh diagram (right).

The local equivalent pore diameter (Fig. 12 left) together with the Kitagawa-Haigh [5] diagram (Fig. 1 right) allows computing the local fatigue strength for 10^7 load cycles. The safety factor against dynamic failure is computed using the local fatigue strength and the existing stress amplitude.

As explained previously the pores within the defined regions are distributed randomly. This means for the computation of the safety factor that if only one computation is done the most critical pore could lie in a critical (low safety factor) or uncritical area (high safety factor). Therefore it is necessary to run multiple times through the assignment of the pores to cover all possible scenarios. The resulting distribution of the safety against cyclic failure for all three qualities is shown in Fig. 13.

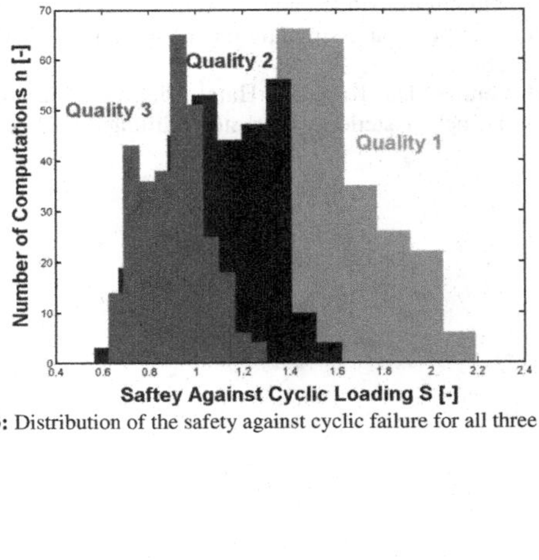

Fig. 13: Distribution of the safety against cyclic failure for all three qualities.

Conclusion and Outlook

An improved computation of the safety against cyclic failure requires the knowledge of the pore distribution in the component. A basic porosity model derived from a simple hpdc plate has been applied on a complex hpdc component. The results showed that the model derived from the plate can be transferred – with certain restrictions – to a complex geometry.
The investigations showed that the solidification pressure has a major influence on the global porosity in the component. The local pore distribution in the component is defined by the temperature distribution. One other major influencing has not been captured by the model – the trapped air due to poor venting. Nevertheless the entrapped air should not have an influence on the fatigue proof design of a hpdc components utilizing the presented model, as the casting engineer should take care that the critical areas are do not get weakened by entrapped air.
The presented work flow for the fatigue proof design of hpdc components will be validated using the results of a component test. The transferability of the approach is currently tested using the magnesium alloy AZ91hp.

Acknowledgment

Financial support by MAN Nutzfahrzeuge AG, Nürnberg and the Austrian Federal Government (in particular from the Bundesministerium für Verkehr, Innovation und Technologie and the Bundesministerium für Wirtschaft und Arbeit) and the Styrian Provincial Government, represented by Österreichische Forschungsförderungsgesellschaft mbH and by Steirische Wirtschaftsförderungsgesellschaft mbH, within the research activities of the K2 Competence Centre on "Integrated Research in Materials, Processing and Product Engineering", operated by the Materials Center Leoben Forschung GmbH in the framework of the Austrian COMET Competence Centre Programme, is gratefully acknowledged.

References

[1] W. Aichberger, H. Riener, H. Dannbauer:"Regarding Influences of Production Processes on Material Parameters in Fatigue Life Prediction", SAE 2007 World Congress, Detroit
[2] F. Klein, F. Plattenhardt: "Influence of die filling on the density and volume of pressure die castings", Giessereiforschung 45, p. 115–124, 1993
[3] R.C. Atwood, P.D. Lee: "Simulation of the three-dimensional morphology of solidification porosity in an aluminium-silicon alloy", Acta Materialia 51, p. 5447–5466, 2003
[4] R. Viets:"Integration der Fertigungssimulation bei der Audi AG", MAGMASOFT®-Expertenforum, Aachen, 2008
[5] C. Oberwinkler, H. Leitner:"Das Kitagawa-Haigh Diagramm für die Berechnung der Lebensdauer von Aluminiumdruckgussteilen", Werkstoffprüfung 2008, Berlin, 2008

Shape Casting: The 3rd International Symposium
Edited by: John Campbell, Paul N. Crepeau, and Murat Tiryakioğlu
TMS (The Minerals, Metals & Materials Society), 2009

Advanced Cast Aluminum Alloys

Alan P. Druschitz[1], John Griffin[1]

[1]University of Alabama at Birmingham,
1530 3rd Avenue South, Birmingham, AL 35294-4480, USA

Keywords: casting, aluminum alloy, high strength

Abstract

A recent advancement in cast aluminum alloys has demonstrated that complex shapes can be cast from a microalloyed Al-Cu alloy in dry sand molds with chills and that these castings can be heat treated to produce mechanical and physical properties nearly comparable to wrought 2519 aluminum alloy. Given this initial level of success, further research has been focused on improving this microalloyed Al-Cu alloy so that the mechanical properties consistently meet or exceed those of wrought 2519 alloy. Further, new research has been initiated on ultra-high strength, microalloyed Al-Zn-Mg-Cu alloys with the goal of producing complex castings with properties significantly better than wrought 2519 aluminum alloy and equivalent to or better than the best 7000 series wrought alloys. The development of the appropriate chemistries, casting practices and heat treatments are described in this paper.

Introduction

Wrought aluminum alloy RSA 708 [1] is the highest strength commercially available aluminum alloy and is produced by rapid solidification (melt spinning) followed by extrusion. This production route has demonstrated that aluminum alloys with yield strengths in excess of 690 MPa with good elongation (reportedly 8%) are possible. Wrought 7055 aluminum alloy is the highest strength conventionally processed, commercially available, wrought aluminum alloy [2]. The yield strength of this alloy is less than the rapidly solidified alloy but still about 50% higher than wrought 2519 aluminum alloy. However, the entire 7000 series of aluminum alloys have poor-to-fair general corrosion resistance and poor-to-good stress corrosion cracking resistance. Wrought 2519 aluminum alloy has good strength, good ballistic performance, good stress corrosion cracking resistance but only fair general corrosion resistance. Despite the fair general corrosion resistance, wrought 2519 aluminum alloy is currently used for General Dynamic's amphibious Expeditionary Fighting Vehicle [3]. Wrought 5083 aluminum alloy is widely used for lightweight military armor applications, has good general corrosion resistance but low strength. Wrought 7039 aluminum alloy is starting to be used for lightweight military armor applications, has good stress corrosion cracking resistance but poor general corrosion resistance. BAC of VA, LLC developed a modified version of wrought 2519 aluminum alloy called BAC 100™, a casting production process and thermal mechanical treatments that produce shaped components nearly comparable to the strength and ballistic performance of wrought 2519 aluminum alloy [4]. Preliminary studies with cast aluminum alloys containing zinc, magnesium and copper have demonstrated that high strength is possible but the tensile ductility has been unacceptably low and needs to be significantly improved. Table 1 is a comparison of the properties of the above mentioned materials.

Table 1. Property Comparison of Some Aluminum Alloys

Alloy	Form	Yield Strength, MPa	Hardness, BHN	General Corrosion Resistance	Stress Corrosion Cracking Resistance
5083	plate	198-280 typ	81-93 typ	excellent	good
A356-T6	casting	210 min	90 min	good	na
High Toughness Al-Cu alloy	casting	330 typ	110-130 typ	na	good (~207 MPa)
7039-T64	plate	380 typ	133 typ	poor	good
High Strength Al-Cu alloy	casting	400 typ	130-140 typ	na	good (~275 MPa)
2519-T87	plate	400 min	130 min	fair	good
Al-Zn-Mg-Cu alloy	casting	>493	>160	na	na
7055-T7751	plate	614 typ	na	fair	poor (103 MPa)
RSA 708 T6	extrusion	700 typ	230 typ	na	na

Wrought aluminum alloys (such as 5083, 2519, 7039, 7055, etc) can provide a desirable combination of properties, but, wrought alloys are only available in plate or billet form. Extensive machining of a plate or billet, which is time consuming, costly, and generally restricted to relatively simple shapes that do not have internal passageways, is required to produce a structural component from these alloys. Advanced aluminum casting alloys with enhanced mechanical, physical and ballistic properties would solve this problem. The inherent design flexibility of the casting process would allow for near-net shape structural components to be manufactured with significant cost and weight savings over traditional wrought aluminum alloys. In addition, the ability to cast complex shapes would allow the integration of a number of parts into a single component and thus eliminate expensive weldments and assemblies.

Microalloyed Aluminum-Copper Alloys

Alloy Concept. A new family of microalloyed aluminum-copper alloys was developed in 2005 [4] with improved mechanical properties and improved resistance to hot tearing compared to aluminum alloys 201 and 206. During the development of this new alloy, laboratory experiments were performed to determine the effects of seven potential alloying elements (Cu, Ag, Cr, Mg, Mn, V, Zr). Concurrently, trials were run at a production foundry to determine castability and hot tearing tendency. Both high toughness and high strength variants of this alloy were developed.

Experimental Methods. Twenty-three, 1.1 kg (2.5 lb) heats of microalloyed Al-Cu alloys were made with P1020 ingot (commercially pure aluminum), Al-50%Cu master alloy, Al-20%Cr master alloy, Al-25%Mn master alloy, Al-5%V master alloy, Al-5%Zr master alloy, pure Mg ingot and pure Ag ingot. The metal was melted in a crucible, grain refined with either Al-5%Ti-1%B or Al-3%Ti-1.5%C and poured into Y-blocks that had a copper chill for the base. A spectrometer sample was also poured from each heat. The Y-blocks were hot isostatically pressed (HIP) at 510-524C (950-975°F) and 103 +/- 3.4 MPa (15,000 +/- 500 psi) for 2-3 hours at a commercial HIP'ing service center. The Y-blocks were then sectioned and the samples heat treated to produce the T4, T6 or T7 temper. The solution treatment was 510-516°C (950-960°F) for 2-4 hours followed by 529-535°C (985-995°F) for 16-20 hours then quench in warm water at

60-82°C (140-180°F). For the T4 temper, samples were allowed to age at room temperature for a minimum of 7 days. For the T6 temper, samples were aged at room temperature for a minimum of 24 hours followed by artificial aging at 160-166°C (320-330°F) for 30 hours. For the T7 temper, samples were aged at room temperature for a minimum of 24 hours followed by artificial aging at 196-202°C (385-395°F) for 24 hours. The samples were then machined into tensile bars and tested in accordance with ASTM E-8. Samples were also prepared using standard metallographic techniques and then examined on a Philips 515 scanning electron microscope equipped with an energy dispersive X-ray spectrometer. Semi-quantitative elemental analysis was performed on the particles present.

To evaluate castability in a production environment, a complex 4.3 kg (9.5 lb) seat frame casting was produced. Molds were made from chemically bonded lake sand. Insulated riser sleeves and steel chills were incorporated into the molds. Production quantity heats of 275 kg (600 lbs) were produced from selected alloys, degassed with argon for 10-12 minutes, grain refined with either Al-5%Ti-1%B or Al-3%Ti-1.5%C and molds poured at 721-754°C (1330-1390°F). Castings were examined for hot tears and then the castings were HIP'ed, heat treated and sectioned for determination of mechanical properties. The mold and castings are shown in Figures 1a&b.

(a) seat frame casting being poured (b) seat frame castings after shake-out

Figure 1. (a) Seat frame casting being poured in a production foundry and (b) castings after shake-out.

Results & Discussion. The laboratory experiments revealed the individual effects of seven elements. Cu did not have strong effect on mechanical properties. Ag had a strong positive effect on UTS and YS in the T6 and T7 tempers (as expected) but no effect on mechanical properties in the T4 condition. Cr and Mg had a negative effect on mechanical properties in all tempers. Mn had a positive effect on UTS and YS in the T4 temper and the unusual effect of increasing UTS and elongation while decreasing YS in the T6 and T7 tempers. V, in general, had a negative effect on mechanical properties except for improving elongation in the T6 and T7 tempers. Zr had a positive effect on all mechanical properties in all tempers. The results of the laboratory experiments are shown in Table 2.

Table 2. Results of Laboratory Experiments to Determine the Individual Effects of Seven Elements on the Mechanical Properties of an Al-Cu Alloy in the T4, T6 and T7 Tempers.

		T4			T6			T7		
Element	Range, wt%	UTS	YS	El	UTS	YS	El	UTS	YS	El
Cu	4.5-6.7						-		+	-
Ag	0-0.40				+	+	-	+	+	
Cr	0-0.50	-		-	-	-	-	-	-	-
Mg	0.1-0.80	-		-	-		-	-	-	-
Mn	0.1-0.65	+	+		+	-	+	+	-	+
V	0-0.25	-	-	-	-		+	-	-	+
Zr	0-0.25	+	+	+	+	+	+	+	+	+

Basically, this new alloy is based on the Al-Cu system, which is known to produce high strength and high toughness, coupled with dispersoid strengthening concepts which improve yield strength without reducing ductility. Further, undesirable alloy interactions were accounted for and minimized. Two variants of this alloy were developed, high toughness and high strength. A typical chemistry for this alloy is listed in Table 3. The high toughness variant was produced by reducing the Cu content to less than about 5.60 wt% and eliminating the Ag addition. For the high strength variant, Cu content could be higher and up to 0.40 wt% Ag was added.

Table 3. Typical Chemistry for Microalloyed Al-Cu Alloy.

Element	Cu	Mg	Mn	Ti	V	Zr	Ag	Fe	Si
Wt%	5.73	0.22	0.32	0.07	0.09	0.20	0.20	0.12	0.01

The microalloyed Al-Cu alloy with the chemistry of Table 3 had a liquidus temperature of 640°C (1184°F), a small arrest at 552°C (1026°F) and a solidus temperature of 530°C (986°F). The freezing range for this alloy was 110°C (198°F), which is considered "long". The cooling curve for the microalloyed Al-Cu chemistry listed in Table 3 is shown in Figure 2.

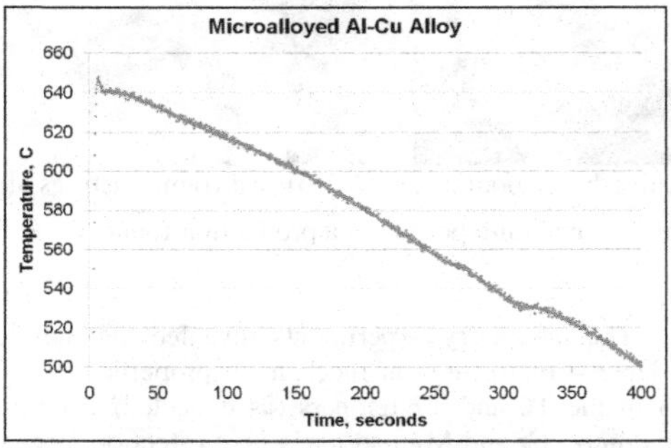

Figure 2. Cooling curve for microalloyed Al-Cu alloy showing a liquidus temperature of 640°C (1184°F), a small arrest at 552°C (1026°F) and a solidus temperature of 530°C (986°F).

The seat frame casting revealed that hot tearing tendency was a strong function of Cu content. Castings with a Cu content less than 5.3 wt% exhibited hot tearing and castings with a Cu content in excess of 5.3 wt% exhibited no hot tearing. A206 alloy castings (Cu content < 5.3 wt%) were also poured and all of these castings exhibited severe hot tearing.

The microstructure of the microalloyed Al-Cu alloys may contain "large" particles (up to 50 μm long) after heat treatment. The largest particles were $CuAl_2$ and the smaller particles generally contained Cu, Fe and Mn. The microstructure of this alloy is shown in Figures 3a&b.

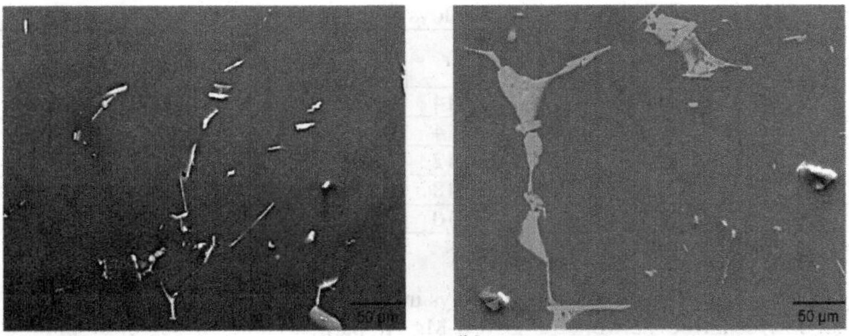

(a) (b)

Figure 3. Microstructure of microalloyed Al-Cu alloy after HIP'ing and heat treatment (T6 temper). (a) shows the generally small size of the particles present after heat treatment and (b) shows a cluster of large, interdendritic $CuAl_2$ particles that were not completely dissolved during the solution heat treatment.

Aluminum-Zinc-Magnesium-Copper Alloys

<u>Alloy Concept.</u> The 7000 series aluminum alloys have the highest strength for wrought aluminum alloys. Building upon the success with the Al-Cu alloys, a program was initiated to determine the feasibility of producing an aluminum casting alloy with significantly better mechanical properties based on the aluminum-zinc-magnesium-copper system. In this study, the ratio of Zn to Mg was chosen to maximize strength (high zinc content) and minimize excess magnesium. Assuming that the strengthening precipitate is Zn_2Mg [5], the calculated "ideal" Zn to Mg weight ratio was 5.39. The ratio of Zn to Mg for commercial 7000 series aluminum alloys was 1.43-4.28, which indicated an excess of Mg. The ratio of Zn to Mg for rapidly solidified commercial aluminum alloys was 2.20-4.78, which also indicated an excess of Mg. The addition of copper has been reported to increase strength but decrease general corrosion resistance if >3 wt% [5]; so, a target maximum of 2 wt% copper was chosen. According to the equilibrium phase diagram [6], an addition of 2 wt% Cu should go completely into solid solution above ~425°C (800°F) and then reprecipitate during low temperature aging at 120°C (250°F). Past experience with zirconium demonstrated improved strength and ductility in Al-Cu alloys [4], so a target of 0.15-0.25 wt% Zr was chosen.

<u>Experimental Methods.</u> Five, 9 kg (20 lb) heats of Al-Zn-Mg-Cu alloys were made with P1020 ingot (commercially pure aluminum), ZA27 ingot, Al-50%Cu master alloy, Al-10%Zr master alloy and pure Mg ingot. The alloys were melted in a SiC crucible, degassed with nitrogen for 2-3 minutes, grain refined with an addition of 1.8 grams of Al-3Ti-1B per kg of alloy and then poured at 704-718°C (1300-1325°F). A thermal analysis sample and a spectrometer sample were poured from each heat. The liquidus, solidus and chemistry of each heat are listed in Table 4.

The alloys were cast in chemically bonded sand Y-block molds with a steel chill for the base. The cast samples were solution treated at 454°C (850°F) for up to 24 hours, quenched in warm water, aged at room temperature for 24 hours and then artificially aged at 120°C (250°F) for 24 hours. Samples were prepared using standard metallographic techniques and then examined on a Philips 515 scanning electron microscope equipped with an energy dispersive X-ray spectrometer. Semi-quantitative elemental analysis was performed on the particles present.

Table 4. Chemistry* of Cast Al-Zn-Mg-Cu Alloys (wt %).

Heat No.	Zn	Mg	Cu	Zr	Liquidus, °C	Solidus, °C	Length of Solidus Reaction, seconds
1	10.93	1.46	1.93	0.14	627	462	7.25
2	10.57	2.13	1.87	0.14	623	467	20.25
3	11.02	2.77	1.41	0.12	620	470	39.50
4	8.95	2.30	0.92	0.18	628	471	18.25
5	7.97	1.76	1.58	0.10	629	467	11.75

* determined by NSL Analytical, Cleveland, OH

Results & Discussion. The Al-Zn-Mg-Cu alloys investigated had a very long freezing range and formed an interdendritic network of Zn-Cu-Mg-Al particles due to segregation of alloying elements during solidification. Microporosity was also present in all of the samples because of the long freezing range and lack of isothermal solidification. Surprisingly, the liquidus and solidus temperatures did not show a large variation despite the Zn-content varying from 8-11 wt%, the Mg-content varying from 1.5-2.8 wt% and the Cu-content varying from 0.9-1.9 wt%. However, the length of the solidus reaction varied significantly (from 7.25-39.5 seconds). The best correlation between chemistry and the length of the solidus reaction was the Mg content (higher Mg content produced long solidus reaction time). Suppressing the segregation of alloy elements would assist in achieving good mechanical properties. Cooling curves for the Al-11.0Zn-2.8Mg-1.4Cu and Al-10.9Zn-1.5Mg-1.9Cu alloys, which had long and short lengths of solidus reaction respectively, are shown in Figures 4a&b.

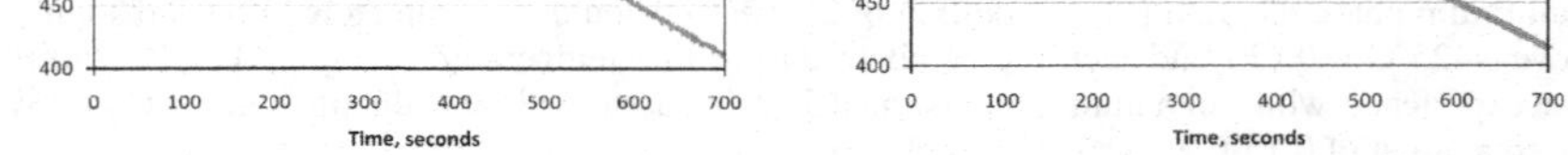

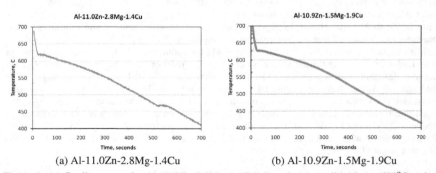

(a) Al-11.0Zn-2.8Mg-1.4Cu (b) Al-10.9Zn-1.5Mg-1.9Cu

Figure 4. (a) Cooling curve for Al-11.0Zn-2.8Mg-1.4Cu alloy showing a liquidus at 620°C and a solidus at 470°C and a "long" solidus reaction time of 39.5 seconds. (b) Cooling curve for microalloyed Al-10.9Zn-1.5Mg-1.9Cu alloy showing a liquidus at 627°C, a solidus at 462°C and a "short" solidus reaction time of 7.25 seconds.

The as-cast microstructure of the Al-Zn-Mg-Cu alloys was similar to the as-cast microstructure of 7000 series aluminum ingots [8]. The heat treatment study revealed that the as-cast interdendritic network was not completely dissolved using a solution treatment temperature of 454°C and a time of 24 hours; additional work is needed to determine the required heat treated to completely eliminate these particles. For the 10.6Zn-2.1Mg-1.9Cu alloy, SEM analysis determined that the predominant intermetallic particles present at the interdendritic boundaries contained Zn, Cu, Mg and Al. There were lesser amounts of two other intermetallic particles; one contained high amounts of Fe and the other contained Al and Cu, presumably $CuAl_2$. After heat treatment, the chemistries of the remaining intermetallic particles were the same, i.e., none of the phases were completely redissolved. For the 9Zn-2.3Mg-0.9Cu alloy, SEM analysis determined that the predominant intermetallic particles present at the interdendritic boundaries contained Zn, Cu, Mg and Al. Also, there was one other type of intermetallic particle that contained high amounts of Fe. The Al and Cu containing intermetallic particles were not present for this composition. After heat treatment, the chemistries of the remaining intermetallic particles were the same, i.e., none of the phases were completely redissolved. Figures 5 and 6 show the as-cast and heat treated microstructures of two of the alloys investigated.

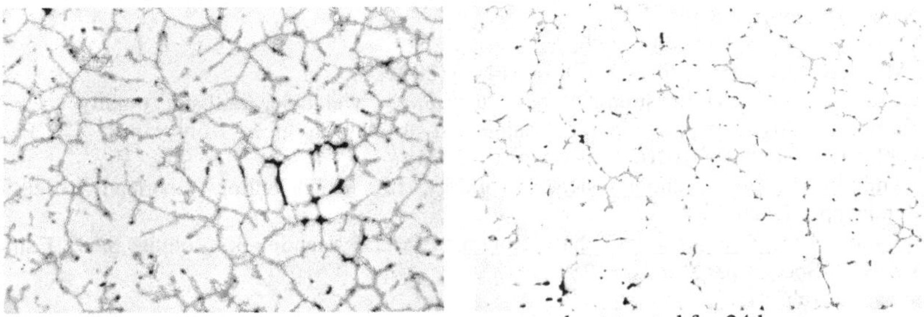

as-cast heat treated for 24 hours

Figure 5. Al-10.6Zn-2.1Mg-1.9Cu as-cast and heat treated microstructures. The Zn-Cu-Mg-Al intermetallic particles present in the as-cast condition were not completely eliminated after 24 hours at 454°C.

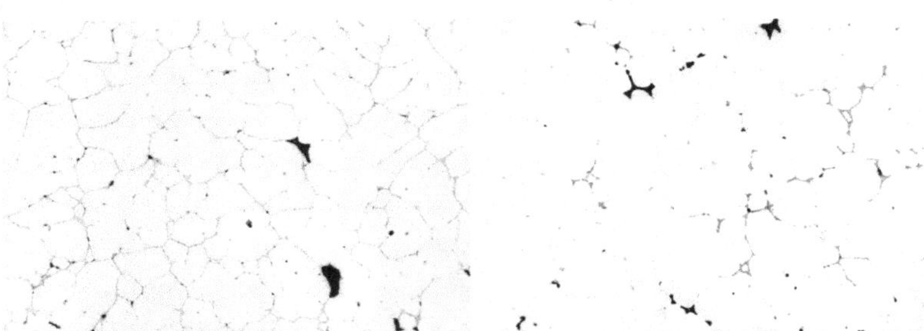

as-cast heat treated for 24 hours

Figure 6. Al-9Zn-2.3Mg-0.9Cu as-cast and heat treated microstructures. The Zn-Cu-Mg-Al intermetallic particles present in the as-cast condition were not completely eliminated 24 hours at 454°C.

Little work has been done on determining mechanical properties since the desired microstructure (no massive intermetallic particles) has not been obtained. However, it has been determined that the yield strength of the Al-10.9Zn-1.5Mg-1.9Cu alloy was greater than 493 MPa and the hardness was 165-170 BHN (500 kg load, 10 mm diameter ball), Table 1. The strength and hardness of this cast alloy were significantly higher than wrought alloys 7039 and 2519. However, the tensile ductility was unacceptably low (essentially zero). Presumably, the reason for the low ductility was the combination of microporosity and Zn-Cu-Mg-Al intermetallic particles.

Future work will include HIP'ing and pressure solidification to eliminate porosity, optimized heat treatment to eliminate the Zn-Cu-Mg-Al intermetallic particles and chemistry optimization to minimize and/or eliminate the Zn-Cu-Mg-Al intermetallic particles in the as-cast condition.

References

1. RSP Technology BV, Metaalpark 2, 9936 BV Farmsum, the Netherlands (www.rsp-technology.com).
2. Alcoa Alloy 7055-T7751 Tech Sheet.
3. AMPTIAC Quarterly, Vol. 8, No. 4, pp. 14-20 (2004).
4. Druschitz, A.P., "High strength, high toughness, weldable, ballistic quality, castable aluminum alloy, heat treatment for same and articles produced from same," US Patent Application No. 20070102071 (filed Nov. 9, 2005).
5. Aluminum: Properties and Physical Metallurgy, J.E. Hatch, editor, American Society for Metals, pp. 51-52 (1984).
6. Metallography, Structures and Phase Diagrams, Metals Handbook, Volume 8, 8th Edition, American Society for Metals, p. 259 (1973).
7. www.matweb.com
8. Aluminum: Properties and Physical Metallurgy, J.E. Hatch, editor, American Society for Metals, pp. 79-82 (1984).

SHAPE CASTING:
3rd International Symposium
2009

Processes

Session Chair:
Alan Druschitz

Shape Casting: The 3rd International Symposium
Edited by: John Campbell, Paul N. Crepeau, and Murat Tiryakioğlu
TMS (The Minerals, Metals & Materials Society), 2009

INFLUENCE OF HYDROGEN CONTENT AND BIFILM INDEX ON FEEDING BEHAVIOUR OF Al-7Si

Derya Dispinar[1], Arne Nordmark[1], Joruun Voje[2], Lars Arnberg[3]

[1]SINTEF, Materials and Chemistry;
7465 Trondheim, Norway
[2]Elkem Aluminium Research;
Kristiansand, Norway
[3]NTNU, Department of Materials Science and Engineering
7491 Trondheim, Norway

Keywords: Aluminium, Feeding, SiBloy,

Abstract

The relationship between 'hydrogen-porosity' and 'Sr modification-porosity' has long been investigated. In this study, this phenomenon has been investigated in terms of bifilm content. A feeding test with deliberately inadequate feeding to promote some degree of shrinkage porosity has been used to compare feeding and porosity of Al-7Si alloy in gravity die casting. A melt with three different hydrogen contents was prepared by degassing first and then upgassing to low, mid and high hydrogen levels (0.1, 0.2 and 0.3 respectively) with Ar-10%H_2 mixture. Reduced pressure test samples were taken for bifilm index calculation and 10 tensile test bars were cast into sand moulds for mechanical testing. The results showed that feedability was increased with B-grain refining (SiBloy alloy) than the conventional Ti-B grain refined. However, pore distribution was scattered in both cases when the alloys were Sr-modified.

Introduction

Casting of pore-free aluminium and its alloys has been the central focus for many years due to the highly critical applications such as automotive industry. Considerable effort has gone into understanding and controlling defects that are formed during casting. Extensive efforts have been made to investigate the feeding characteristics of aluminium alloys with respect to the alloying elements, hydrogen content, pouring temperature and so on. Grain refinement (transition from columnar to equaxied) and modification of Si crystals (transition from needle to fibrous) have been considered to provide benefits for the feeding process [1].

The most common grain refiners used for Al-Si alloys are Al-Ti-B master alloys. There are several theories about the nucleation process. Some researchers suggest that $TiAl_3$ nucleates α-Al grains. There is some evidence that TiB_2 are the favourable sites. It is also been argued that borides are pushed to the grain boundaries if no solute titanium is present, thus no grain refinement effect was observed [2, 3, 4]. There is also a negative aspect of Ti refinement which is known as 'fading'. $TiAl_3$ crystals settle at the bottom of the furnace within 20 minutes at around about 700°C [5, 6].

As an alternative to Ti grain refining, B refining have been investigated [2, 7, 8]. SiBloy is a patented Ti-free alloy based on the AlSi foundry alloy system with a permanent grain refining effect due to AlB_2-particles. The purpose of the present work is to investigate the effect of hydrogen on feeding behaviour in SiBloy, in order to improve our understanding of porosity formation in SiBloy and check out the possibility that bifilm index could assist in describing the porosity observed in SiBloy grain refined and commercial available A356 castings. In addition, the effect of modification with Sr over the feeding characteristics was also investigated.

Experimental Work

A resistance furnace was used to melt 65 kg of the alloy in a crucible at 725°C. The compositions of the alloys are given in Table 1. Both A356 and SiBloy alloys were provided by Elkem Aluminium, Norway.

Table I. Base alloy composition in wt%.

Alloy	Si	Fe	Mn	Mg	Zn	Ti	B	Na	Ca	Sr	P
SiBloy	7.38	0.088	0.017	0.28	0.005	0.0014	0.0175	0.003	0.000	-	0.0005
A356	7.03	0.091	0.008	0.41	0.005	0.11	-	0.0017	0.0014	0.0005	0.0002

The melts were initially degassed with a rotating impeller degas unit operating at 350 RPM using pure Ar (5 l/min). The hydrogen content was measured with ALSPEK-H and the aim was to achieve three different hydrogen levels: 0.1, 0.2 and 0.3 cc/100 grams of Al. The gas content of the melt was controlled by purging Ar-10% H_2 (10 l/min) through the rotary impeller.

At each hydrogen level, a feeding test was carried out where a deliberate inadequate feeding was promoted to form some degree of shrinkage porosity. The geometry (Figure 1) consisted of two cylindrical disks A and B with a diameter of 40 mm and heights of 17 mm and 12 mm, respectively. The disks were fed through two 4x40 mm plates in a V-shape with a common insulated feeder at the end. The moduli of the two disks A and B were 2.5 and 2.0 respectively, times the modulus of the 4 mm thick feeding plate, which resulted in some degree of shrinkage porosity in the cylindrical disks.

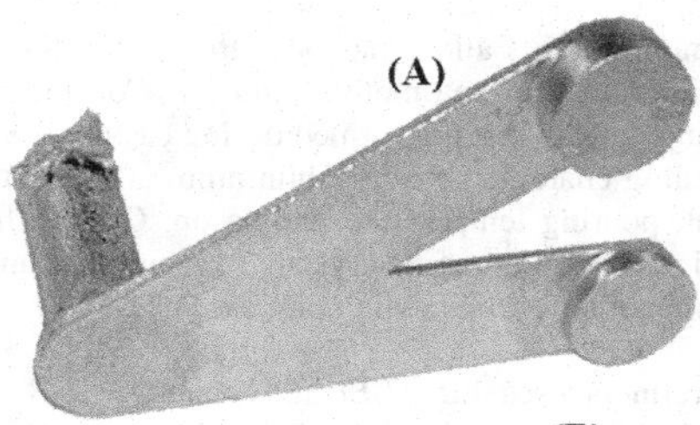

Figure 1. Picture of the casting: (A) 40x17, (B) 40x12

The permanent mould used in the experiments was made of steel and equipped with Ø10 mm cooling channels that were positioned 16 mm from the casting surface. The channels were connected to an oil heating/cooling unit by flexible tubes. In the present experiments, calorific oil was pumped through the system at a set temperature of 320 °C.

The two die halves were heated in a furnace to 180°C before the die surfaces were coated with an insulating coating. The total thickness of the coating was measured to be 200 - 250 μm. Prior to casting, a 25 mm (1") insulating sleeve was mounted in the upper die and was used both during filling and as a feeder during solidification. This sleeve was renewed for each casting.

Six samples were cast with each alloy and each hydrogen level. In the final stage, both alloys were modified with Sr and the feeding tests were completed.

In addition to feeding tests, reduced pressure test samples were collected from the melt with each casting trials to measure the metal quality. A sand mould was also prepared to produce 10 test bars (Ø12x150mm) in order to correlate the mechanical properties with bifilm index [9].

Results

During all the casting experiments, the pouring temperature, mould temperature and cavity temperature was kept constant and under control and was recorded on a datalogger.

For the feeding tests, the cylindrical disks (Fig. 1) were cut from the castings and subjected to density measurement and metallographic examination. Archimedes' principle was used to measure the density of the samples. To assess the overall amount of internal porosity, the difference between the volume associated with the surface sink and internal volume of porosity was measured as depicted in Fig. 2 [10]. The volume of external shrinkage and the volume of internal porosity can be computed from the measured disk volumes and densities by assuming a reference density for the pore-free material (calculated using the ALSTRUC model to 2669 kg/m3 for SiBloy) [11]. Volume fractions can be obtained by sub-dividing these volumes by the volumes of the cylindrical cavities A and B (Fig. 1). The density values are reported as mean values of four repeated measurements with the corresponding standard error in the mean. The external shrinkage includes the shrinkage in solid state from solidification (density 2580 kg/m^3) and to room temperature (density 2669 kg/m^3).

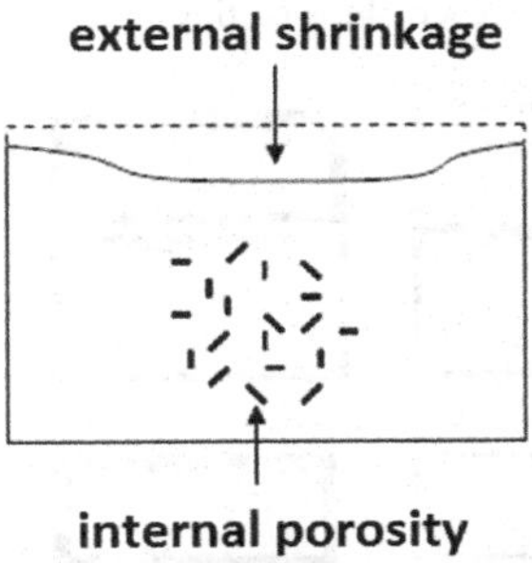

Figure 2. Definition of external shrinkage and internal porosity [10].

Figures 3 and 4 show the external and internal shrinkage together with total shrinkage (i.e. volume fractions) for disks A (17mm) and B (12mm) with the various hydrogen levels. Figure 5 shows the cross section of disks at various hydrogen levels showing the size and distribution of porosity.

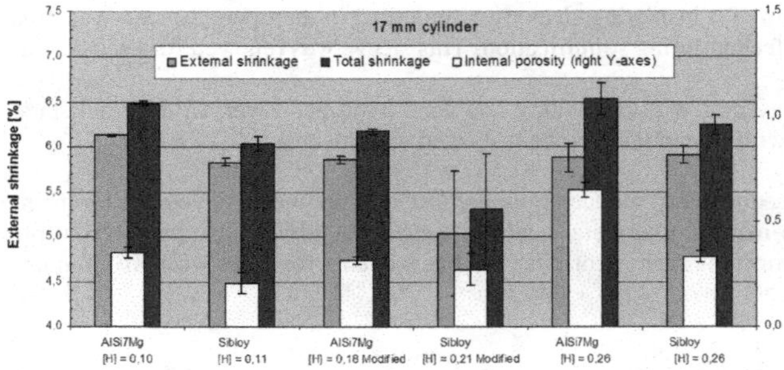

Figure 3. Volume fraction of external shrinkage (left Y-axis) and internal porosity (right Y-axis) and total shrinkage (left Y-axis) for each of the casting series (Disk A in Fig 1 – 17 mm disk)

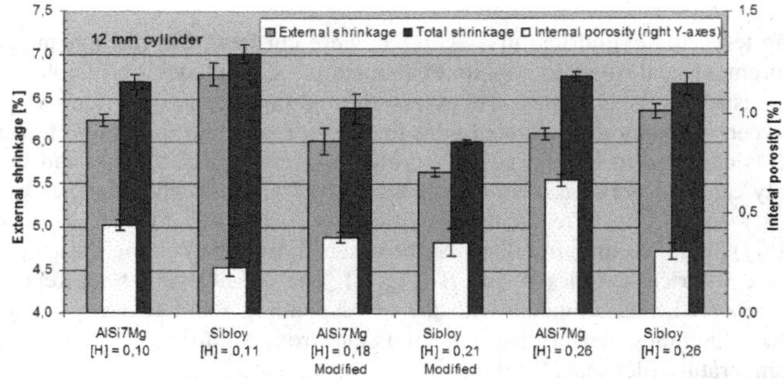

Figure 4. Volume fraction of external shrinkage (left Y-axis) and internal porosity (right Y-axis) and total shrinkage (left Y-axis) for each of the casting series (Disk B in Fig 1 – 12 mm disk)

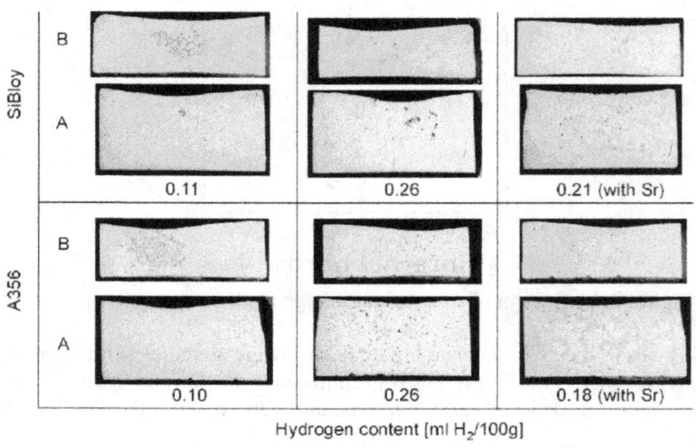

Figure 5. Macrographs revealing the distribution of porosity in cross-sections of randomly selected disks from the casting series

In Fig. 6, the bifilm index change of both SiBloy and reference A356 alloy with hydrogen is given. As seen in figure, SiBloy has lower bifilm index and with less scatter compare to A356. However, the mechanical test results show the other way around. In Fig 7a, the change in the elongation values at fracture are given as a function of hydrogen content. As seen from the figure, SiBloy alloy elongation values are more scattered but the trend shows an increasing value with increased gas content. On the other hand, A356 alloy has slightly higher elongation values but it decreases with increasing hydrogen content. The similar scenario exists for UTS values. As seen in Fig 7b, UTS values of SiBloy alloy increases with increasing hydrogen, whereas it decreases for A356.

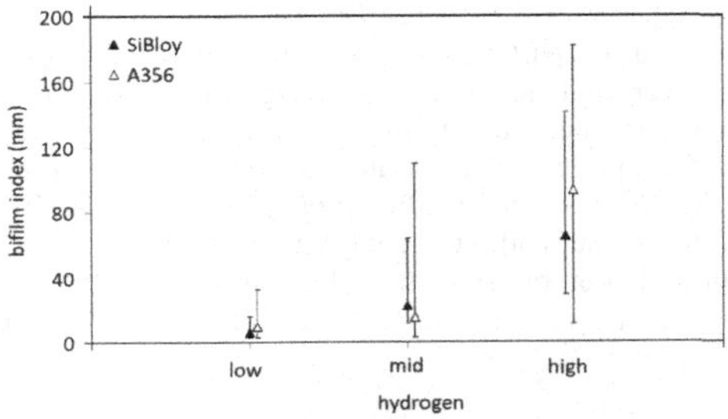

Figure 6. Bifilm index change with hydrogen content

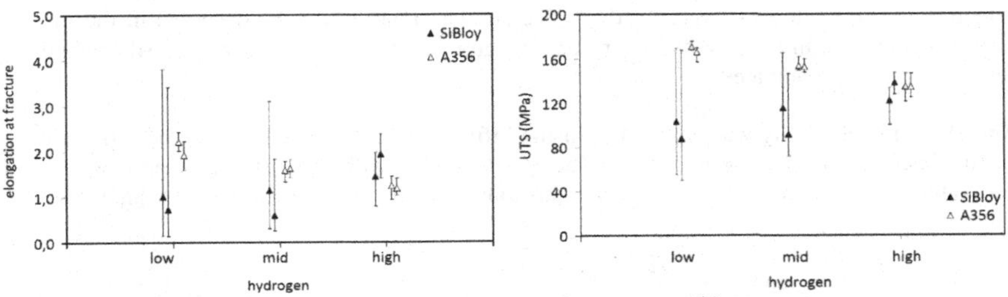

(a) (b)

Figure 7. Change in (a) elongation at fracture, (b) UTS

Discussion

Fig 3 and 4 show the volume fractions for disks A (17mm) and B (12mm) for SiBloy and A356 with the various hydrogen levels. In each case, the internal porosity of SiBloy alloy is lower than the A356 alloy. This can also be seen in the micrographs at the cross section of the castings in Fig 5.

At low hydrogen levels, typical central pores are observed at both castings. However, as the hydrogen content was increased, the amount of localized porosity remained unchanged in SiBloy alloy (Fig 5). On the other hand, the noticeable difference in A356 is the wide spread distribution of pores all over the cross section.

It is important to note that addition of strontium seems to reduce both the internal porosity and the external shrinkage for both alloy systems, which is visible in Fig 5 and in the bar charts in Fig 3 and 4.

The large standard error in the mean value on SiBloy with medium hydrogen content (0.21 ml/100g) is due to the fact that one of the four samples had a very low external shrinkage (3.0%) and a higher internal porosity (0.5%) than the other three samples of that series.

Even though castings of SiBloy show less porosity compared with conventional alloys, the porosity tends to be more concentrated which could cause problems in structural castings subjected to x-ray inspection.

With reference to Fig 3 and 4, apart from Disk B in the lowest hydrogen series, the SiBloy castings always have a lower total shrinkage. Assuming that the method for grain refining does not significantly affect the thermal conditions, which seems reasonable, this implies that the integrated rate of flow of liquid through the plates to feed shrinkage in the cylinders is larger in SiBloy. In other words, SiBloy has better feeding properties than the A356 reference. A plausible explanation can be that the smaller and more compact grains in SiBloy delay dendrite coherency and thus increase the solid fraction range for mass feeding. The measured average grain sizes for SiBloy and A356 was 200 μm and 258 μm respectively with SiBloy having more globular structure than A356.

Why the typical hydrogen porosity is not present in SiBloy is not clear. However, we might speculate if the boron has some influence on the nucleation of the hydrogen pores, *e.g.*, by affecting the formation of oxides/bifilms. The similar situation can be observed in the thinner Disk B (Figure 6). In the low-hydrogen cases (Figure 6), the difference between SiBloy and the reference is less pronounced.

Therefore, metal quality was checked by using bifilm index. As seen in Fig 7, the bifilm index, i.e. oxide content of the casting, has a tendency to increase with increasing time and gas content for both alloys. However, SiBloy alloy appears to have lower scatter than A356 for all cases.

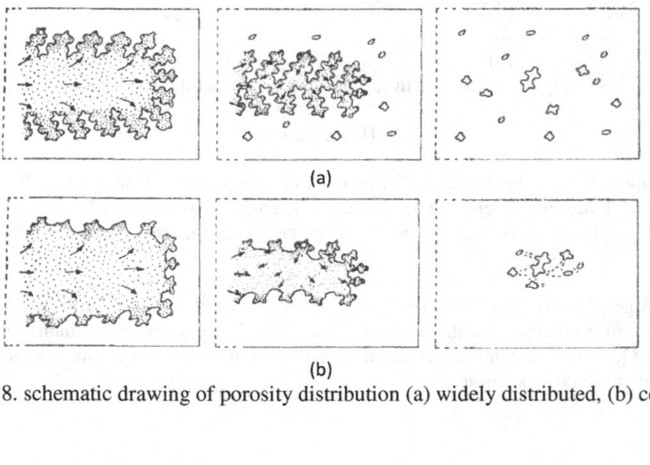

Figure 8. schematic drawing of porosity distribution (a) widely distributed, (b) centred

During solidification, bifilms are being pushed by the dendrites (Fig 8). It is this pushing action that may affect the behaviour of bifilms [12]. Because one of the important features for pore formations in aluminium castings is the opening of bifilms. In the presence of hydrogen with its reduced solubility, hydrogen can diffuse into the bifilms to help the pore to grow.

If there is a coarse and irregular structure with secondary and ternary dendrite arms, the bifilms can find it is way to open themselves. And together with the help of the hydrogen, pores can start to grow inbetween the dendrite arms, which is the situation for A356 alloy (Fig 8a). For SiBloy alloy, the dendrites are more globular and smaller. In this case, bifilms are just pushed away without opening or without being tangled by the secondary arms such that all bifilms are possibly collected in the last region to solidify (i.e. centre of the casting) leading to a localized, centred porosity (Fig 8b). However this lead to the fracture at the grain boundaries as seen on the tensile tests results. There were even premature failures before reaching the yield stress (Fig. 7).

On the other hand, the mechanical test results are somewhat interesting. As seen in Fig 7, SiBloy castings had an increasing tendency of elongation and UTS values, as the gas content was increased. Nevertheless, they have higher scatter and lower values compare to A356. Conversely, A356 castings suffered from decreasing ductility as the hydrogen level was increased.

Moreover, modification is often been associated with an increase in porosity, especially wide dispersed porosity [13]. Boot *et al* [14] also found out that the amount of internal porosity was increased with Sr addition. In the Sr-modified cases in this study, similar results were also observed for both alloys. The permeability of the solid network during the critical stages of alloy solidification will strongly depend on the eutectic nucleation mode. Lu *et al* [15] and Dinnis *et al* [13] suggested that in Sr modification, the eutectic nucleates independently and grow to eventually block the feeding channels (i.e. permeability of the network is deteriorated) that results in dispersed pores throughout the casting. Dispinar [16] had observed similar findings in the unmodified A356 where the permeability was dramatically decreased (and almost came to a stop) when the eutectic reaction began.

Conclusions

SiBloy alloy consists of compact dendrites that delay dendrite coherency leading to better feeding compare to A356.

Porosity in SiBloy tends to be more localised regardless of the hydrogen content but it is widely distributed in A356.

Strontium modification increases the tendency to wide dispersed porosity in both SiBloy and A356.

References

1. Lu, L. and A.K. Dahle, *Effects of combined additions of Sr and AlTiB grain refiners in hypoeutectic Al-Si foundry alloys.* Materials Science and Engineering: A, 2006. **435-436**: p. 288-296.

2. Mohanty, P.S. and J.E. Gruzleski, *Grain refinement mechanisms of hypoeutectic Al---Si alloys.* Acta Materialia, 1996. **44**(9): p. 3749-3760.
3. Mohanty, P.S. and J.E. Gruzleski, *Mechanism of grain refinement in aluminium.* Acta Metallurgica et Materialia, 1995. **43**(5): p. 2001-2012.
4. Guzowski, M., G. Sigworth, and D. Sentner, *The role of boron in the grain.* Metallurgical and Materials Transactions A, 1987. **18**(5): p. 603-619.
5. Schaffer, P.L. and A.K. Dahle, *Settling behaviour of different grain refiners in aluminium.* Materials Science and Engineering: A, 2005. **413-414**: p. 373-378.
6. Limmaneevichitr, C. and W. Eidhed, *Fading mechanism of grain refinement of aluminum-silicon alloy with Al-Ti-B grain refiners.* Materials Science and Engineering A, 2003. **349**(1-2): p. 197-206.
7. Nafisi, S. and R. Ghomashchi, *Boron-based refiners: Implications in conventional casting of Al-Si alloys.* Materials Science and Engineering: A, 2007. **452-453**: p. 445-453.
8. Kori, S.A., B.S. Murty, and M. Chakraborty, *Development of an efficient grain refiner for Al-7Si alloy and its modification with strontium.* Materials Science and Engineering A, 2000. **283**(1-2): p. 94-104.
9. Dispinar, D. and J. Campbell, *Critical assessment of reduced pressure test. Part 2: Quantification.* International Journal of Cast Metals Research, 2004. **17**: p. 287-294.
10. Nordmark, A., et al., *Influence of gas fluxing with argon-hydrogen mixture and grain refinement, on feeding behavior of AlSi7Mg in gravity sand casting.* AFS Transactions, 2006. **113**: p. 06-033.
11. Dons, A., et al., *The alstruc microstructure solidification model for industrial aluminum alloys.* Metallurgical and Materials Transactions A, 1999. **30**(8): p. 2135-2146.
12. Campbell, J., *Castings*. 2nd ed. 2003: Buttonworths.
13. Dinnis, C., et al., *The influence of strontium on porosity formation in Al-Si alloys.* Metallurgical and Materials Transactions A, 2004. **35**(11): p. 3531-3541.
14. Boot, D., et al., *A comparison of grain refiner master alloys for the foundry*, in *Light Metals 2002*, W.A. Schneider, Editor. 2002, Minerals, Metals & Materials Soc: Warrendale. p. 909-915.
15. Lu, L., et al., *Eutectic solidification and its role in casting porosity formation.* JOM Journal of the Minerals, Metals and Materials Society, 2004. **56**(11): p. 52-58.
16. Dispinar, D., et al. *Measurement of permeability of A356 aluminium alloys* in *ICASP-2.* 2008. Graz, Austria.

Shape Casting: The 3rd International Symposium
Edited by: John Campbell, Paul N. Crepeau, and Murat Tiryakioğlu
TMS (The Minerals, Metals & Materials Society), 2009

OXIDE ENTRAINMENT STRUCTURES IN HORIZONTAL RUNNING SYSTEMS

C Reilly[1], N.R Green[2] and M.R. Jolly[1]

[1]School of Mechanical Engineering, University of Birmingham, UK
[2]School of Metallurgy and Materials, University of Birmingham, UK

Keywords: Casting, Oxide, Quality Assessment, Entrainment, Running System

Abstract

During the transient phase of filling a casting running system surface turbulence can cause the entrainment of oxide films into the bulk liquid. Research has shown that these are detrimental to the material's integrity. Common mechanisms for this entrainment include returning waves, arising during filling of the runner bar, and plunging jets, found when pouring into a basin. One of these, the returning wave, has been studied in greater depth, using real-time X-ray and process modelling techniques alongside the application of physical principals. It has been concluded that when developed, returning waves cannot attain the more stable and less entraining tranquil flow regime desirable in the running system of castings.

Introduction

Real time X-ray studies have shown that when a stream of fluid impacts the end of a runner bar a chaotic flow regime is produced for a short period in this locality. This chaotic regime develops into a returning wave. These waves are known to be highly entraining and detrimental to casting integrity as they entrain double oxide films [1, 2]. Such flow phenomena have previously been observed to resemble hydraulic jumps.

It was the original intention of this research to determine the threshold flow conditions that lead to formation of, and gas entrainment by, hydraulic jumps in liquid metals. A threshold has been shown experimentally in water, where the hydraulic jump is relatively well understood and the theory for which is reliant on certain assumptions being made [3]. One of these; the assumption that surface tension effects are negligible may not hold true however for most liquid metals. After detailed experimental investigations, contrary to what had been shown through simulation previously [4], it was not possible to create hydraulic jumps within an open channel running system at liquid metal velocities in excess of 3 ms^{-1}, flow depths of 0.01 m and flow distances of 1 m. Thus it was concluded a 'trigger' is required to initiate a 'hydraulic jump' type structure. It was observed that in this geometrical configuration only the return wave generated at the end of the runner created such a structure.

Returning waves have been known as entraining hydraulic structures but little work has been undertaken thus far on characterising them. This work has concentrated on attempting to understand the trigger, using a combination of experimental, numerical and first principles methods, in order to allow further insight into the constrained return wave found in running systems.

The foundry engineer can calculate the returning wave velocity for a constrained wave in a runner bar of uniform thickness via the principles of conservation of volume [3], as represented in equation 1. The difficulty comes in predicting the height of the returning wave (and thus its velocity) when the constraint is removed.

Past work has shown how casting integrity is greatest when the system fills in a tranquil manner. However, if return waves are formed the persistence of these is critical in determining their overall damage to casting integrity [5, 6]. Although the use of low profile runners, *i.e.* a height of less than the sessile drop height of the fluid has been advocated to stop such waves forming [7] this is not always possible; for example due to manufacturing constraints or lack of flow control for multiple gated systems.

Criterion Development

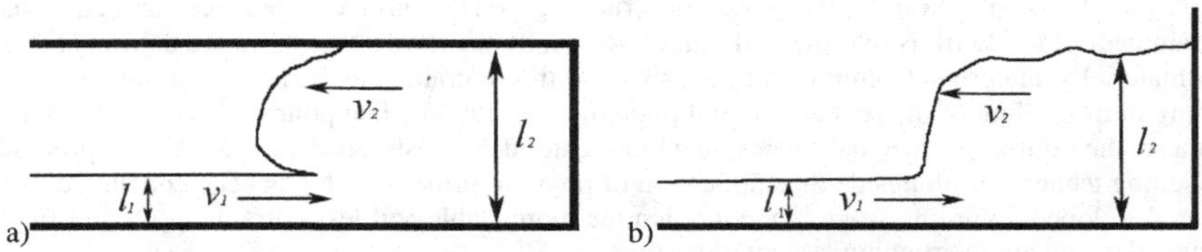

Figure.1 a) Constrained flow schematic b)Unconstrained flow schematic

Constrained Flow

For a constrained return wave the velocity can be calculated from principles of conservation of volume [3] as shown in Equation 1 where v_1 is the inflowing fluid velocity (ms^{-1}), l_1 is the inflowing fluid height (m), and l_2 is the runner bar height (m). This is shown in Figure 1(a).

$$v_2 = \frac{l_1 . v_1}{l_2 - l_1} \tag{1}$$

However, this condition, is only true for the section of the running system between the end of the runner and the in-gate (Section A, Figure 2) To calculate the return wave velocity for the area between the in-gate and downsprue (Section B, Figure 2) it is necessary to account for the fluid volume which is flowing into the casting volume.
For most industrial castings this is extremely difficult to determine from first principles without making assumptions that severely limit the accuracy of the calculation. One difficulty comes from the fact that the volume flowing into the casting varies with time.

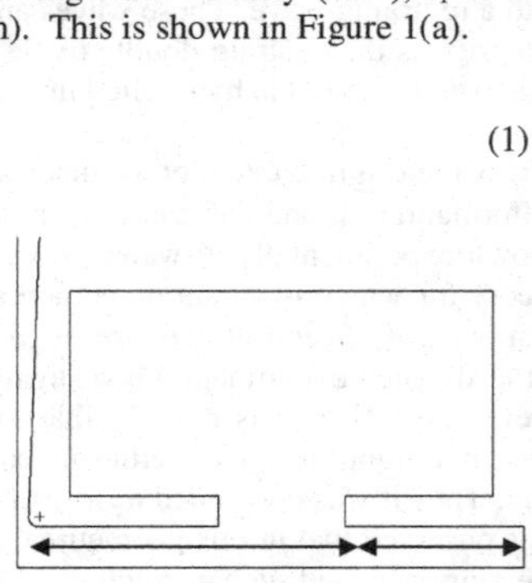

Figure 2. Casting schematic

Unconstrained Flow

For an unconstrained returning wave the velocity for a channel of uniform and unit thickness can be calculated by balancing the energy equation, as shown in the derivation below: (Figure 1b)

Incoming flow kinetic energy	+	Potential energy of fluid height l_1	=	Returning wave kinetic energy	+	Potential energy of fluid height l_2

The potential energy (PE) of the of incoming flow per unit width per unit time = mgh where m is the mass, g the acceleration due to gravity and h is a length

$$m = \rho l_1 v_1 \quad \text{and} \quad h = \frac{l_1}{2}, \text{ therefore PE} = \frac{1}{2}\rho l_1^2 v_1 g \tag{2}$$

where ρ is the fluid density (kgm^{-3}). The kinetic energy (KE) of the incoming flow per unit width per unit time = $\frac{1}{2}mv_1^2$

$$\text{As } m = \rho v_1 l_1 \text{ from (2) then KE} = \frac{1}{2}\rho l_1 v_1^3 \tag{3}$$

Potential energy of return wave per unit width per unit time = mgh

$$m = \rho(l_2 - l_1)v_2 \quad \text{and} \quad h = \left(\frac{l_2 - l_1}{2} + l_1\right) \text{ therefore PE} = \frac{1}{2}\rho(l_2 - l_1)v_2\left(\frac{l_2 - l_1}{2} + l_1\right)g \tag{4}$$

Kinetic energy of return wave per unit width per unit time $= \frac{1}{2}mv_2^2$; where wave depth $= l_2 - l_1$,

$$m = \rho(l_2 - l_1)v_2, \text{ therefore KE} = \frac{1}{2}\rho(l_2 - l_1)v_2^3 \tag{5}$$

Balancing the energies of incoming and returning flows (energy flux) gives;

$$\frac{1}{2}\rho g l_1^2 v_1 + \frac{1}{2}\rho l_1 v_1^3 = \frac{1}{2}\rho(l_2 - l_1)v_2\left(\frac{l_2 - l_1}{2} + l_1\right)g + \frac{1}{2}\rho(l_2 - l_1)v_2^3 \tag{6}$$

$$g l_1^2 v_1 + l_1 v_1^3 = 2(l_2 - l_1)v_2\left(\frac{l_2 - l_1}{2} + l_1\right)g + (l_2 - l_1)v_2^3 \tag{7}$$

For this equation to be of use it needs to be solved independently of $v_2$or l_2. Substituting v_2 from equation 1 and simplifying gives;

$$g l_1^2 v_1 + l_1 v_1^3 = l_1.v_1((l_2 - l_1) + 2l_1)g + (l_2 - l_1)\left(\frac{l_1.v_1}{l_2 - l_1}\right)^3 \tag{8}$$

Rearranging;

$$0 = \left(g l_1^2 v_1 + l_1 v_1^3\right) - \left(l_1.v_1((l_2 - l_1) + 2l_1)g + (l_2 - l_1)\left(\frac{l_1.v_1}{l_2 - l_1}\right)^3\right) \tag{9}$$

Equation 10 was solved in Matlab using the 'solve' function. The solutions are 0 and equation10;

$$l_2 = \frac{1}{2g}\left(v_1^2 + 2gl_1 \pm \sqrt{-4gl_1v_1^2 + v_1^4}\right) \tag{10}$$

Upon determining l_2 the velocity, v_2 of an unconstrained wave can be calculated by substitution into equation 1 giving the results plotted in Figures 3 through 6.

Experimental Design

Moulds were cast in a real-time X-ray flow imaging facility, the principles of which have been reported previously [8]. Castings (Figure 3) were poured in resin bonded silica sand moulds (AFS grade 60 sand) using aluminium alloy 2L99. A range of sprue heights whose design conforms to that defined by Campbell [1] were used to adjust the metal flow velocity and mass flow rate. The cast weight was 11 kg, the pouring temperature 760 °C and the pouring was controlled using a robotic system. Filling was viewed using real-time X-ray radiography to allow the qualitative assessment of the transient flow. Flow images were captured at a rate of 100 s^{-1} and resolution 800×600 pixels over a field of view of approximately 200×150 mm.

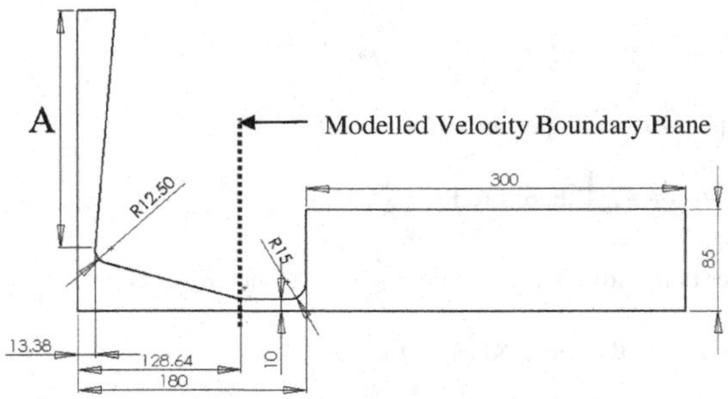

Trial	A (mm)
1	70
2	105
3	120
4	200
5	360

Figure 3. Experimental mould (all dimensions in mm)

<u>Process Modelling</u>

Flow-3D [9], (a commercially available CFD code) was used to model the casting experiments. A velocity boundary condition was imposed on the plane shown in Figure 3. Velocities of 1.02, 1.1, 1.18, 1.25 and 1.37 ms^{-1} measured from the experimental results for v_1 from Trials 1 through 5 respectively were used as the boundary condition. Only the section of the mould to the right of the velocity boundary plane shown in Figure 3 was modelled. This allowed a more accurate comparison with the theoretical results as v_1 could be kept constant. However, it should be noted that with this approach the simulations may not fully match the real time X-ray results as during the transient filling phase when the downsprue is backfilling the pressure (and therefore velocity) of the fluid flowing through this section would not be expected to have reached the stable equilibrium as modelled.

Results

The results shown in Table 1 and Figures 4 and 5 show how, for a given set of inlet parameters there are two alternative flow regimes; rapid and tranquil flow. The exception for this is where the energy is at a minimum at the 'nose' of the returning wave depth curve. This is analogous to the energy nose observed in hydraulic jump structures [3].

Rapid flow is when the flow takes the high velocity and low wave height parameter. This regime is normally considered much more turbulent and highly entraining than the tranquil regime were the fluid takes the alternative parameters, namely a deep but slow moving flow. A much reduced, if at all, entraining regime.

Figures 6 to 8 show examples of the experimental results. These show how initially the returning wave (circled in Figure 7) takes the rapid flow form before immediately jumping to try and attain the tranquil regime.

Experimental results also show the returning wave periodically retreating along the runner bar, giving an insight of the potential magnitude of turbulent energy losses caused by the shearing in the region of the wave front. Figure 6b, 0.2-0.4 s shows clearly that the wave has progressed very little in 0.2 seconds when compared with the distance travelled between 0.4 and 0.6 s.

Figure 7 shows two different but clearly defined examples of the initial wave of low height followed by a jump. This initial wave size matches the theoretical values calculated for Trials 1 through 5, where l_1 equals 10 mm ±2 mm and v_1 ranges from 1.02 ms^{-1} to 1.35 ms^{-1}. This gives l_2 a range of values between 20.6 to 21.2 mm; equation 10. This is obviously beyond the accuracy available with the real-time X-ray equipment of ±2 mm at a turbulent free surface. However, this flow form can be seen in all experimental trails, with the theoretical height of the initial wave lying within the experimental accuracy of the equipment. The modelled results gave l_2 values of 20-22 mm frequently throughout all modelled trials. However, with such waves being unstable the accurate definition of this can be troublesome for many cases. For example Figure 6 b) 0.6 s where the nose of the wave can be seen to be of uniform angle with l_2 values rising from 15 to 35 mm with no clear definition of the 21mm initial wave.

It can be seen in Figure 8 that even with an unconstrained wave there is entrainment of oxide films and bubbles present at the wave front, even for this relatively low energy condition. Further work is required to define if an entrainment threshold exists for a stable wave or whether the rapid regime is entraining for all inlet conditions.

The modelled flow results showed good agreement with the experimental and first principles models with respect to the flow profiles observed. Figure 9 shows example images of the returning wave form that initially advances in the rapid form before immediately attempting to jump to the tranquil form.

Figures 10 and 11 summarise experimental and *Flow 3D* results which correlate favourably. In reviewing these it must be remembered that;

i) There are large errors associated with extracting data from both the experimental results and modelled results. The error within the results is significant due to the following factors;

- The wave has not reached equilibrium, meaning results vary heavily depending on the time frame(s) chosen for analysis. Future work should look at using longer channels to allow a stable wave to form, giving more accurate analysis.
- The bulking effect of air being entrained into the liquid changes the local fluid density; this effect it not considered in either the theoretical calculations or the model.
- The measuring of the incoming fluid velocity v_1 is not accurate as the initial jet is unrepresentative of the steady state condition and the effective head height is constantly varying as the sprue and basin back fill during this initial transient period.
- Difficulties in assessing experimental fluid depth; accuracy being ±2 mm.
- Parallax error from the camera.

ii) The wave height plotted in these figures is not a measure of the initial rapid wave height, but the height the wave attains when trying to reach the tranquil flow regime.

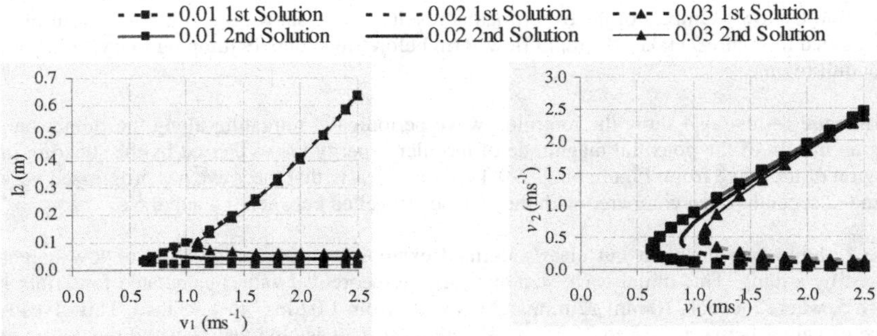

Figure 4. v_1 Vs l_2 (l_1 of 0.01, 0.02 and 0.03 m) Figure 5. v_1 Vs v_2 (l_1 of 0.01, 0.02 and 0.03 m)

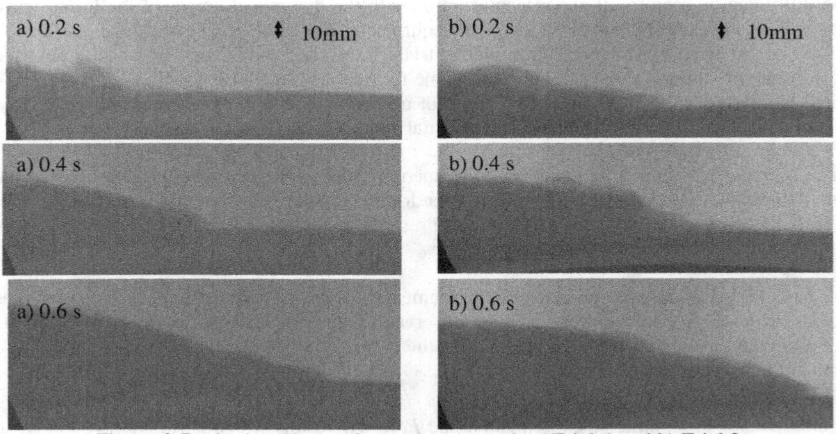

Figure 6. Real –time X-ray of returning waves in a)Trials1 and b) Trial 2

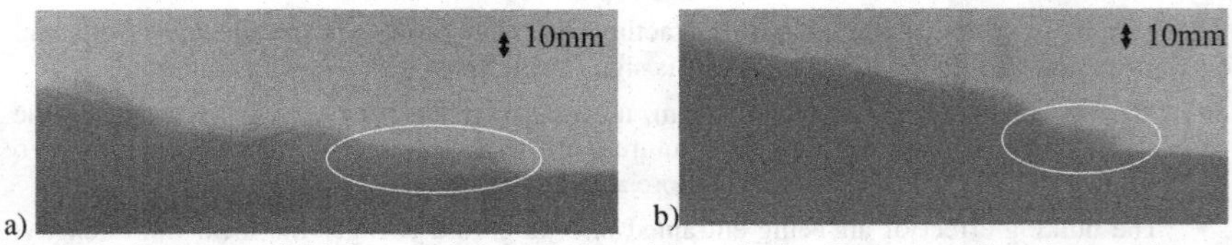

Figure 7. Images of initial wave height followed by a hydraulic jump. Ringed areas denote regions where the initial wave heights were measured.

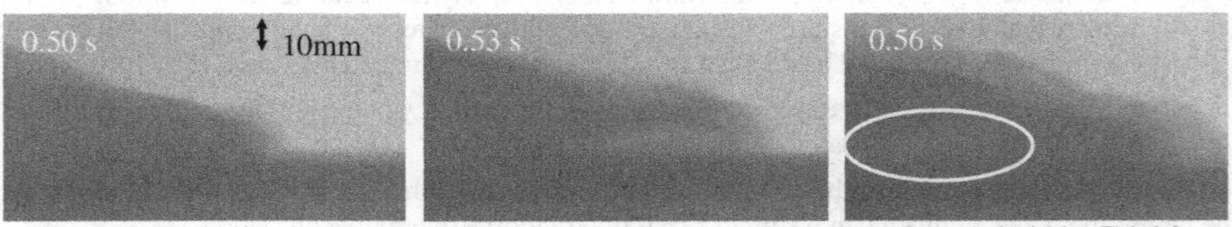

Figure 8. Instability in advancing flow front leading to entrainment of an air bubble; Trial 2

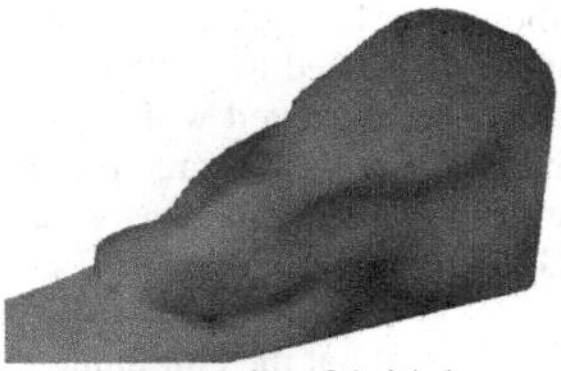

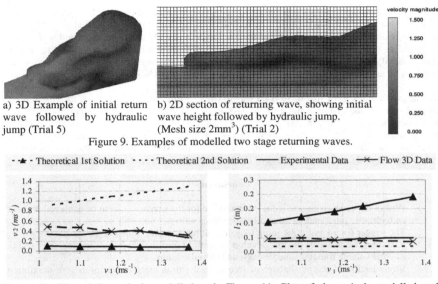

a) 3D Example of initial return wave followed by hydraulic jump (Trial 5)

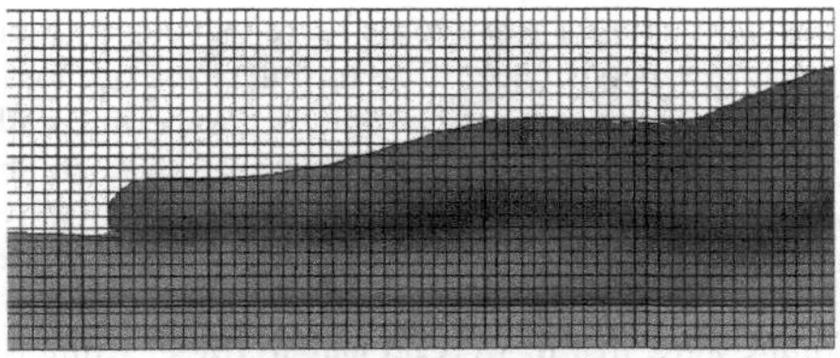

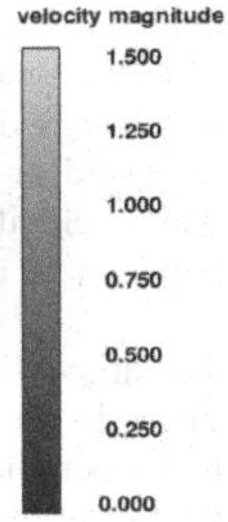

b) 2D section of returning wave, showing initial wave height followed by hydraulic jump. (Mesh size $2mm^3$) (Trial 2)

Figure 9. Examples of modelled two stage returning waves.

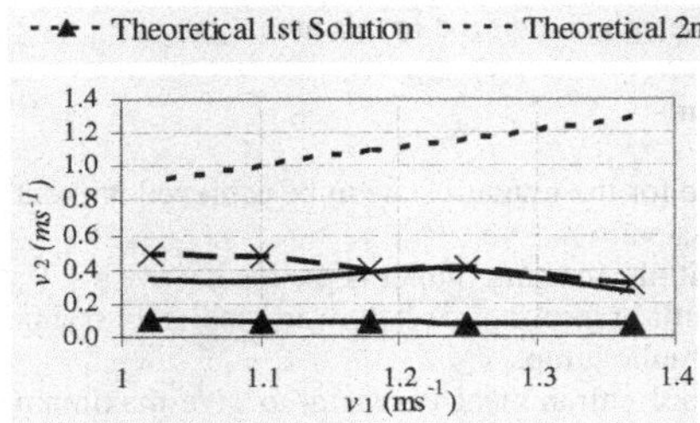

Figure 10. Plot of theoretical, modelled and measured return wave velocities.

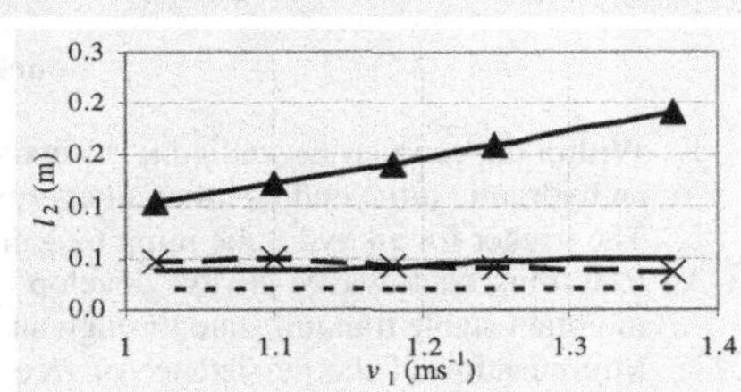

Figure 11. Plot of theoretical, modelled and measured return wave heights.

Discussion

It should be noted that the velocity v_2 and height l_2 of the returning wave for both constrained and unconstrained waves, are dependent on both the approach velocity v_1 and fluid depth l_1, but independent of the physical characteristics of the fluid, for example density, surface tension and viscosity (Equation 8). In this study the well structured incoming flow had a Froude number in the range 3.3 to 4.4. In many casting systems even higher values could reasonably be expected. As stated in the introduction, the flow of liquid aluminium alloys under such inlet conditions has not been observed to initiate a hydraulic jump in a continuous open channel flow. The reflected wave and the shearing interface are essential components of the flow to initiate entrainment.

The presence of two stable flow depths suggested by the simple energy balance is analogous to that derived for an hydraulic jump [3]. In this series of experiments the greater (tranquil) calculated flow depth has not been observed and the predicted height remains somewhat surprising. It also appears to be the case that the thin jet forms ahead of the main returning flow and thus such systems will always be prone to severe entrainment. Therefore reduction of the persistence of the entrainment event is key to achieving optimum casting integrity [6]. One approach applied successfully in industry is the application of reticulated foam filters.

The observed returning flow depths suggest significant energy loss, due principally to shearing at the wave front, bulk turbulence and the bulking affect of air entrainment. It should also be noted that the tranquil depth of flow is far greater than that which could be accommodated in the runner bar of a casting design within normal parameters. Therefore the flow becomes constrained, causing the return wave to increase its velocity along the runner bar with high levels of entrainment but low persistence. The turbulent energy loss can be quantified by calculating the energy difference between the flow obtained experimentally and that associated with either the rapid or tranquil flow regime using Equation 7. Calculation based upon the average return wave velocity and height derived from experimental data shows energy losses in the system of between 50 and 73%. This is clearly well described within the RNG turbulence model.

The above findings show that there is no way to fill a casting runner bar without the entrainment of oxide films if the system is not a single pass design [1]. Single pass systems have been shown to be beneficial to casting integrity but are not appropriate for systems with multiple gates or where large flow rates have to be used because of the casting geometry aspect ratios. Where single pass designs are not possible it appears that running system geometry should concentrate on dissipation of the return wave energy. Further research is required in this area.

Conclusions

1. Within the geometries studied it is impossible for the tranquil state to be achieved without an hydraulic jump and its inherent energy dissipation.
2. The trigger for an hydraulic jump to occur within a casting runner is the back wave.
3. Returning back waves always develop an initial rapid regime before immediately trying to obtain stable tranquil state through an hydraulic jump.
4. Minimisation of the persistence of free surface entrainment is crucial to give maximum casting integrity.
5. Correlation between the theoretical model proposed for an unconstrained wave and experimental data is good. Further data are required for definitive validation

Acknowledgments

The authors would like to acknowledge the help of The School of Mechanical Engineering, The University of Birmingham for sponsoring the PhD of CR, the support of Flow Science Inc and the EPSRC support of NG's chair (EP/D505569/1).

References

1. J. Campbell, *Castings 2nd Edition* (Butterworth Heinemann, 2003) 71.
2. N.R Green and J. Campbell, "Influence in Oxide Film Filling Defects on the Strength of Al-7si-Mg Alloy Castings", *Transactions of the American foundry society,* 114 (1994) 341 -347.
3. B. S. Massey, *Mechanics of Fluids 6th Edition* (Chapman & Hall, 1992).365-369.
4. F.-Y. Hsu, "Further Developments of Running Systems for Aluminium Castings", (PhD Thesis, The University of Birmingham, 2003) 137-138.
5. C. Reilly, "Surge Control Systems for Gravity Castings", (Final Year Project, The University of Birmingham, 2006) 45-72.
6. C. Reilly, N. R. Green, M. R. Jolly and J. C. Gebelin, "Using the Calculated Fr Number for Quality Assessment of Casting Filling Methods", *Modelling of casting, welding and advanced solidification process XII.* (2009).
7. J. Campbell, *Castings Practice: The 10 Rules of Casting* (Butterworth Heinemann, 2004) 41-47
8. M. Cox, R. A. Harding and J. Campbell, "Optimised Running System Design for Bottom Filled Aluminium Alloy 2199 Investment Castings", *Materials science and technology* 19 (2003) 613-625
9. *Flow3D*, www.flow3d.com.

Shape Casting: The 3rd International Symposium
Edited by: John Campbell, Paul N. Crepeau, and Murat Tiryakioğlu
TMS (The Minerals, Metals & Materials Society), 2009

DEGASSING: A CRITICAL STAGE IN THE MANUFACTURING OF Al-Si-Cu ALLOYS FOR AUTOMOTIVE CASTINGS

E. Velasco[1], R. Valdes[2], J. Niño[2]

[1]Texas State University, Dept. of Technology, 601 University Drive
San Marcos, TX 78666, USA
[2]NEMAK, Arco Vial Km 3.8, Garcia N.L. CP 66000, México.

Keywords: liquid metal treatment, degassing, reduced pressure test (RPT), fluidity, inclusion content, porosity, mechanical properties

Abstract

In the manufacturing of aluminum blocks for automotive engines, porosity requirements on sealing surfaces and other critical areas are so stringent that special controls and special processes are required to remove hydrogen and inclusions from the molten aluminum. Several trials and adjustments in the degassing parameters and melting practices were evaluated at a production foundry. To evaluate the degassing process, density measurements were performed and to evaluate the inclusion content, fluidity and PreFil™ measurements were performed. Additionally, the mechanical properties of the casting in the heat treated condition were determined. The relationship between the density as determined by the reduced pressure test (RPT) and actual hydrogen content as determined by an AlScan™ unit was established for A319 aluminum alloy. The results of the mechanical property tests showed a direct relation with the metal quality of the alloy.

Introduction

High technology components for the automotive industry, such as cylinder blocks and cylinder heads, have to meet exacting requirements, such as high mechanical properties, good high cycle fatigue characteristics and special finished surface conditions. In these components, the quality of the base liquid metal has a big impact on the final mechanical properties, such as tensile strength and fatigue life. These properties are directly related to the hydrogen content and the inclusion content.

Hydrogen is soluble in liquid aluminum and the solubility has a strong dependence with temperature [1]. The solubility of hydrogen in a liquid aluminum-silicon alloy may be expressed by an equation of the general form:

$$\log_{10} S = -\frac{A}{T} + B$$

where S is the solubility in cc of hydrogen at standard conditions per 100 grams of metal and T is temperature in °K.

For Al-8%Si [1] we have:

$$\log_{10} S = -\frac{3050}{T} + 2.95$$

According to this equation and for this alloy at 1382°F (750°C) the maximum solubility increases 43% compared at 1292°F (700°C). However, as Figure 1 shows, the main problem associated with hydrogen porosity in castings appears during the solidification process when the dissolved hydrogen in excess of the solubility limit may precipitate in molecular form, resulting in the formation of voids.
The precipitation of hydrogen follows the laws of nucleation and growth, and if inclusions are present in liquid metal, these particles act as preferential nucleation sites and facilitate void formation.

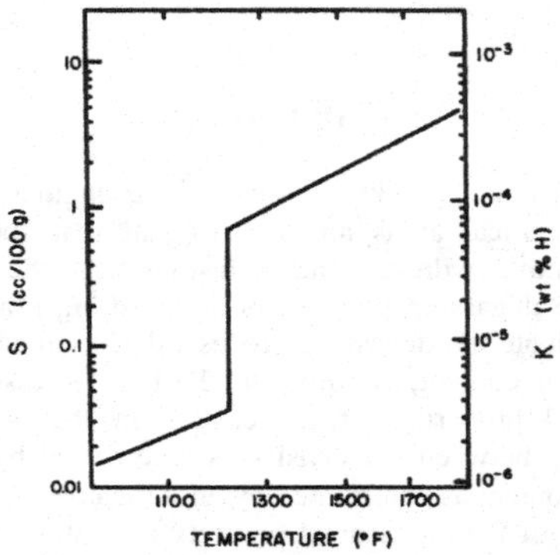

Figure 1. The solubility of hydrogen at one atmosphere pressure in pure aluminum.

Additionally, the solidification of complex geometrical shapes, such as cylinder blocks and cylinder heads, with varying walls and section thicknesses creates conditions under which internal porosity may form due to poor feeding during the natural contraction of liquid to solid. If the degassing is not adequate enough to remove most of the hydrogen and inclusions present in the liquid, the final result will be a casting with porosity as a result of poor metal treatment practices and solidification conditions [2].

Factors related with the porosity such as dissolved hydrogen content and non-metallic inclusion content can be measured and controlled during the metal treatment stage. For decades, the reduced pressure test (RPT) has been a simple, low cost, reliable method to assess the propensity of a melt to form microporosity during solidification. Today, the reduced pressure test is, by far, the most commonly used technique to evaluate the level of dissolved hydrogen and the cleanliness of liquid aluminum alloys [3].

Fluidity is a characteristic associated with liquid metal and is indicated by the distance through a mold passages or a constant cross-sectional area that the metal will flow before it solidifies. Fluidity is a complex property and depends upon many factors related to the alloy (chemical composition, heat of fusion, solidification range), mold interactions and the thermal properties at the liquid metal-mold interface. If most of these factors are constant, the fluidity of the molten metal will be a function of impurities (such as oxides or intermetallic particles) and the pouring temperature.

D. Groteke showed in a classical technical paper on filtration practice and fluidity testing, that inclusions play an important role in fluidity, Figure 2. His work showed that reducing the inclusion and oxide content by filtering, the fluidity is highly improved [4].

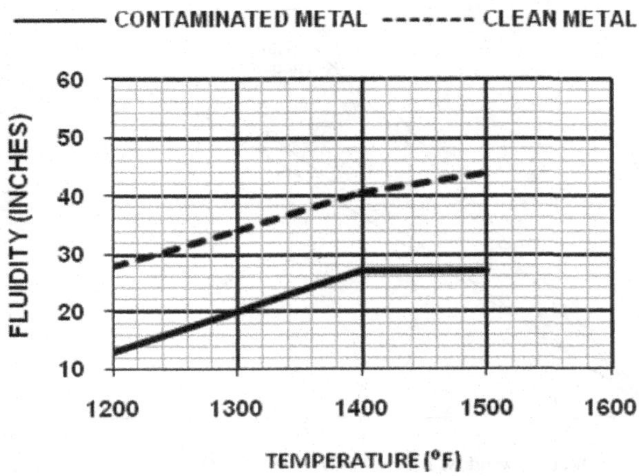

Figure 2. The effect of filtration on fluidity of die casting alloys [4].

Experimental Procedures

In order to determine the factors associated with internal porosity in an automotive casting, a series of reduced pressure test (RPT) samples were taken during five days from the output well of a four metric ton, gas fired furnace. The molten metal held in this furnace was A319 aluminum alloy and its average composition is shown in the Table 1. No adjustment in the chemical composition was carried out during the RPT sampling period.

Table 1. Average chemical composition (wt %) of the A319 aluminum alloy sampled.

Si	Cu	Fe	Mg	Mn	Zn	Ti	Cr	Pb	Sr*	Na*	Ca*
7.5	3.25	0.66	0.30	0.42	0.64	0.09	0.05	0.03	100	6	2

* ppm

Each RPT sample (with replica) was taken at 26 +/- 1 inches of Hg with a GasTech™ tester. At the same time, the hydrogen content was measure using an AlScan™ ABB unit.

Using a digital scale FMA3001 Model with integrated software based on the Archimedes principle, the density of each RPT sample was determined and a correlation curve relating RPT densities to the actual hydrogen measurements obtained by the AlScan™ was obtained.

Fluidity and metal quality obtained during production were compared with reference values for fluidity and PreFil™ curves. The reference values correspond to 319 aluminum alloy that were produced with fluxing with chlorine gas injection to remove inclusions. Figure 3 shows the

PreFil™ curve with an operative window for clean metal. Turnings and chips were added to the molten metal and the cleanliness of the resulting liquid metal was evaluated. The details of the relationship between fluidity and PreFil™ curves for this alloy are describe in previous works [5,6].

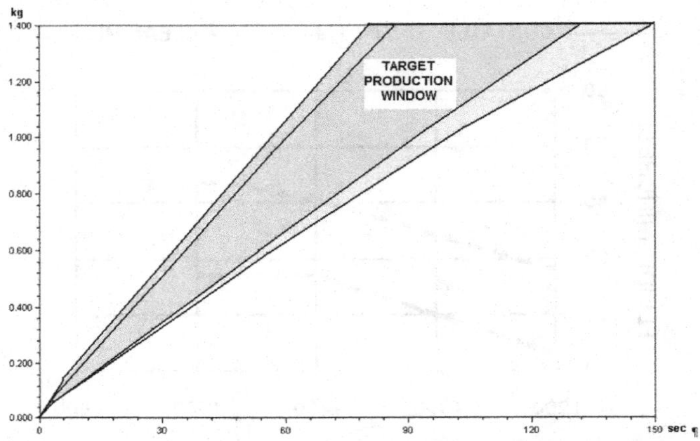

Figure 3. Target production window for 319 alloy. The center area corresponds an adequate metal in terms of cleanliness [5].

A set of ten automotive precision sand casting were taken from a low pressure production line verifying when high RPT density and fluidity values were reached. The quality of the metal was then evaluated by PreFil™ testing. The fluidity was determined using a Ragone Tester. Every two hours, fluidity samples were taken in the holder furnace with a Pyrex glass tubes preheated at 100°C during 60 minutes. The vacuum used in the Ragone Tester for sampling was 100 psi. The fluidity reported during production was the average of five samples.

Mechanical properties and percent porosity were determined in a chilled area of production castings in order to evaluate the effectiveness of the metal treatment process on the soundness of the casting. Similarly, castings that were scrapped during production due to low RPT density were taken to evaluate their mechanical properties.

Results and Discussion

Figure 4 shows the correlation of RPT densities to actual hydrogen measurements obtained by using an AlScan™. As can be seen in the curve, the simple density measurement has a good correlation with actual hydrogen content at values below 2.73 g/cc; poor correlation was found between 2.74 - 2.75 g/cc, values typically not easily reached with conventional degassing equipment (shaft and rotor system).

The lack of correlation at low levels of hydrogen could be associated to RPT factors such as sample pouring temperature, variations of chamber pressure, mold temperature or molten metal cleanliness [7].

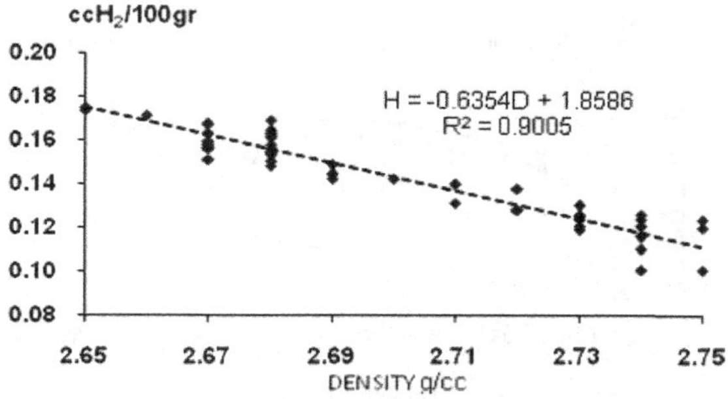

Figure 4. Correlation curve for RPT densities and actual hydrogen content values obtained by AlScan™ in the production foundry for A319 aluminum alloy.

The density values recorded for the period of time during production of castings are shown in Figure 5. The main holding furnace maintained higher density values than the receiver holding furnace. These values are typical of processes where the gas fired burners are used in the holding and reverberatory furnaces, where filtering, rotary degassing and frequent skimming of melt surface are carried out [8]. The receiver holder showed lower densities than the main holder due to turbulence during metal transfer and less cleanliness caused by splashing during the ladle transfer process.

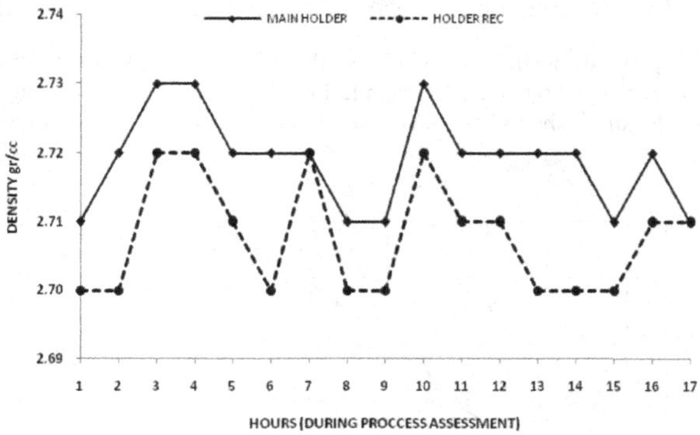

Figure 5. RPT density measurements for the main holder and the receiver holding furnace for 17 hours of production.

In order to assess the molten metal quality, a series of pre-established fluidity curves were taken as reference; these fluidity curves are shown in Figure 6. The best condition (high fluidity) for this alloy corresponds to chlorine gas degassing with the addition of flux. Inclusions and aluminum oxide films in the liquid metal significantly decreased the fluidity. Good references of this effect can be found in the technical casting literature [4,8,9,10,11].

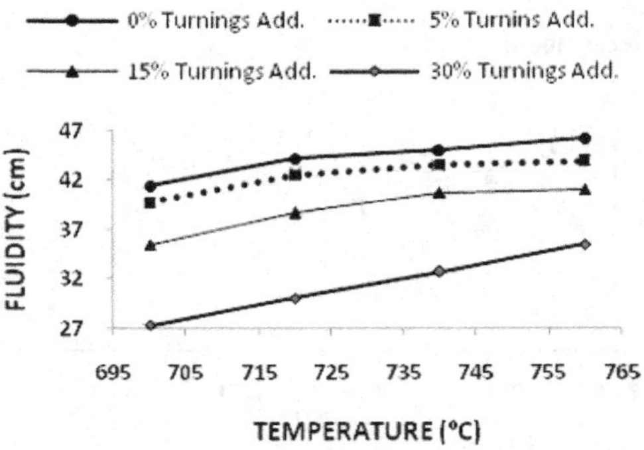

Figure 6. Effect of using turnings as a charge material on the fluidity of 319 aluminum alloy.

Temperature, inclusions and silicon content are the main factors associated with the fluidity of the Al-Si alloys, but during production, the factors influence the fluidity; are temperature and the inclusions generated during melting, degassing and transfer. Table 2 shows the temperature in the well furnace, near to the pouring station. This metal has been previously filtered and the fluidity values can be considered as relatively clean metal.

Table 2. Fluidity measurements of filtered metal in a production casting line.

Temperature (°C)	743	742	741	742	742	745	750
Fluidity (cm)	37.5	38.7	39.5	42	38.7	37.5	43

In addition to fluidity measurements, the quality of the metal was evaluated by PreFil™ sampling; a few samples were taken in the main holder furnace, prior to the grain refiner and modifier additions. Figure 7 shows that the curve obtained is inside the target production window.

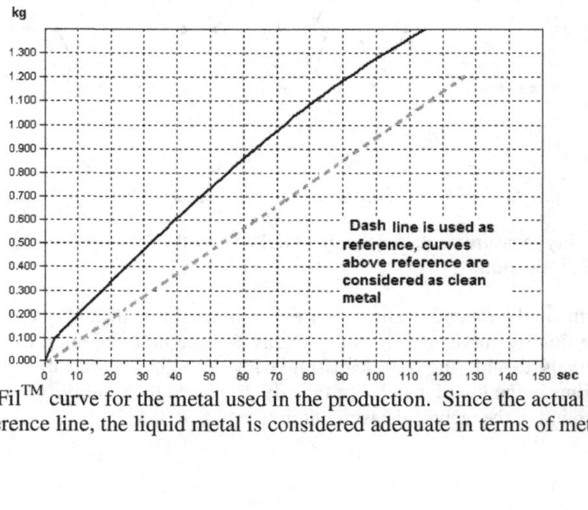

Figure 7. PreFil™ curve for the metal used in the production. Since the actual data is above the reference line, the liquid metal is considered adequate in terms of metal quality.

The mechanical properties of the final product were determined. Tensile specimens were taken from areas near to chilled zone in the casting. Table 3 shows the tensile values along with some microstructural characteristics associated with the castings.

Table 3. Average of metallographic evaluation and mechanical properties of ten automotive castings after T7 heat treatment.

Average Porosity (Area %)	Maximum Porosity (Area %)	Maximum Pore Size (µm)	DAS (µm)	UTS (MPa)	YS (MPa)	Elong. (%)	BHN
0.03	0.14	37	18	328	248	2.9	100
Average of six automotive castings with oxide inclusions and porosity on the fracture surface of the tensile specimen.							
0.13	0.15	110	24	280	220	1.4	92.6

Figure 8 shows oxides and porosity at the failure origin of a tensile specimen. Porosity and oxide inclusions are related to the metal treatment process.

Figure 8. Surface of tensile specimen with low mechanical properties; oxide inclusions are apparent.

According to the ASM Handbook, Typical mechanical properties for separately cast test bars of alloy 319.0 alloy in the T6 condition [12]:

UTS = 280 MPa
YS = 185 MPa
Elongation = 3.0 %
BHN = 95

High values of mechanical properties are the result of many factors, such as a suitable metal treatment processes, heat treatment and low DAS (dendrite arm spacing). Inappropriate control in these processes can reduce the mechanical properties by as much as 20% [13,14,15].

Conclusions.

Basic process measurements, such as the RPT density and fluidity can be a reliable method to assess and control the quality of liquid treatment processes.

Good mechanical properties can only be achieved if the casting is produced with low porosity and high cleanliness; a critical point is metal quality control.

References

1. W. R. Opie and N. J. Grant, "Hydrogen solubility in aluminum and some aluminum alloys," Journal of Metals (AIME Transactions) 188:1237-1241, 1950.
2. R. Monroe, "Porosity in Castings," AFS Transactions 2005, pp1-28.
3. F. Chiesa and J. Mammen, "Correlating Microporosity to Local Solidification Conditions and RPT in Aluminium A356 Castings" AFS Transactions 1999, pp 379-387
4. D. E. Groteke, "A Production Filtration Process for Aluminum Melts," Trans. Soc. Die Cast Eng. 1983.
5. E. Velasco and J. Proulx. "Metal quality of secondary alloys for Al castings," Light Metals 2006, TMS, pp 721 – 724.
6. E. Velasco et. al. "Effect of temperature and charge borings in the fluidity of an aluminum-silicon alloy". Light Metals 2008, TMS, pp 609 -612.
7. W. LaOrchan, "Quantified Reduce Pressure Test," AFS Transaction 1995, pp 277 – 286.
8. G. Crum, et, al. "Control of hydrogen in 319 aluminum alloy for commercial scale production of premium quality sand castings," Aluminum Conference AFS. pp 843 – 858.
9. M. Di Sabatino, L. Arnberg, S. Rørvik and A. Prestmo, "The influence of oxide inclusions on the fluidity of Al–7 wt.%Si alloy," Materials Science and Engineering: A. Vol. 413-414, December 2005, pp 272-276 International Conference on Advances in Solidification Processes.
10. M. Di Sabatino et. al. "Fluidity evaluation methods for Al–Mg–Si Alloys," International Journal of Cast Metals Research, Vol. 19 No. 2. Feb. 2006. pp 94 -97
11. Timelli, G.; Bonol.lo, F. "Fluidity of aluminium die castings alloy," International Journal of Cast Metals Research, Vol. 20, No. 6, Dec. 2007 , pp. 304-311.
12. ASM Specialty Handbook, "Aluminum and Aluminum Alloys" ASM International 1996, pp.714-715.
13. M. C. Flemmings, *et.al.* "Dendritic Arm Spacing in Aluminum Alloy" AFS Transactions 1991, pp 501-506.
14. G.W. Mugica, *et. al.* "Effect of Porosity on the Tensile Properties of Low Ductility Aluminum Alloys" Materials Research, Vol.7, 2004, pp. 221 -229.
15. Z.Li, et.al. "Parameters controlling the performance of AA319-type alloys".Part I. Tensile properties. Materials Science and Engineering A 367 (2004) 96–110

Shape Casting: The 3rd International Symposium
Edited by: John Campbell, Paul N. Crepeau, and Murat Tiryakioğlu
TMS (The Minerals, Metals & Materials Society), 2009

HEAT TREATMENT OF A356.2 ALUMINUM ALLOY: EFFECT OF QUENCH RATE AND NATURAL AGEING

Manickaraj Jeyakumar, Mohamed Hamed, Sumanth Shankar

Light Metal Casting Research Centre (LMCRC), Department of Mechanical Engineering, McMaster University, Hamilton, ON, Canada L8S 4L7

Keywords: A356.2 alloy, Mg_2Si, co-clusters, precipitation hardening, mechanical properties.

Abstract

A356.2 aluminum alloy is a popular commercial alloy used for structural shaped castings in automotive applications. The heat treatment of this alloy is critical to obtain the desired mechanical and performance properties. The three stages of heat treatment include, solutionizing, quenching and artificial ageing. In this study, the effect of quenching in water at 80 °C and forced air at 37.4 ms^{-1} velocity is quantified. Further, the effect of natural ageing treatment prior to the artificial ageing is also quantified. It is observed that the mechanical properties of the castings are significantly affected by both the rate of quenching and the natural ageing treatment. Mechanical properties and hardness values are presented for standard uniaxial tensile test specimen cast by gravity permanent mold process and subjected to various heat treatment conditions.

Introduction

A356 aluminum alloy is used in automobile industry due to its good castability and the high strength to weight ratio. Most importantly, it is one of the heat treatable aluminum alloys and age-hardening response of the alloy is very significant to attain the optimum mechanical properties and performance. The strength of the alloy is increased by the nucleation and growth of Mg_2Si precipitates, a phenomenon occurring during artificial ageing. Our current understanding is that during heat treatment the solutionizing and ageing temperatures, and the duration of these treatments are critical in dissolution and homogenization of the solute elements and subsequent precipitation of specific phases in the primary Al phase matrix for strengthening, respectively. However, it has been shown recently that in Al wrought alloy, the rate of quench at the end of solutionizing treatment and the duration of natural ageing before the artificial ageing treatment are also critical in influencing the morphology and distribution of the strengthening precipitates in the primary phase matrix [1-7].

The rate of quenching of Al-Si-Mg alloys will directly influence the behavior of solute atoms during the natural ageing process. This study aims to evaluate any effect on the precipitation of the strengthening phase during artificial ageing from the rate of quenching and natural ageing process.

Background

Quenching rate is one of the important factors in the heat treatment process to preserve the solid solution in the matrix. Residual stresses developed in the sample during quenching depend on the rate of quenching. Many researchers [1,8-9] have studied the effect of quenching rate on the mechanical properties and it is observed that quenching rate affects the solute and vacancy concentration in the matrix and it changes the ageing sensitivity. Byczynski et al [10] studied A319 alloy and observed that increasing the quenching rate increases the macro surface hardness and the Ultimate Tensile Strength (UTS) but decreases elongation. Liu et al [2] studied the quenching rate on 7XXX series alloys and concluded that decreasing the quenching rate decreases the mechanical properties due to loss of solute concentration and void concentration. Cavazos et al [3] observed precipitates in sample quenched at a slow cooling rate (<10 °Cs^{-1}), which in turn increased the micro-hardness after solutionizing.

Natural ageing plays a crucial role in the subsequent artificial ageing. Cooling rate during quenching has a direct impact on the precipitation events during natural ageing [4]. In Al-Si alloys, co-clusters are formed during natural ageing and consequently affect the kinetics of precipitates during artificial ageing [5-6]. Iskandar et al [7] observed that increasing the natural ageing time increases the yield strength and ductility of the air quenched samples as compared to the water quenched samples.

Literature analysis shows that quench rate sensitivity and the role of natural ageing in subsequent artificial ageing affects the mechanical properties and its magnitude of the effect depends upon the solute contents in the alloy. In A356.2 alloy, the solubility of Mg and Si in the aluminum matrix increases with temperature [11-12]. During solutionizing, coarsening and spheroidizing of the Si particle and dissolution of the Mg and Mg_2Si depends upon the process time and temperature. The main objective of the present study is to analyze the effect of quenching rate and delay time before artificial aging on the mechanical properties of A356.2 alloy. Two different quenching media, water at 80°C and high velocity air were chosen for the study; the aging condition ranges between immediate artificial ageing, materials stored at room temperature (25 °C) and (-15) °C for 8 h and 3 days, respectively before artificial ageing. The mechanical properties studied include micro-hardness of the primary Al phase matrix, UTS, yield strength (YS) at 0.2% strain and % of elongation (%el).

Materials and Procedure

A356.2 alloy was used and its composition (weight %) measured by ICP* was Al-7.5 Si-0.11 Fe-0.06 Cu-0.11Mn-0.46 Mg-0.04Zn-0.17 Ti-0.02 Sr. The melt was degassed by ultra-high purity argon gas using rotary impeller degasser. ASTM B108a standard cylindrical tensile bars were cast using gravity permanent mould process.

In Table I, the first alphabet 'x' in the Sample ID represents the isothermal time in hours for the artificial ageing process. 'x' is an integer between 0 and 10. The second alphabet in the Sample ID is 'W' for water quench medium and 'A' for air quench medium used immediately after solutionizing treatment. The third and last numeral in the Sample ID is '1', '2' or '3' for 0 h natural ageing, 8h natural ageing at room temperature and 3 days natural ageing at (-15)°C, respectively.

All samples were solutionized at 540 °C and held for 12h isothermal time, *t* using an electric resistance furnace. Subsequently, the samples were quenched by two methods: one was by water maintained at 80 °C and the other was by air at 37.4 ms^{-1} velocity using pressurized air blower (diameter of the exit nozzle was 20 cm). The velocity distribution of the air was measured using a hot wire Thermo-Anemometer at 25 cm away from the nozzle exit, where the sample was placed during quenching. A total of 15 samples were quenched by each medium at one time. The samples were quenched in water maintained at 80 °C followed by water at room temperature to arrest all precipitation reactions and the sample in the air quenching process was placed such

* Inductively Coupled Plasma Spectrometry.

that the air velocity was uniform in all samples and the direction of air velocity was parallel to the length of the tensile test bar. Sample temperatures were measured during solutionizing and quenching using data acquisition system with K- type thermocouple embedded at three different locations in a reference tensile bar sample placed along with the experiment samples.

Table I. Sample identification and experiment conditions. RT represents room temperature and all times during solutionizing and ageing are isothermal holding times.

Sample ID	Quench Medium	Solutionizing	Natural Ageing	Artificial Ageing
xW1	Water at 80 °C	540 °C for 12h	None	155 °C for x h. 0<x<10h.
xW2			8h at RT	
xW3			-15 °C for 3 days	
xA1	Air		None	
xA2			8h at RT	
xA3			-15 °C for 3 days	

Artificial ageing at 155 °C was carried out in all samples after three unique natural ageing processes. For each quench medium, five samples were artificially aged immediately after quenching, five samples were artificially aged after 8h natural ageing (room temperature) and five samples were artificially aged after storing them for 3 days at (-15) °C. The artificial ageing was carried out for 0h to 10h isothermal times and samples were obtained at every hour.

Uniaxial tensile tests were carried out at room temperature by using an INSTRON 8800 testing machine at a rate of elongation of 1 $mm.min^{-1}$ and an extensometer was used to measure percentage elongation. Tensile tests were carried out on all the six experiment conditions shown in Table I after 5h and 8h artificial ageing times (x=5 and x=8). Clemex intelligent microscopy micro-hardness testing machine was used to measure the micro-hardness (Vickers scale, HV) with a 25 g load applied at the centre of primary Al phase matrix in the sample microstructure. The arithmetic mean of ten micro-hardness values was recorded for each experiment condition in Table I.

Results and Discussion

Figure 1(a) shows a photograph of the reference tensile bar sample fitted with three thermocouples. Two thermocouples are placed in the grip sections and one in the middle of the gauge length. The thermocouples were placed at equal distances (50 mm) from each other. The gauge length of all the samples was 50 mm. Figure 1(b) shows the thermal profile during a typical water quenching process. Thermal profiles from all the three thermocouples in Figure 1(a) were identical during water quenching. Figure 1 (c) shows the temperature profile as a function of time for the three thermocouples in Figure 1 (a) obtained during a typical air quenching process. Figure 1 (c) shows no significant difference in temperature gradient among the three thermocouples in Figure 1 (a). This was also confirmed by negligible variations in macro surface hardness measurements at the three thermocouple locations and across the gauge length.

Figures 2(a) and 2(b) show typical microstructures after heat treatment for the water and air quenched samples, respectively. Figure 2(b) also shows typical micro-hardness indentations in the primary Al phase matrix. It can be observed in Figure 2 that the microstructures of both the water and air quench samples are identical with the same level of spherodization of the eutectic Si phase during solutionizing.

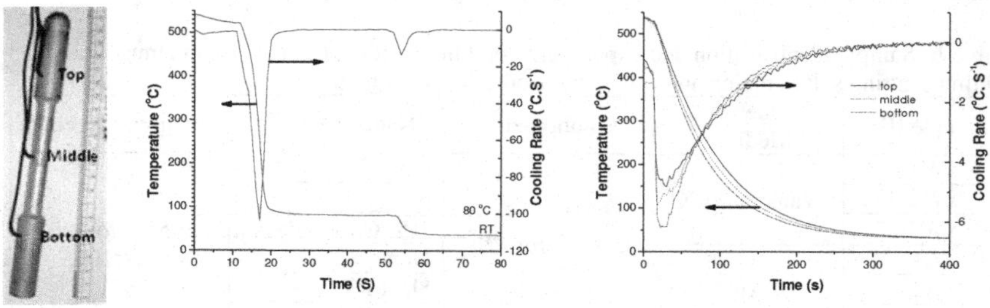

(a) (b) (c)

Figure 1. Reference sample and typical thermal data during quenching. (a) reference tensile bar sample with three thermocouples, (b) water quenching, and (c) air quenching.

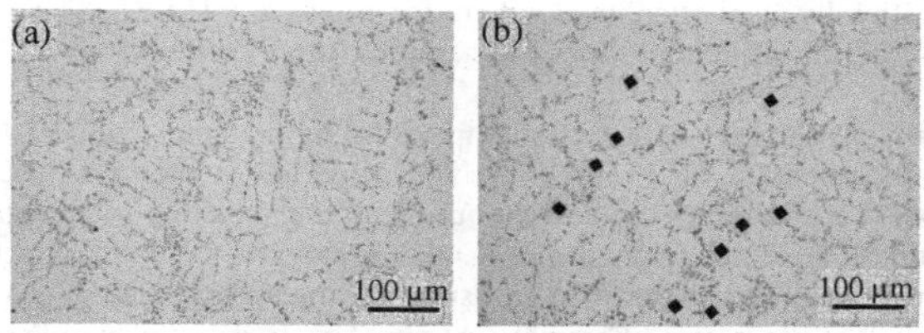

Figure 2. Typical microstructures from optical microscopy (a) water, and (b) air quenched.

Figure 3 shows the results of the micro-hardness measurements for all experiment conditions. An error bar is only shown for the xW1 sample to aid better visualization of data, and the error bars were of similar magnitude for the other samples as well.

Typically there are five major stages during the formation of Mg_2Si by precipitation reaction [13-14].

- Stage I. Formation of independent *self-clusters* of Mg and Si atoms, respectively. This takes place between (-30) °C and 70 °C.
- Stage II. Disintegration of *self-clusters* of Mg and Si atoms.
- Stage III. Formation of *co-clusters* that contain Mg and Si atoms in a ratio of 1:1. This takes place at a temperature around 70 °C along with Stage II.
- Stage IV. Transformation of the co-clusters to β" precipitates which forms as more Mg atoms join the co-clusters such that the ratio between Mg and Si atoms vary from 1:1 to 2:1. This takes place at the artificial ageing temperature.
- Stage V. Formation and growth of β-Mg_2Si phase when ratio of Mg and Si atoms reach 2:1.

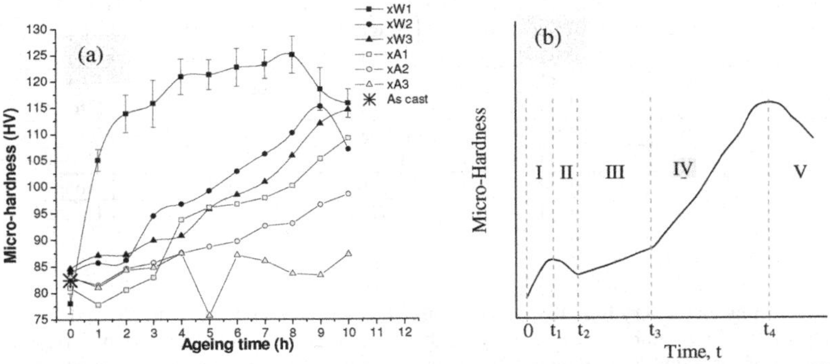

Figure 3. Relationship between micro hardness (Vickers scale) data as a function of artificial ageing time and the five stages in the precipitation hardening reaction. (a) experiment data and (b) schematic showing various stages.

Edwards et al [14] claim that in Stage II, the Si self-clusters do not disintegrate and only the Mg self-clusters disintegrate and join the Si self-clusters by diffusion to form the co-clusters. The kinetics of all Stages is directly proportional to the vacancy concentration present at the each stage [15] which influences the mobility of the Mg and Si atoms. β" precipitates is the major contributor to the increase in the strength and micro-hardness of the alloy sample and it has been proposed that the β-Mg_2Si phase is formed during the peak and over aged conditions [16-17]. The vacancy concentration in the water quenched samples at the end of quenching will be higher than in the air quenched samples because of the significantly higher cooling rates as shown in Figures 1(b) and 1(c). At higher vacancy concentration there will be higher mobility of Si and Mg atoms in the matrix during the precipitation reaction.

The hardness data in Figure 3(a) reflect the various stages of the precipitation reaction in all the experiment conditions. Figure 3(a) also shows that the rate of quenching after solutionizing treatment and natural ageing process prior to artificial ageing has significant effect on the precipitation reaction mechanism and kinetics as reflected by the hardness data of the various experiment conditions. Figure 3(b) presents a schematic of the various stages of the precipitation reaction as observed by variation in the micro-hardness values as a function of time. Hardness value for a homogeneous distribution of Si and Mg atoms in the primary Al matrix will be the lowest as reflected by time, t=0h in Figure 3(b) for each experiment condition, respectively. The hardness value increases during Stage I, decreases at Stage II and increases again at Stage III such that the hardness in Stage III will be higher than that in Stage I (refer to Figure 3(b)). Between Stage III and Stage IV the hardness value continuously increases and reach a maximum at the beginning of Stage V and continues to decrease as the β-Mg_2Si phase grows to result in an over-aged condition.

Table II presents the duration of the five stages in each of the six experiment conditions listed in Table I and shown in Figure 3(a). The following discussion on the precipitation reaction sequence (Stages I to V) for each experiment condition listed in Table 1 will be carried out in reference to Figures 3(a) and 3(b).

Table II. Duration of the five stages in the six experiment conditions as given in Figure 3.

Sample ID	Stage I (t_0 - t_1), h	Stage II (t_2 - t_1), h	Stage III ($t_3 - t_2$), h	Stage IV ($t_4 - t_3$), h	Stage V (>t_4), h
xW1	<1	<1	<1	1 to 8	8 to 10
xW2	at 0	<1	<1 to 2	2 to 9	9 to 10
xW3	at 0	<1	<1 to 4	4 to 10	None
xA1	at 0	0 to 1	1 to 3	3 to 10	None
xA2	at 0	0 to 1	1 to 6	6 to 10	None
xA3	at 0	0 to 1	Unclear	Unclear	Unclear

<u>xW1</u>

Stage I takes place at a very high rate and so does Stages II and III because the Si and Mg solute atoms has very high mobility due to the high vacancy concentration in the matrix and the temperature of the sample is immediately increased to the artificial ageing temperature without delay. Hence, the hardness curve for xW1 increases at a very high rate initially and reaches the highest maximum value amongst the samples evaluated. It may be that the Stage I and Stage II may not even exist in xW1 sample because of high vacancy concentration and instant increase in temperature to greater than 70 °C. More work is being carried out to determine the existence of Stages I and II in xW1 condition.

<u>xW2</u>

In xW2, the Stage I will already exist at t=0h due to the 8h natural ageing process at room temperature wherein the self-clusters were formed. The vacancy concentration in xW2 is still high but not as high as in xW1 because some of the vacancies would be utilized and disappear during the 8h natural ageing process at room temperature. Stage II takes place at a high rate due to the high vacancy concentration and hence not discernable at 1h. Between 0 and 2h, Stage III along with Stage II will occur and hence show a marginal increase in hardness value in Figure 3(a). Upon completion of Stage III, the hardness value increases at a high rate during the transformation of co-clusters to β". Due to the lower vacancy concentration in xW2 than xW1, the time taken to reach maximum peak harness is longer in xW2 and a lower peak harness value is achieved as well. Transformation of β" to β-Mg_2Si begins at 9h.

<u>xW3</u>

Compared to xW1 and xW2, xW3 sample will have lesser vacancy concentration and the longer natural ageing time at (-15 °C) will ensure a higher initial hardness at t=0h due to a higher concentration and/or size of self-clusters of Mg and Si as shown in Figure 3(a). Stage II takes places between 0 and 1h and this is immediately followed by the Stage III wherein the co-clusters form. Due to lower vacancy concentration, the rate of co-cluster formation is slower in xW3 than in xW1 and xW2. Further, the time to obtain the peak maximum hardness is longer in xW3 than xW2 and xW1 because of lower vacancy concentration.

<u>xA1</u>

Due to the slow cooling rate of the air quench process, the possibility of formation of Stage I (self-clusters) is high in xA1 during quenching. This can be observed by the higher hardness value of xA1 at t=0 than xW1. Further, the vacancy concentration in xA1 is far less than any of the water quenched samples. Hence, Stage II can be observed in the hardness curve between 0 and 1h time period wherein the hardness value decreases during the disintegration of the self-clusters. Stage II is followed by the formation of the self-clusters in stages III and IV between 1

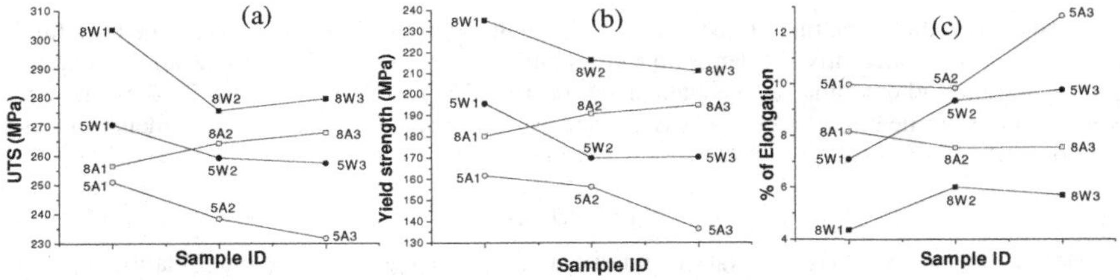

Figure 4. Mechanical properties for experiment conditions shown in Table 1. (a) Ultimate Tensile Strength, (b) yield strength, and (c) % of elongation.

and 3h, and 3 and 10h, respectively. Due to the low vacancy concentration, Stage IV extends to beyond 10h and the hardness value at any time is lower than the water quenched samples.

<u>xA2</u>

The vacancy concentration in xA2 is lower than xA1 due to the natural ageing process for 8 h at room temperature. The natural ageing process will enable a higher concentration and/or size of the self-clusters than in xA1 and this is reflected by the higher hardness value at t=0h. Subsequently, Stages II, III and IV take place during the time periods shown in Table II. Stage V is not reached in XA2 because of low vacancy concentration and the hardness value is always lower than xA1 and the water quenched samples.

<u>xA3</u>

The vacancy concentration is lowest among all the sample conditions. At t=0h, the hardness value of xA3 is higher than xA2 because of the longer ageing time of 3 days at (-15) °C which results in a larger concentration and/or size of the self-clusters. Stage II wherein the disintegration of self-clusters occur takes place between 0 and 1h time period. However, the time periods for Stages III, IV and V are unclear as seen by the erratic behavior of the sample hardness value between 1 and 10h time periods. This may be attributed to the very low vacancy concentration resulting in a very low mobility of the Si and Mg atoms during precipitation reaction. There seem to be an anomalous behavior in the hardness value of xA3 between 4 and 10h time period wherein the hardness value is equal to or less than that observed at t=4h. Stage II may be occurring between t = 1h and 4h.

Figure 4 shows the results of the tensile test samples for the water and air quenched samples after 5h and 8h artificial ageing times. It can be observed that variation of the quenching rates and the natural ageing process results a significant variation of the mechanical properties. The value of yield strength is directly proportional to the harness value at the respective artificial ageing times and the elongation is inversely proportional to the hardness value. This can be notable observed in the anomalous behavior of the xA3 sample in which the hardness value decreases drastically at t=5h and reflected by the lowest YS and highest %el values in Figures 4(b) and 4(c), respectively. Sample 8W1 shows the highest hardness value among all samples and this is reflected by the highest YS and lowest %el values in Figures 4(b) and 4(c).

Conclusions

The study showed that the rate of quenching after solutionizing and the extent and type of natural ageing process significantly influences the mechanical properties of the A356.2 alloy. Further study is required to examine the mechanism of precipitation reaction during artificial ageing and propose optimum heat treatment process for specific mechanical property requirements of cast components.

References

1. Deschamps and Y. Brechet, "Nature and distribution of quench-induced precipitation in an Al–Zn–Mg–Cu alloy," *Scripta Materialia*, 39 (11) (1998), 1517–1522.
2. S.D. Liu, X.M. Zhang, M.A. Chen and J.H. You, "Influence of aging on quench sensitivity effect of 7055 aluminum alloy," *Materials characterizations*, 59 (2008), 53-60.
3. J.L. Cavazos and R. Colas, "Quench sensitivity of a heat treatable aluminum alloy," *Materials Science and Engineering A*, 363 (2003), 171–178.
4. S.T. Lim, S.J. Yun and S.W. Nam, "Improved quench sensitivity in modified aluminum alloy 7175 for thick forging applications," *Materials Science and Engineering A*, 371 (2004), 82–90.
5. M. Murayama, K. Hono, M. Saga and M. Kikuchi, "Atom probe studies on the early stages of precipitation in Al–Mg–Si alloys," *Materials Science and Engineering A*, 250 (1998), 127–132.
6. M. Murayama and K. Hono, "Pre-precipitate clusters and precipitation process in Al-Mg-Si alloys," *Acta materialia*, 47 (5) (1999), 1537-1548.
7. M. Iskandar, D. Reyes, Y. Gaxiola, E. Fudge, J. Foyos, E.W. Lee, P. Kalu, H. Garmestani, J. Ogren and O.S. Es-Said, "On identifying the most critical step in the sequence of heat treating operations in a 7249 aluminum alloy," *Engineering Failure Analysis*, 10 (2003), 199–207.
8. E. Cerri, M. Cabibbo, P. Cavaliere and E. Evangelista, "Mechanical behaviour of 319 heat treated thixo cast bars," *Materials Science Forum*, 331–337 (2000), 259– 264.
9. J.E. Hatch, "Microstructure of alloys, in: Aluminum: Properties and Physical Metallurgy" *ASM International*, (1984), 58–104.
10. G.E. Byczynski, W. Kierkus, D.O. Northwood, D. Penrod and J.H. Sokolowski, "The effect of quench rate on the mechanical properties of 319 Aluminium alloy castings," *Materials Science forum*, 217-222 (1996), 783-788.
11. D. Apelian, S . Shivkumar and G. Sigworth, "Fundamental Aspects of Heat Treatment of Cast Al-Si-Mg Alloys," *AFS Transactions*, 97 (1989), 727-742.
12. S. Shivkumar, S. Ricci, Jr. and D. Apelian, "Effects of solution parameters and simplified supersaturation treatments on tensile properties of cast Al–Si–Mg alloys," *AFS Transactions*, 98 (1990), 913–922.
13. C.D. Marioara, S.J. Andersen, J. Jansen and H.W. Zandbergen, "The influence of temperature and storage time at RT on nucleation of the β" phase in a 6082 Al–Mg–Si alloy," *Acta Materialia*, 51 (2003), 789–796.
14. G.A. Edwards, K. Stiller, G. L. Dunlp and M. J. Couper, "The precipitation sequences in Al-Mg-Si alloys," *Acta materialia*, 46 (11) (1998), 3893-3904.
15. J.T Staley, T.H Brown and R. Schmidt, "Heat treating characteristics of high strength Al–Zn–Mg–Cu alloys with and without silver additions," *Metallurgical Transaction*, 3 (1) (1972), 191–199.

16. Gaber, M.A. Gaffar, M.S. Mostafa and E.F. Abo Zeid, "Precipitation kinetics of Al–1.12 Mg2Si–0.35 Si and Al–1.07 Mg2Si–0.33 Cu alloys," *Journal of Alloys and Compounds*, 429 (2007), 167–175.
17. S. J., Anderson, H.W. Zandbergen, C. Traeholt, U. Tundal and O. Reiso, " The crystal structure of the β" phase in Al-Mg-Si alloys,"*Acta Materialia*, 46 (1998), 3283-3298.

Shape Casting: The 3rd International Symposium
Edited by: John Campbell, Paul N. Crepeau, and Murat Tiryakioğlu
TMS (The Minerals, Metals & Materials Society), 2009

PROCESS PARAMETERS STUDY FOR NET-SHAPE STEEL CASTING

S. N. Lekakh, D. Kline, K. Chandrashekhara, J. Chen, and V. Richards

Missouri University of Science & Technology, Rolla, MO 65409, USA

Keywords: casting, steel, ceramic shell, properties, modeling

Abstract

The objective of this research was the experimental study and computational modeling of process parameters of a net-shape steel casting process. Burst pressure (pressurized water), mechanical and physical properties at room and high temperatures for ceramic shells were evaluated. The experimental data was used to simulate stresses in the ceramic shell, mold filling, and solidification using ABAQUS, FLUENT, and MAGMA software. Models were verified by pouring steel into ceramic shell molds and monitoring with electrical sensors and thermocouples connected to a high-speed DAQ. The results will be used for the optimization of an industrial process.

1. Introduction

Increasingly, rapid manufacturing (RM) is being applied to automotive, motor sports, military components, jewelry, dentistry, orthodontics, medicine and collectibles. Metal casting has a unique possibility to produce complicated three-dimensional shapes from RM patterns with minimal energy and material consumption at short lead times. RM is based on solid freeform fabrication (SFF) which is a collection of techniques for manufacturing solid objects by the sequential delivery of energy and/or material to specified points in space to produce that solid. SFF is sometimes referred to as rapid prototyping, rapid and layered manufacturing. For metal casting processing, different variations of these techniques can be used for mold or pattern preparation. For example, a RM process was recently developed by MS&T[1] based on rapid prototyping (layer by layer freezing) of ice patterns for an investment casting process. That process was unique because of ice's volume contraction during melting compared to the volume expansion of melting wax patterns. Unfortunately, application of this and other known cast RM processes is limited by geometry and sizes of castings.

The objective of this study was to develop a RM process for large scale steel precision castings with complicated geometry. This RM process is based on CAD design and CNC machining of light-weight foam patterns, special ceramic shell development, pattern burn-out during shell firing, and, finally, pouring melt into the shell. RM requires intensive modeling of all stages of processing. Modeling was done by using the specific process parameters and properties in commercial software (MAGMA, FLUENT, ABAQUS). The experiments in this project were performed for measurement and optimization of foam pattern and shell properties for the specific application in the designed process.

2. Experimental Measurements Of Material Properties

Experiments were performed for collecting physical properties of the materials involved in this studied metal casting process, in particular, thermal expansion and burning of different plastic foam patterns and mechanical and thermal properties of the ceramic shell at different temperatures.

2.1 Plastic Foam Pattern

CNC machining (or foaming from beads or liquid reactants in a master mold) is used to produce a light-weight plastic foam pattern. The pattern is used for shell building and will burnout during ceramic shell sintering. The specific properties of the pattern material, such as thermal expansion and foam decomposition, are important for process development.

Thermal expansion of the pattern could create cracks in the weak "green" ceramic shell as well as produce casting distortion during pattern melt-out and burnout. A laser-based large scale dilatometer with sample length up to 100 mm and $\pm$ 1 micron precision was used for measurement of thermal expansion of pattern materials in free standing and mechanically pre-loaded conditions. A variety of EPS (expanded poly-styrene) based foams with differing densities (1.3 – 3 lb/ft^3) and a rigid high density foam produced for investment casting were tested. An example of linear changes during heating of a foam sample 80 mm in length with 1.6 lb/ft^3 density, under 0.9 kPa pressure, is given in Fig. 1a. This material showed a high coefficient of thermal expansion (CTE $=6.38*10^{-5}$ K^{-1} at 20°C – 60°C). Applying pressure significantly decreased thermal expansion of the foam (Fig. 1b) and under a particular pressure the pattern could have near zero expansion (Fig. 1c). The type of foam influenced thermal expansion; in particular, the high density foam had significantly larger thermal expansion (Fig. 1d).

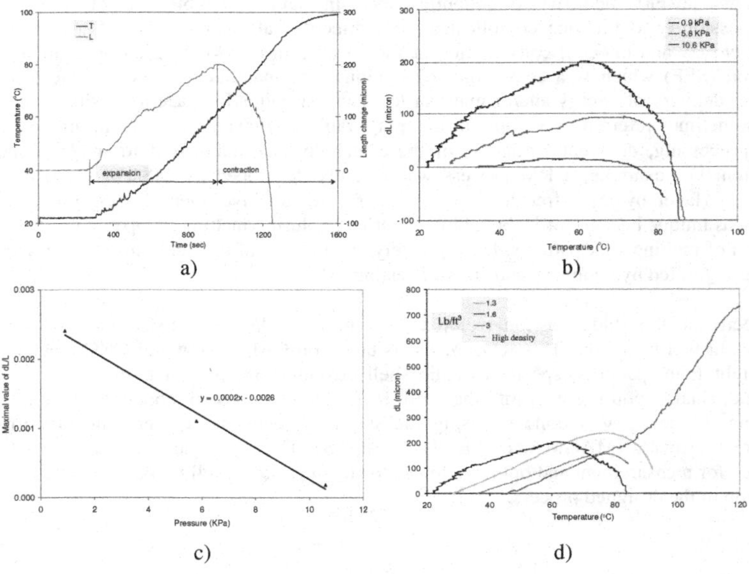

Figure 1. Thermal expansion of foam with density 1.6 lb/ft^3 (a), effects of applied pressure (b, c), and density (d).

Thermal gravimetric analysis (TGA). In order to insure reproducible carbon content of steel castings, full foam decomposition (without residual carbon) during shell firing is required. A large scale TGA apparatus (up to 300 grams sample capacity with 10^{-4} gram precision) was used for the study of thermal decomposition of different types of foams in air. In addition, different materials used for pattern processing (wax for surface quality improvement, glue for pattern assembly) were tested. The majority of foam materials were almost completely decomposed below 500°C while removing residual carbon required up to 900°C when heating at 10°C/min (Fig. 2). The rigid, high density foam had a significantly higher decomposition temperature when compared to general EPS foams. Holding time decreased the upper decomposition temperature limit from 900°C to within the range 820°C-850°C. The decomposition results are summarized in Table I.

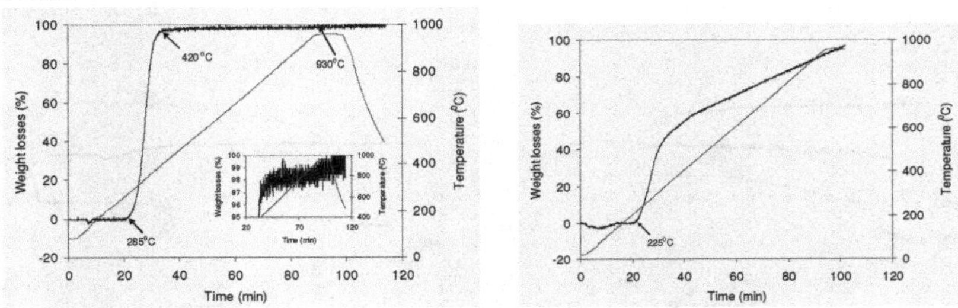

a) b)

Figure 2. TGA analysis (red – temperature, black – weight) of different foam: a) EPS foam with density 1.6 lb/ft^3 and b) high density foam.

Table I. Decomposition (Burning) Temperatures (°C) of Pattern Materials.

Material	Start burning	Finish burning	Carbon deposit burn-out
EPS foam (1.3 lb/ft^3)	285	420	930
EPS foam (1.6 lb/ft^3)	295	450	920
30%EPS+ 70%PMMA Foam	270	420	-
High density foam	225	900	920
Wax	200	415	690
Glue	280	480	570

2.2 Ceramic Shell

Different processes take place during ceramic shell development on foam pattern, firing, and metal pouring. The physical and mechanical properties of a ceramic shell play an important role in casting quality. In this project, a special high strength ceramic shell was developed. The shell was made using colloidal silica binder with silica-zircon flour for all coats. -30+50 mesh fused silica was used for stuccoing of the 7 back-up layers.

Thermal expansion. Different structural processes including sintering and devitrification of amorphous silica could occur during firing of the "green" ceramic stucco shell. These processes are responsible for changing shell dimensions during firing and cooling. The laser assisted large scale dilatometer was used for measurement of the thermal expansion/contraction of the shell during two thermal cycles. The first cycle imitated shell firing at different temperature-time conditions (heating at 10°C/min, holding from 10 min to 60 min at 800°C, 950°C, or 1200 °C and cooling in furnaces). After that, all shells were tested during a second heating cycle at

1250°C which imitated steel pouring. The results are given in Fig. 3 where positive lineal change is expansion and negative is contraction.

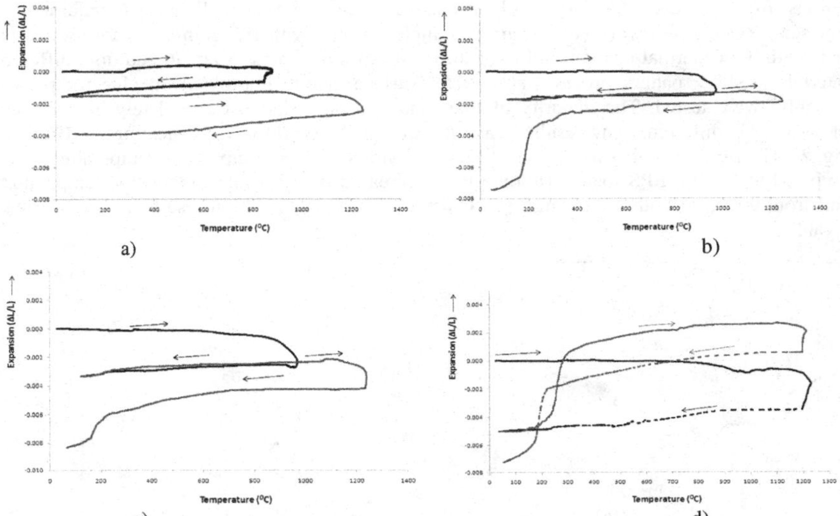

Figure 3. Changing shell linear dimensions during firing (black) a) 850°C, 10 min., b) 950°C, 10 min., c) 950°C, 60 min., and d) 1200°C, 30 min., cooling and second heating at 1250°C (red).

The "green" shell had minimal expansion during first heating from room temperature to 700°C, which later changed to contraction at temperatures above 800-900°C. Contraction was a result of sintering and became significant during holding the shell at 1200°C. When the fired shell was cooled, contraction with a slightly higher CTE took place when compared to the heating cycle. As a result of sintering, the length of a cooled sample after the first firing cycle decreased by 0.15% for a 850°C firing temperature and by 0.5% for a 1200°C firing temperature. The second heating cycle, imitating steel pouring, could increase the total dimensional changes by additional contraction during firing at high temperature as well as by partial crystallization of originally amorphous silica binder. Cooling the crystalline silica transformation product causes a large volume change which could intensify micro-cracks formation. Controlled by firing temperature, cristabolite transformation could be helpful for detaching the shell from the casting and removing internal cores.

Mechanical properties. A material's response to loading is an important factor in predicting possible ceramic shell failures. Stress-strain curves for previously fired shells were created using a 3-point flexural test. The flexural test was performed at room and elevated temperatures up to 1200°C with 300°C increments. Temperature dependence of the maximum flexural stresses and strains are displayed in Fig. 4. In most cases, Stress-strain curves exhibited non-linear behavior which could be affected by different mechanical properties of prime and back-up layers. The ceramic shells were able to support the most stress at 600°C and 900°C while they exhibited the large strain upon failure at higher temperature. Experimental data provided quantitative information about the thermal behavior of ceramic shells and will be used for process parameter optimization. According to the preliminary analysis, an optimal firing process could provide high strength of shell with the minimal dimensional changes.

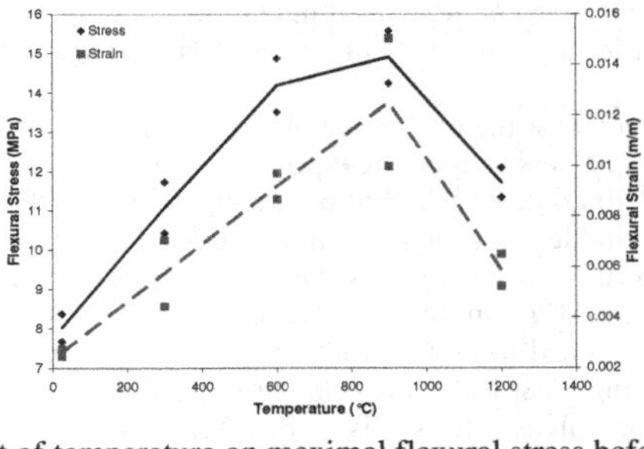

Figure 4. The effect of temperature on maximal flexural stress before failure of shell.

Coefficient of thermal conductivity (k) of a ceramic shell plays an important role during casting solidification. The modified "Hot wire" experimental-computational method was suggested for measurement of *k* for a ceramic shell at different temperatures. A heater of platinum wire (0.4 mm diameter, 62 mm length) had two outside leads connected to a programmable power supply. Two internal leads were connected to a 24-bit data acquisition system. Significantly larger power and heating rate were used in this modified method as compared to ASTM C1113-99 recommendations. This increase was necessary for correct measurement of thin ceramic shell at elevated temperatures. A special computational procedure using FLUENT software was used for inverse calculation of the thermal properties based on fitting experimental data to computed heating rates. The results are given in Fig. 5.

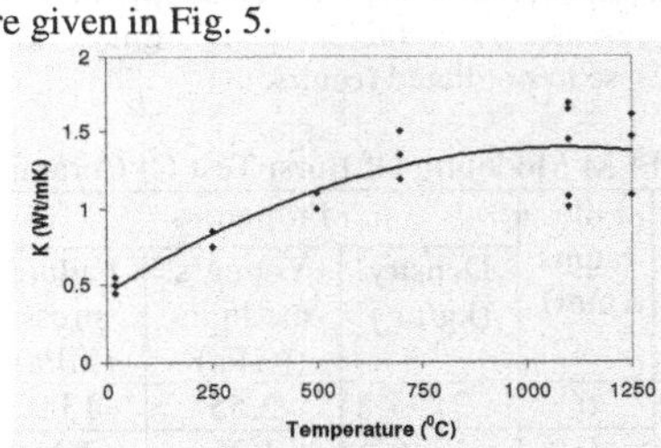

Figure 5. Thermal conductivity of ceramic shell.

3. Modeling and Experimental Verification

3.1 Stress Shell Modeling and Burst Test Verification

A comprehensive nonlinear finite element model was developed to predict the behavior of ceramic shells subjected to internal pressure. The model was formulated in a three-dimensional frame with material nonlinearity taken into account. A brittle material model[2] was used to investigate the mechanical behavior of the ceramic shell during loading. The brittle behavior of ceramic is associated with cleavage, shear and mixed mode fracture mechanisms that are observed under tension and tension-compression states of stress. The brittle material model represents the discontinuous macro-crack brittle behavior in ceramics. The presence of cracks affect the stress and material stiffness associated with each material calculation point. The material model was input in the commercial finite element code ABAQUS/Explicit which is suitable for problems with material complexities or for meshes with a large number of elements. Due to symmetry, a quarter of the ceramic shell is modeled in ABAQUS/CAE. A brick element with eight nodes was used to mesh the shells with different shapes. To avoid the shear-locking problem associated with this element, a reduced integration strategy with hourglass control[3] was used. The material properties used in the simulation are listed in the Table 2.

The stress modeling was done for the evaluation of material properties and shape affect on shell mold integrity. Square prisms versus cylindrical prisms, with bottom, were compared. The effect of fillet radius also was defined. In this study, room temperature MOR (modulus of rupture) test results were used for the modeling (Table 2). It was done for the possibility of experimental verification of modeled results. Modeling was done for a square prism with bottom having 3.7 mm wall thickness and 8x30.5x70 mm internal dimensions and for a cylindrical prism with 30.5 mm diameter, 70 mm length and the same wall thickness. The results of stress modeling were experimentally verified using burst tests. The burst test apparatus included a water pump, digital pressure gages, and needle valves. The shells were glued to the device and internally applied water pressure was increased up to shell fracture. The modeling results showed that the shape of the shell had a large effect on shell integrity and fillet radius plays an important role in burst pressure.

Fig. 6 shows the stress distribution of ceramic shells with different shapes. For a square prism shell, the maximum stress concentrates at the corner of the wall. The crack occurs at the top of the corner and propagates along the corner and will result in shell failure. For a round prism shell, the maximum stress occurs at the corner where the wall connected with the bottom. The failure occurs first at the bottom of the shell. The predicted burst pressure of ceramic shell with different shapes is shown in Table 2. In comparison to the square prism, the round prism can withstand just over three times the pressure. Also, the round prism with a 5mm fillet at the corner can further increase burst pressure by a factor of 1.7 times, as shown in Table II. The experimental data were very close to predicted results.

Table II. FEM Modeling Of Burst Test Of Ceramic Shell

Shell shape	Fillet radius (mm)	Properties			Burst pressure (kPa)	
		Density (kg/m^3)	Young's modulus (GPa)	Failure stress (MPa)	Predicted	Measured
Square prism	0	2195	2.55	13.9	360	380
Industrial square prism (7.5 mm thickness)	0	1816	1.25	2.41	175	190-220
Round prism	0	2195	2.55	13.9	1125	969
Round prism	5	2195	2.55	13.9	1925	N/A

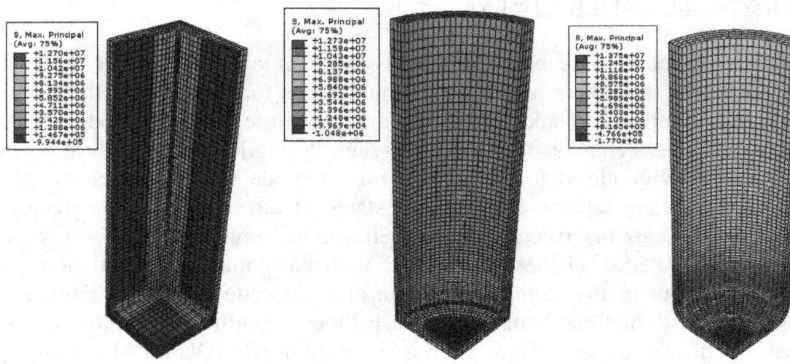

a) b) c)

Figure 6. Stress distribution at failure in square prism without fillet (a), round prism without fillet (b) and with 5 mm fillet radius (c).

3.2 Mold Filling Modeling and Experimental Verification

Re-oxidation of steel during mold filling can damage the quality of high strength steel. To avoid this effect, different gating systems for pouring a 127x152x30.5 mm vertical plate were modeled with MAGMA software. The recommended bottom fill gating system design for minimizing of the "fountain" effect and melt-air turbulences in the casting is shown in Fig. 7a. For experimental verification of melt flow, steel wires (18gage) were inserted through the plate with equal spacing forming a 7x8 array and connected to a high speed data acquisition system. Also, another wire was placed in the gating which was connected to a voltage source and served as an electrode. Contour plots of the simulated and actual filling times across the plate are displayed in Fig. 7. The profile of contour lines was approximated for 0.1 sec increments using the discrete values of measured time when electrical circuits were formed by the steel, pins and electrode placed in the gating system. Constant melt velocity between neighboring pins was assumed. It was shown that the tested gating system practically eliminated the "fountain" effect during pouring in the ceramic shell.

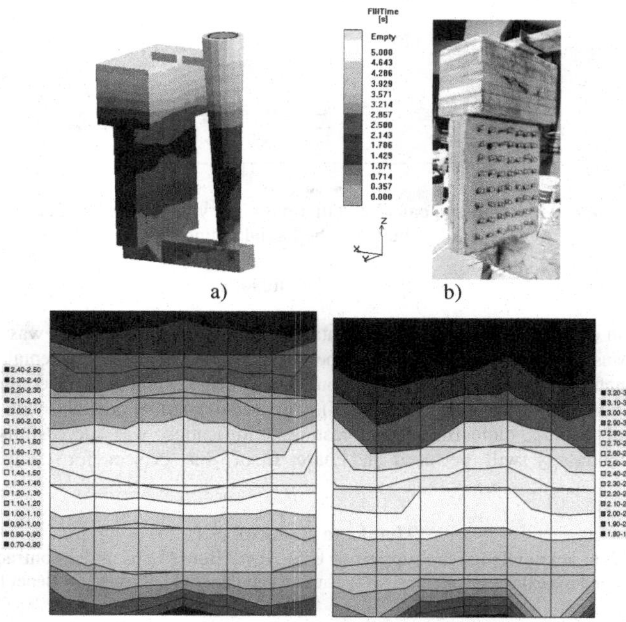

Figure 7. Gating system computational (a) and experimental (b), simulated (c) and experimental (d) fill patterns, for a vertical plate. Each color band represents a 0.1 second time interval.

3.3. Casting Solidification Modeling and Experimental Verification

The comparison was done for 20 mm thickness vertical plate (100x100 mm) from 4340 steel (0.4%C. 0.7%Cr, 0.7%Mn. 0.3%Mo, 1.8%Ni) poured into room temperature shell mold. Three thermocouples were placed in the center of the casting (S-type), next to the prime coat (S-type), and near the external shell surface (K-type). Fluent software was used for solidification modeling with known thermal properties of materials under a variety of casting-shell boundary conditions.

Solidification temperature and enthalpy of 4340 steel were calculated with FACTSAGE software. When the coupled boundary conditions without any additional thermal resistance at the mold-metal interface were used, the computed solidification time was equal to that experimentally measured (Fig. 8). Casting-shell and shell-sand boundary temperatures were calculated directly at interfaces, while thermocouples reading were restricted to approximately 1 mm inside the shell. Modeling showed that the small changes in thermocouple position had an effected on the absolute value of measured temperature. The thermal behavior of calculated and experimental curves (Fig. 8) are identical (position of maximum) while the experimental values were inside the calculated temperature interval. The main part of latent heat was accumulated by the shell rather than transferred to the sand. This result confirmed that shell thermal properties (heat capacity, density and heat conductivity) play an important role in thin wall steel casting solidification because heat transfer occurred under non-steady state thermal conditions.

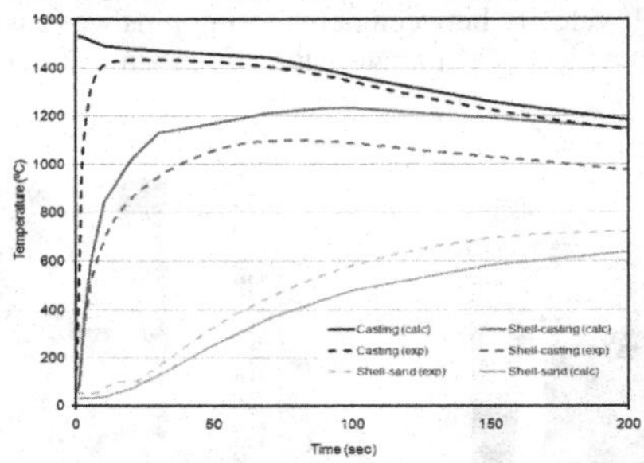

Figure 6. Measured and calculated thermal history in the center of casting, casting-shell and shell-sand boundaries.

Conclusions

The combination of a vast variety of experimental and computational methods was used for the study of a process for production of net-shape steel castings. The experimentally measured properties of materials involved in the shell building process were used in conjunction with software (FLUENT, MAGMA, and ABAQUS). The verification of models, which described mechanical shell loading, mold filling, and casting solidification, was successfully done. The developed methodology will be used in Phase II of this Project for complex process optimization.

Thanks to sponsors

The authors wish to acknowledge the support of U.S. Army Benet Labs under contract W15QKN-07-2-0004. The opinions expressed are those of the authors and not of U. S. Army or Benet Labs.

References

1. Chun-Ju Huang, Ming C. Leu, Von L. Richards, "Investment Casting With Ice Patterns And Comparison With Other Types Of Rapid Prototyping Patterns", *Shape Casting: The 2nd International Symposium*, TMS, 2007.
2. A. Hillerborg, M. Modeer, and P.E. Petersson, "Analysis of Crack Formation and Crack Growth in Concrete by Means of Fracture Mechanics and Finite Elements," *Cement and Concrete Research*, 6 (1976), 773-782.
3. ABAQUS, ABAQUS Theory Manual, ABAQUS Inc., Providence, RI, 2008.

Shape Casting: The 3rd International Symposium
Edited by: John Campbell, Paul N. Crepeau, and Murat Tiryakioğlu
TMS (The Minerals, Metals & Materials Society), 2009

IMPROVING BUILD SPEED IN RAPID FREEZE PROTOTYPING THROUGH INCREASE OF HEAT TRANSFER

Ming C. Leu [1], Sriram P. Isanaka [1], Von L. Richards [2]

[1]Dept. of Mechanical and Aerospace Engineering, Missouri University of Science and Technology,
400 W. 13th St., Rolla, MO 65409, USA
[2]Dept. of Materials Science and Engineering, Missouri University of Science and Technology,
223 McNutt Hall, 1400 N. Bishop, Rolla, MO 65409- 0330, USA

Keywords: Rapid Freeze Prototyping, Chilling Plate, Fluent Analysis

Abstract

The heat transfer in the Rapid Freeze Prototyping (RFP) process has been significantly increased for improvement of build speed. RFP is a solid freeform fabrication process in which water droplets are deposited and solidified layer-by-layer to form three-dimensional ice patterns for investment casting. Mechanisms have been devised to cool the substrate to as low as -140 °C. Chilling plates were developed to enable effective transfer of heat with the aid of conduction. To ensure that the deposited water does not freeze to the chilling plate, various surface coats were investigated. The most effective interface material was identified using contact angles measured with high resolution digital photography. The experimental results were substantiated with simulations performed using Fluent. The improvements in build speed after incorporating the above changes were measured to verify the trends predicted from the simulations.

Introduction

The Rapid Freeze Prototyping (RFP) apparatus at the Missouri University of Science and Technology consists of a substrate table capable of X-Y travel and a nozzle head capable of motion in the Z direction [1]. The entire setup including the axes is housed inside a freezer in a controlled environment continuously maintained at -20°C as shown in Figure 1. The required part geometry is traced by the X-Y table, while fine water droplets are being deposited through the nozzle head mounted on the Z axis. The water droplets freeze on contact with the substrate in a layer-by-layer manner, thereby forming a three-dimensional geometry. The three-dimensional parts can be used as replacements for wax patterns in investment casting [2, 3, 4]. Based on research we conducted in the past [5, 6, 7] one of the factors that greatly affects the build speed is the temperature prevalent in the system during part fabrication.

This paper describes our attempt to lower the temperature near the part during the deposition process to increase the heat transfer and reduce the build time. Increase in heat transfer is achieved by

1. Conduction – This is done by lowering the temperature of the substrate by inclusion of slots and by using chilling plates to facilitate higher heat conduction.
2. Convection – This is done by using fans to create forced convection in the system.

Predictive studies by means of heat transfer simulations using Fluent were performed and compared with experimental results.

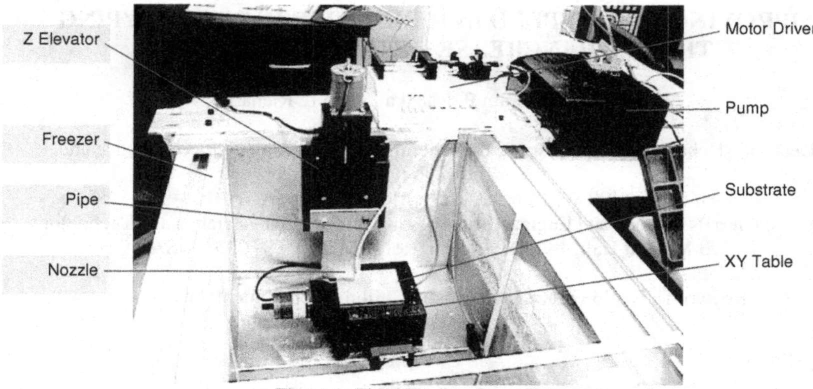

Figure 1: The RFP experimental setup.

Modification of Substrate

The substrate initially used in the RFP system was a solid block of aluminum which could be maintained at a temperature of -20°C. The original substrate design was modified to incorporate slots into which liquid nitrogen could be input. The logic behind this idea was to increase the surface area of contact between liquid nitrogen and the aluminum substrate, thereby reducing the temperature of the substrate considerably. Using the new substrate design in conjunction with liquid nitrogen we have been able to reach substrate temperatures down to -140°C. Figures 2 and 3 show the original substrate and the modified substrate, respectively.

Figure 2: Original substrate design.

Figure 3: New substrate design.

Temperatures that could be achieved using the two substrates were measured. Figures 4 and 5 are the plots of the temperature data of the original and new substrate design, respectively. To obtain the temperature data from the two substrates, holes were drilled into the substrates just under the surface. This was done to facilitate the insertion of thermocouples at nine different points along the surface of the substrates. Because the temperatures measured by the nine thermocouples are very similar, Figures 4 and 5 show the temperatures obtained from only two thermocouples. With the original substrate design the thermocouples indicate a downward temperature trend (Figure 4), which was expected because of the use of liquid nitrogen. However, even with considerable use of liquid nitrogen the minimum temperature achievable was only -40°C, after which the temperature leveled out and remained a constant. However, with the new substrate

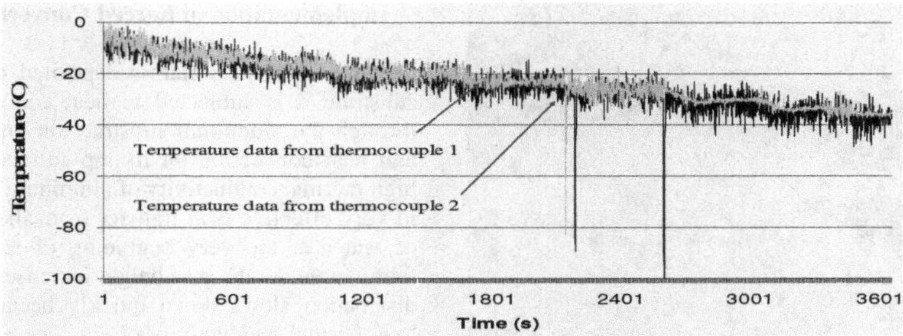

Figure 4: Temperature during part build (with original substrate design).

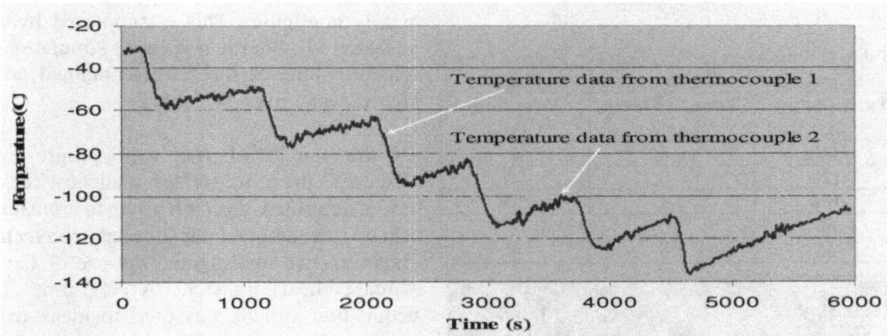

Figure 5: Temperature during part build (with new substrate design).

design the downward trend continued to progress till about -140°C before leveling out. Also the amount of liquid nitrogen used to achieve this very low temperature was less than one half of that used for the original substrate.

Simulations performed in Fluent [8] to predict the time taken for a layer to freeze showed a significant reduction in water droplet freezing time with decrease in the substrate temperature. The total heat loss and the liquid fraction in the system were simulated in Fluent to identify the wait time between layers. The results indicate that with the temperature around -140°C, the wait time between layers can be shortened to around 5 seconds. One of the successful parts built using this method is shown in Figure 6.

These simulation results were compared with the experimental results wherein test parts were built with reduced wait time between layers. From Table 1 we conclude that the lower the substrate temperature, the lower the wait time between layers. During the practical implementation of this concept we observed the occurrence of a bowing phenomenon in the built ice part. At temperatures lower than -70°C we noticed the advent of the bows, and the severity of the bows increased with decreasing temperature. Due to this phenomenon we restricted ourselves to fabricating parts at temperatures above -70°C. With the new substrate design we could reduce the wait time between layers to about 15 seconds and still produce parts with good accuracy. The smallest possible wait time with the older substrate was 40 seconds. Thus we have achieved a reduction of over 60% in the wait time between layers when using the new substrate.

Figure 6: Part built with lowered substrate temperature.

Table 1: Simulation results of time to freeze a layer of water at various temperatures with Fluent.

Temperature (°C)	Time to Freeze (s)	Wait time (s)
-140	0.8	5
-120	0.9	5.5
-100	1.1	6.4
-80	1.0	7
-60	1.5	8

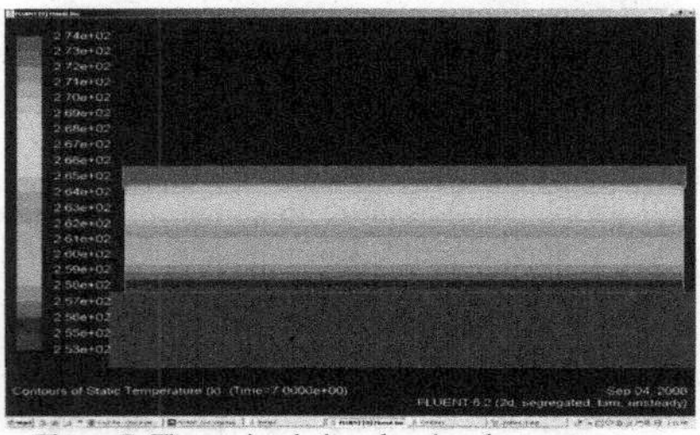

Figure 7: Fluent simulation showing the temperature distribution across the height of an ice wall at 7 seconds after depositing a layer of water.

Table 2: Simulated values of time to freeze and wait time for a water layer deposited at a particular height.

Height * (mm)	Time to Freeze (s)	Wait Time (s)
0	2.8	5.5
1.0	5.8	9.7
2.5	8.4	17.5
5.0	10.1	28.3
10.0	12	42
* 0 mm = on substrate surface		

Implementation of Forced Convection

When a layer of water is deposited on the substrate it is subjected to heat conduction through the aluminum substrate on one side and heat convection on its top surface. The high thermal conductivity of aluminum leads to very effective heat transfer from the layer of water at the very beginning of ice part fabrication. As the part height increases, heat dissipation slows down quickly because of low thermal conductivity of ice. The amount of heat dissipated through convection is nearly negligible. This is supported by past research [5]. We ran a series of simulations in Fluent. Some of the results obtained are in Figure 7 and given in Table 2.

As the height of the wall progressively increases, there is greater tendency for the newly deposited layer of water to transfer its heat to the ambient air through convection. Therefore, we investigated the use of fans to increase heat transfer. A real-time data acquisition system was used to measure the temperature in the RFP apparatus using two thermocouples, one placed at 10 mm from the part build surface and the other at 1 meter away of part fabrication. The results obtained are shown in Figure 8. From this figure we can see that the heat transfer is concentrating in the neighborhood of the deposited water. This is evidenced from the fact that the thermocouple that was away from the part build surface exhibited nearly constant temperature throughout the build cycle. Liquid nitrogen was utilized to reduce the temperature around the part. The effect of liquid nitrogen is indicated by the sharp downward spikes in Figures 8 and 9.

When a similar investigation was conducted with the introduction of forced convection by using a fan to blow liquid nitrogen cooled air into the build area, the cooling of the part was found to be much faster than without forced convection. Using this method, the ambient temperature in the range of -25 °C to -30 °C could be easily maintained in the build region.

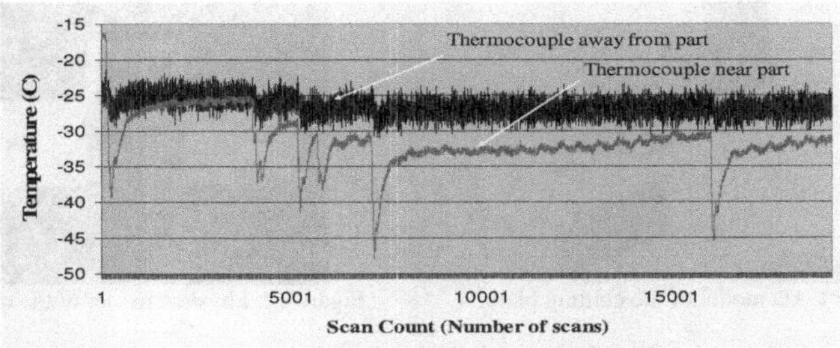

Figure 8: Temperature vs. Time for part build without forced convection.

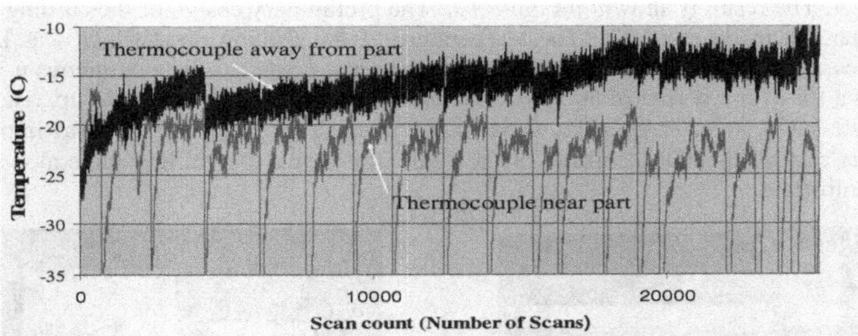

Figure 9: Temperature vs. Time during part build with forced convection.

Successful parts were built with the wait time of 20 seconds in this case vs. a 40 second wait time in the case without forced convection, hence a significant reduction of 50% in the part build time has been achieved. Although this cooling method provided considerable advantages, it came along with the problem of frosting. As can be seen in Figure 10 excessive amounts of frost appear on the surface of the ice part. Therefore, despite the reduction in wait time, this method was not seen as a viable option for reducing the build time for ice part fabrication.

Figure 10: Wall built using forced convection.

Concept of Chilling Plate

Another idea we investigated was utilizing pre-cooled plates to freeze deposited water more effectively. Besides cooling, the chilling plate provided a surface which could be used to flatten the newly deposited layer of water. As conduction is a more effective mode of heat transfer than convection, it was conjectured that this plate could reduce the build time significantly. Preliminary experiments conducted while trying to freeze a layer of water between the ice substrate and the chilling plate indicated a tendency of the water freezing onto the chilling plate. During separation of the ice part from the chilling plate the ice part can break easily. Therefore, identification of an interface material that would prevent the water from freezing onto the chilling plate became critical to the success of this technique.

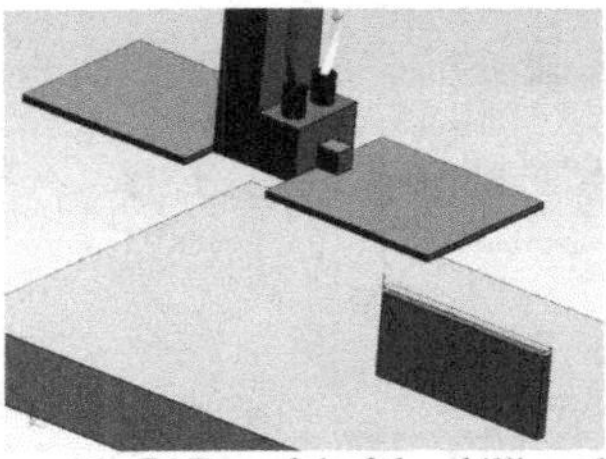

Figure 11: CAD model of the chilling plate.

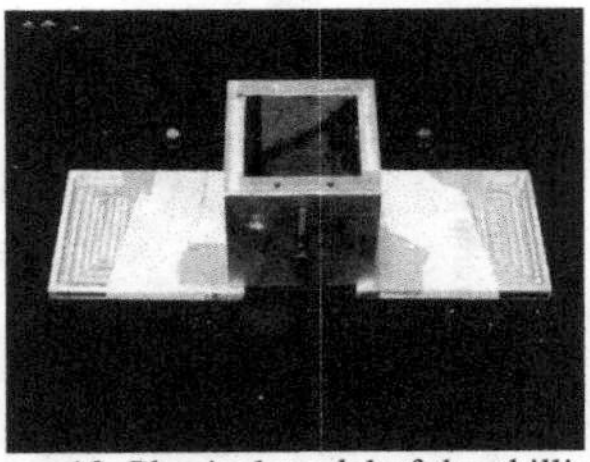

Figure 12: Physical model of the chilling plate.

Fluent was used to simulate the temperature, which is an indicator of heat transfer, and the liquid fraction, which is an indicator of the time taken to freeze a layer of water housed on the chilling plate concept. The result is shown in Figure 13. The preliminary design of the chilling plate was a 3×3" square of thickness 0.2". The temperature of the chilling plate which was initially at -20°C, increases to -8°C after 6 seconds. Fluent simulation predicted that continuous usage of the chilling plate for over a period of 50 seconds would result in residual heat build up, rendering the chilling plate ineffective. Hence, the design for the substrate plate with slots to improve heat transfer was also used for the chilling plate. Figure 14 shows the chilling plate along with the aluminum substrate.

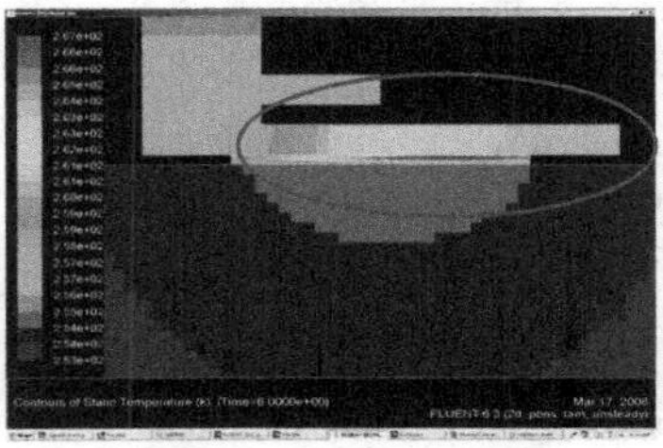

Figure 13: Temperature at 6 seconds. (Area of residual heat buildup circled.)

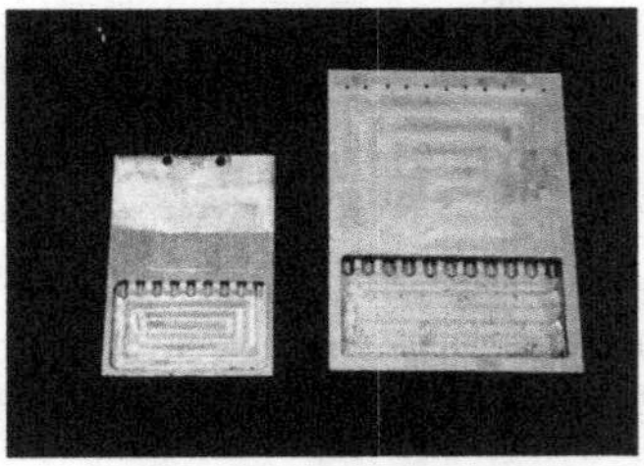

Figure 14: The chilling plate (left) and aluminum substrate (right).

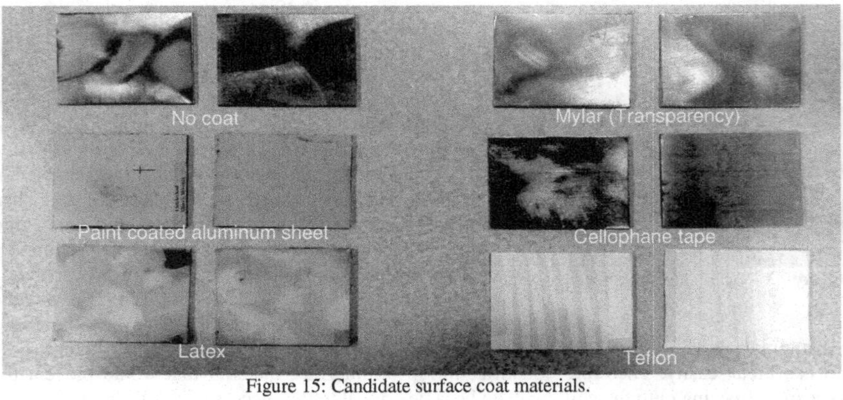

Figure 15: Candidate surface coat materials.

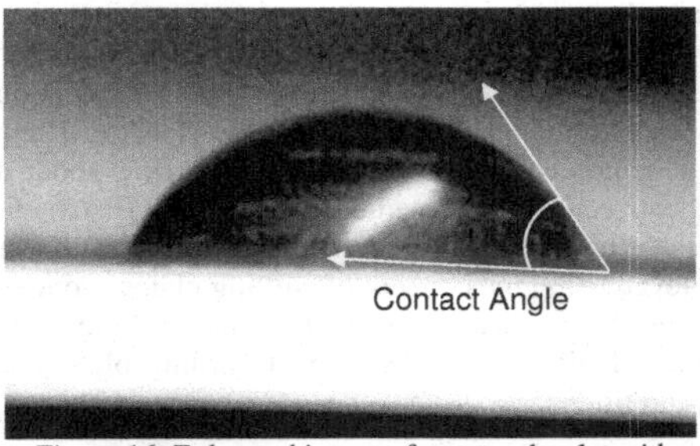

Figure 16: Enhanced image of a water droplet with contact angle shown.

A major problem in the chilling plate approach was finding the right surface coat that could help prevent water from adhering to the chilling plate. Materials including paint coated aluminum sheets, mylar, cellophane tape, latex, and Teflon, as shown in Figure 15, were tested to find the extent of their water repulsive properties at -25°C. The best material was chosen by examining the contact angle between a water droplet and the surfaces of varying coat materials. The higher the contact angle, the greater is water's tendency to repel that surface. Measurement of contact angle was achieved by taking high resolution digital photographs of a water droplet on the surfaces of these materials. With ImageJ software [9], we measured the actual contact angle between a water droplet and a surface coat as shown in Figure 16. Table 3 shows the measured contact angles for different coat on the chilling plate.

Table 3: Contact angles of water droplets with different coat materials.

Coat material	Contact angle of water droplet (degrees)
No coat (unpolished face)	35
No coat (polished face)	50
Paint coated aluminum sheet	44
Mylar (Transparencies)	70
Cellophane tape	65
Latex	80
Teflon	105

From the measurements we found that the highest contact angle occurs when the water droplet was in contact with Teflon. This is also confirmed by measurements done by other researchers [10]. Therefore, Teflon was identified as the best surface coat. The chilling plate coated with a Teflon film of thickness 0.1mm was then utilized to produce ice parts with short wait time between layers. This method produced ice patterns better than using forced convection because of the advantage of no frost formation. However, the greatest advantage gained by utilizing this method is the reduction in wait time. The two parts in Figures 17 and 18 were built with wait time of 10 seconds between layers. This translates to 75 percent reduction in part build time.

Figure 17: Wall of height 3mm.

Figure 18: Wall of height 10 mm.

Conclusion

Three techniques were investigated to increase heat transfer and reduce build time in fabrication of ice parts by the RFP process. Modifications made to the substrate by the use of slots for filling with liquid nitrogen to decrease the surface temperature were successful in bringing about a reduction in part build time of 62.5%. Forced convection implemented with the use of fans to circulate cold air produced desirable reduction of build time but with undesirable formation of frost. High reduction in part build time was achieved using the concept of chilling plates. Contact angle measurements and Fluent heat transfer simulations were used to help the chilling plate design and identify the ideal surface coat for the RFP system. The use of chilling plate was shown to bring about a 75% reduction in part build time.

References

1. Leu, M.C., Zhang, W., and Sui, G., 2000, "An Experimental and Analytical Study of Ice Part Fabrication with Rapid Freeze Prototyping", CIRP Annals, Vol. 49/1, pp. 147-150.
2. Richards, V. L., Druschitz, E., Isanaka, S. P., Leu, M. C., Cavins, M., and Hill, T., "Rapid Freeze Prototyping of investment cast thin-wall metal matrix composites I – Pattern build and molding parameters", TMS 2008 Annual Meeting & Exhibition, New Orleans, Louisiana, U.S.A, March 9-13, 2008.
3. Liu, Q., Leu, M. C., and Richards, V. "Investment casting with ice patterns from rapid freeze prototyping", 2004 NSF Design, Service and Manufacture Grantees and Research Conference, Dallas, U.S.A, January 5-8, 2004.
4. Liu, Q., Sui, G., and Leu, M. C., 2002, "Experimental study on ice pattern fabrication for the Investment Casting by Rapid Freeze prototyping (RFP)," Journal of Computers in Industry, 48(3), pp. 181 – 197.
5. Sui, G., and Leu, M. C., 2003, "Thermal Analysis of Ice walls built by Rapid Freeze Prototyping," Journal of Manufacturing Science and Engineering, Volume 125, Issue 4, pp. 824-834.
6. Sui, G., and Leu, M. C., 2003, "Investigation of Layer Thickness and Surface Roughness in Rapid Freeze Prototyping," Journal of Manufacturing Science and Engineering, Volume 125, Issue 3, pp. 556-563.
7. Leu, M. C., Liu, Q., and Bryant, F.D., 2003, "Study of Part Geometric Features and Support Materials in Rapid Freeze Prototyping," Annals of CIRP, 52 (1), pp. 185 – 188.
8. Fluent Inc., 2003, Fluent Online Users Guide, last referred on 19th September, 2008 (http://bubble.me.udel.edu/wang/teaching/MEx81/fluent-help/html/ug/main_pre.htm).
9. ImageJ, Users documentation, last referred on 10th September, 2008 (http://rsbweb.nih.gov/ij/docs/index.html).
10. Teflon contact angle reference, last referred on 20th September, 2008 (http://www.lib.umich.edu/dentlib/Dental_tables/Contangle.html).

Shape Casting: The 3rd International Symposium
Edited by: John Campbell, Paul N. Crepeau, and Murat Tiryakioğlu
TMS (The Minerals, Metals & Materials Society), 2009

COOLING PROPERTIES OF FROZEN SAND MOLDS FOR CASTING OF LEAD FREE BRONZE

Hiroyuki Nakayama[1], Shuji Tada[1], Toshiyuki Nishio[1], Keizo Kobayashi[1]

[1]AIST (National Institute of Advanced Industrial Science and Technology);
2266-98, Anagahora, Shimoshidami, Moriyama, Nagoya, 463-8560, JAPAN

Keywords: Effset process, microstructure, temperature transition

Abstract

A frozen mold is an advanced sand mold produced by freezing a mixture of sand and water. Cast lead free bronze plates, which had dimensions of 30x150x5 mm, were produced using the frozen mold casting process, also called the "*Effset Process*" [1]. The cooling behavior of the cast plates were measured at three points (near the gate, the center and the far side) by a high speed recorder with a sampling time of 1 ms. The cooling behavior of the bronze castings was divided into 3 stages; (1) rapid cooling, (2) plateau region and (3) slow cooling. The plateau temperature decreased and the plateau time shortened as the water content of the mold was increased. This data demonstrated that the cooling properties of the frozen mold were increased as the water content was increased. As a result, a finer microstructure was obtained by increasing the water content of the mold and by decreasing the pouring temperature.

Introduction

The "*Effset Process*" was developed in the United Kingdom in the1970's [1]. In this process, molds are produced by freezing a mixture of water and sand. The comparison of the *Effset process* and the conventional bonded sand casting process are schematically shown in Figure 1.

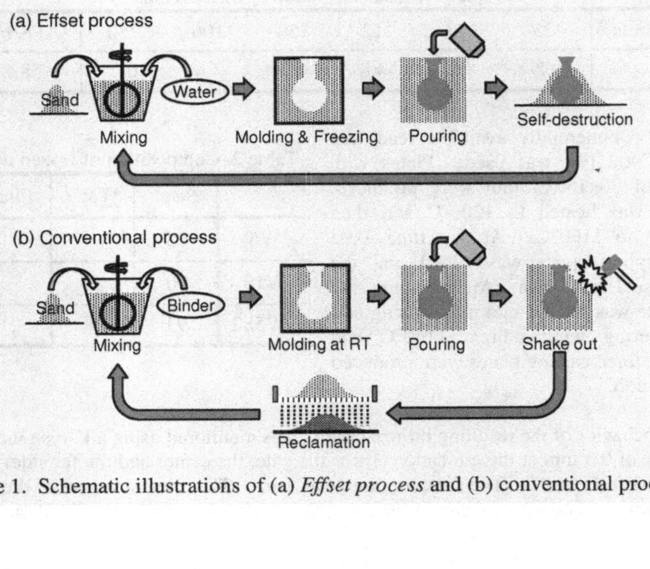

Figure 1. Schematic illustrations of (a) *Effset process* and (b) conventional process.

Since the sand is bonded with ice, a frozen mold does not need any chemical binders for its forming. Therefore, the *Effset process* is an environmental friendly process. In addition, the mold spontaneously decomposes when the ice thaws (self-decomposing). This means no shake-out process is required. Therefore, problems such as noise, vibration and dust, which are inevitable in the conventional bonded sand casting process, are reduced. At the same time, the *Effset process* has disadvantageous. The production cost of a frozen mold used to be very high, because the production of frozen molds required a large volume of liquid nitrogen. Therefore, the *Effset process* was not practical. Recently, a new method to produce a frozen mold with a lower cost and shorter processing time has been developed [2], thus the *Effset process* is receiving new attention.

The cooling rate of the molten metal just after pouring affects the microstructure of the casting. In the case of the *Effset process*, the water near the inner surface of the mold, which contacts the molten metal, is immediately evaporated. Thus, the cooling properties of a frozen mold change during casting cooling. For the *Effset process*, the details of the cooling behavior of the casting have not been studied. In this study, the effects of the frozen mold production variables (sand/water ratio and additives) on the casting cooling behavior and microstructure were investigated.

Experimental Procedure

Silica sand and water were used for the mold. The size distribution of the sand is shown in Table 1. In some molds, fine silica particles (colloidal silica) with an average diameter of 10 - 20 nm were added to enhance the mold strength [3]. The silica sand, water and colloidal silica were mixed together, poured into a mold and the mold was frozen by placing it in a freezer held below -30°C for 12 hrs. The compositions and dimensions of the frozen molds used in this study are listed in Table 2 and shown in Figure 2, respectively. In Figure 3 shows the frozen mold used in this study and the cast bronze produced by *Effset process*.

Table 1. Size Distribution of Silica Sand.

Micron	425	300	212	150	106	750	AFS-FN
Mass %	0.2	5.6	54.8	34.8	4.4	0.2	58.7

For casting, commercially available lead free bronze, CAC902 [4], was used. Plates with dimensions of 30x150x5 mm were produced. The bronze was heated to 1200°C and then cooled down to 1160°C. At that time, 0.02 mass% phosphor-copper was added and the bronze was held for 4 min. After holding, the molten bronze was poured into molds using two different pouring temperatures: 1140°C and 1100°C. The three casting plates were produced in each condition.

Table 2 Composition of frozen molds.

	Sand	Water	Fine Silica
W5	95	5	-
W10	90	10	-
W5S2	93	5	2

The cooling behavior of the resulting bronze casting was monitored using a K-type thermocouple with diameter of 0.3 mm at three locations (near the gate, the center and the far side) as denoted by solid circles in Figure 2. The sampling rate was 1 ms. The microstructure of the cast plates was examined. The cast plate was cut along the longitudinal direction and then metallographically polished. The surface was etched in a solution of 4 mass% $FeCl_2$ – 9 mass% HCl – 87 mass% H_2O.

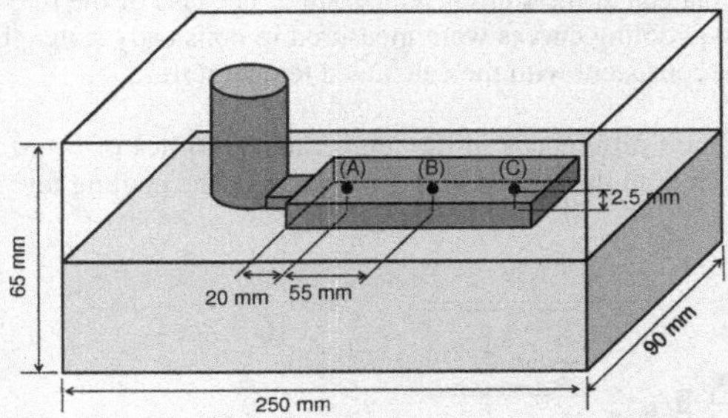

Figure. 2 Drawing of the mold used in this study. Temperature measurement points are denoted by solid circles.

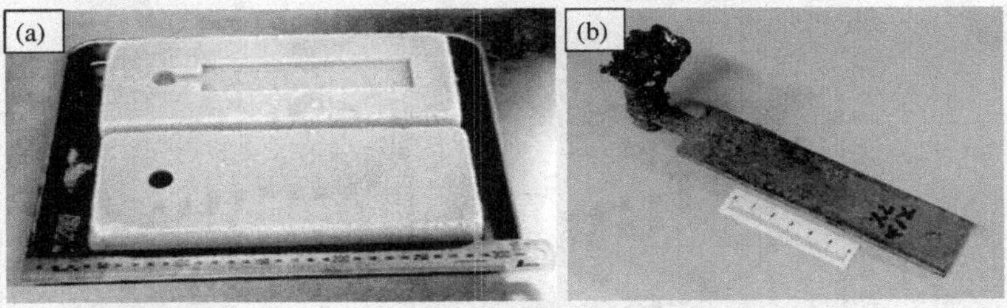

Figure. 3 (a) Photograph of the frozen mold used in this study and (b) appearance of the cast bronze produced by *Effset process*.

Results and Discussion

Figure 4 shows the temperature profile at each location in the bronze casting poured at 1140°C. Enlarged graphs are also shown on the right side of the Figure and are denoted (a'), (b') and (c'). For comparison, the temperature profile of a bronze casting poured in a green sand mold containing 2.6% water is also shown. The cooling profiles of the frozen mold are similar to that of the green sand mold. The temperature profile shortly after pouring at location (A), which is close to the gate, had fluctuations, see Fig. 3(a'). These temperature fluctuations may have been due to interrupted flow. In contrast, the temperature did not fluctuate at the points (B) and (C). The cooling behavior at the points (B) and (C) can be divided into three stages; (1) rapid cooling, (2) plateau and (3) slow cooling as shown in Fig. 3(b'). In stage 1, the moisture existing at the inner surface of the mold would immediately vaporize when contacted by the molten bronze. The plateau seen in stage 2 was caused by the thermal balance between the cooling capacity of the mold and the release of the latent heat of solidification of the molten metal. In the stage 3, solidification was complete and the bronze was cooling. In this stage, the contained water at the inner surface of the mold contacting the cast plate was already vaporized, so that the cooling was done by sand and air. This means that the cooling capacity of the mold was less compare to the stage 1. The cooling rate in this stage was lower than that of stage 1. The calculated liquidus and solidus lines of the bronze used in this study were about 1040°C and 900°C, respectively. In the steady state, the cooling curve exhibits the small hump between the liquids and solidus temperatures and the plateau at the solidus temperature, because of the release of the latent heat. However, in this study, cooling curves were measured in nonsteady state. Therefore, the plateau temperatures were not consistent with the calculated temperatures.

There were no significant differences in the temperature profiles of bronze cast in green sand, W5 and W5S2 molds in both the stage 1 and 2. In stage 3, the cooling rate of the bronze poured

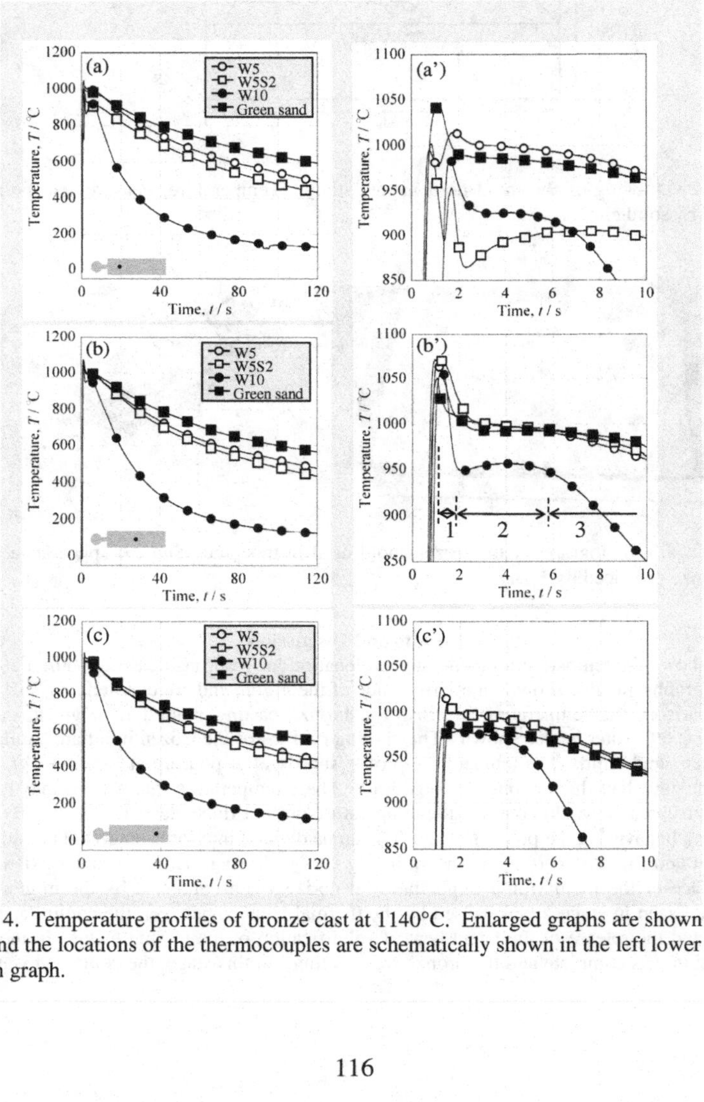

Figure 4. Temperature profiles of bronze cast at 1140°C. Enlarged graphs are shown on the right and the locations of the thermocouples are schematically shown in the left lower corner of each graph.

in the W5 and W5S2 molds were higher than that poured into the green sand mold. The frozen molds had a larger cooling capacity than the green sand mold, because the water was solid before pouring and the water content was higher than that of the green sand mold. Also, the bronze poured into the W5 mold was similar to that poured into the W5S2 mold. The cooling rate in stage 3 for the bronze poured into the W10 mold was higher than that of the other molds. This was due to the greater amount of water in the W10 mold. The increasing water content increases the cooling capacity of the mold. In addition, the higher permeability of the W10 mold might increase the cooling capacity. However, the detailed investigation for the permeability effect will be necessary.

The temperature profile at the points (B) and (C) exhibited similar behavior. The temperature plateau of the bronze poured into the W10 mold occurred at a lower temperature than that into the W5 and W5S2 molds, and the length or the plateau was shorter than that poured into the W5 and W5S2 molds. These results demonstrate that the W10 mold had greater cooling capacity

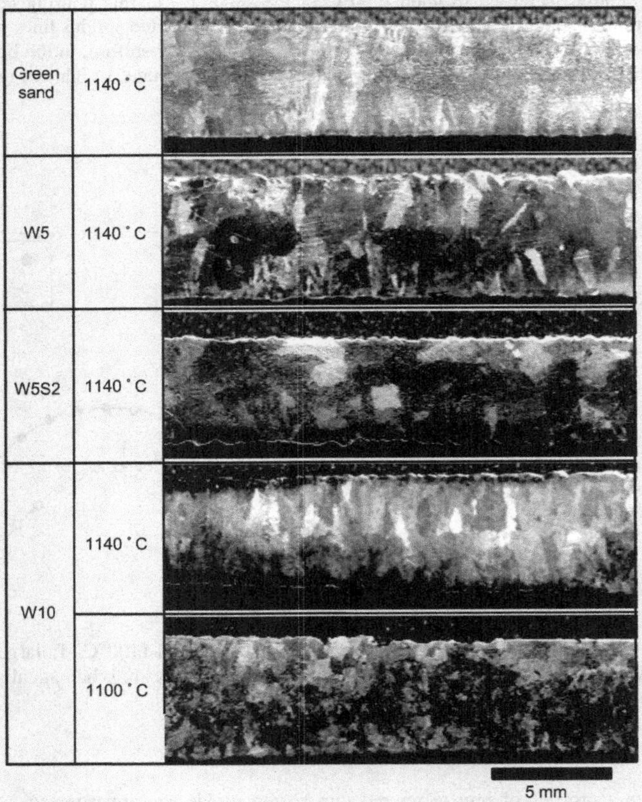

Figure 5. Photographs of bronze cast in frozen and green sand molds.

than the W5 and W5S2 molds. The W10 mold contained 10% water, therefore, the amount of heat absorbed by the contained water was larger than that of the W5 and W5S2 molds.

The microstructures of the bronze castings produced in this study are shown in Figure 5. The images are taken from the center of the plates. For a comparison, the microstructure of the bronze poured into the green sand mold containing 2.6% water at 1140°C is also shown. In the case of the bronze poured at 1140°C, all of the observed microstructures are similar excepting the bronze poured into the W5S2 mold. The microstructure exhibits the columnar structure at the surface of the plate. The rapid cooling behavior was observed in the bronze poured into the W10 mold compared to the W5, W5S2 and green sand molds, but grain refinement was not observed. In contrast, the decreasing pouring temperature from 1140°C to 1100°C led to the formation of an equiaxed and finer microstructure.

Figure 6 shows the temperature profiles at location B in the plate poured at 1100°C using the W10 mold. For comparison, the temperature profile in the plate poured at 1140°C is shown in the same figure. The temperature profiles were similar to the plate poured at 1140°C, but the temperature rapidly decreased to about 900°C in the stage 1. In this pouring condition, the microstructural refinement was observed. The calculated liquidus and solidus lines of the bronze used in this study were about 1040°C and 900°C, respectively. Therefore, in the bronze poured at 1100°C, the solidification was almost completed at the end of stage 1. This suggests that the frozen mold has the potential to produce a finer microstructure

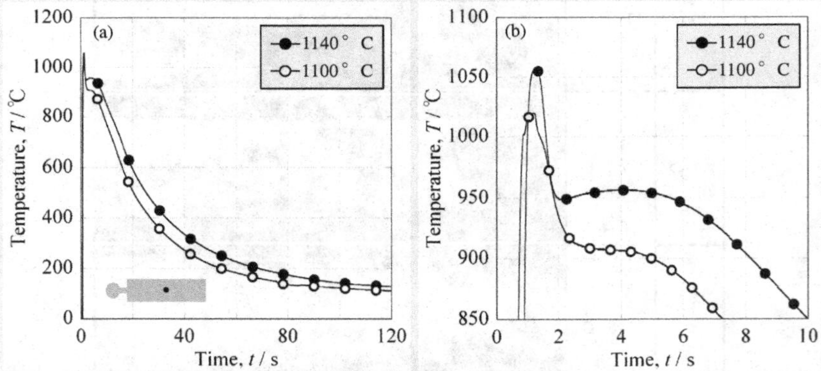

Figure 6. Temperature profiles for bronze poured at 1140°C and 1100°C. Enlarged graph of (a) is shown on the right and the locations of the thermocouples are schematically shown in the left lower corner of (a).

Conclusion

The temperature profiles of bronze poured into frozen molds was investigated. The cooling behavior could be divided into three stages, which were (1) rapid cooling, (2) plateau and (3) slow cooling. The water content and permeability of the frozen molds may have increased the cooling capacity of the molds. In particular, higher water content effective in decreasing the plateau temperature and reducing its time. A finer microstructure was produced in the bronze poured into the frozen mold containing 10% water at 1100°C. With these casting conditions, the temperature of the cast bronze rapidly fell to the solidification temperature.

References

1. C. Moore and D. Beet, "Effset – Metallurgy, Sand Technology and Economics," *Foundry Trade Journal*, May (1979), 1049-1063

2. Hideto Matsumoto, Yoshiyuki Maeda and Yoshinobu Fukuda, "Introduce Casting Plant for Frozen Molding Process," Journal of Japan Foundry Engineering Society, 80 (2008), 370-374

3. Shuji Tada, Yuya Makino, Toshiyuki Nishio, Keizo Kobayashi and Ken Aoyama, "Compressive Strength of Frozen Mold with Colloidal Silica Addition," *Journal of Japan Foundry Engineering Society*, 80 (2008), 531-535

4. Japanese Industrial Standards (JIS): JISH5120

SHAPE CASTING:
3rd International Symposium
2009

Characterization

Session Chairs:
Sumanth Shankar
Srinath Viswanathan

Shape Casting: The 3rd International Symposium
Edited by: John Campbell, Paul N. Crepeau, and Murat Tiryakioğlu
TMS (The Minerals, Metals & Materials Society), 2009

OXIDE FILM AND POROSITY DEFECTS IN MAGNESIUM ALLOY AZ91

Liang Wang[1], Hongjoo Rhee[1], Sergio D. Felicelli[1,2], Adrian S. Sabau[3], John T. Berry[1,2]

[1]Center for Advanced Vehicular Systems, Mississippi State University, Mississippi State, MS 39762, USA

[2]Mechanical Engineering Department, Mississippi State University, Mississippi State, MS 39762, USA

[3]Materials Science and Technology Division, Oak Ridge National Laboratory, Oak Ridge, TN 37831, USA

Keywords: Magnesium alloy, AZ91, Oxide film, Defect, Porosity

Abstract

Porosity is a major concern in the production of light metal parts. This work aims to identify some of the mechanisms of microporosity formation in magnesium alloy AZ91. Microstructure analysis was performed on several samples obtained from gravity-poured ingots in graphite plate molds. Temperature data during cooling was acquired with type K thermocouples at 60 Hz at three locations of each casting. The microstructure of samples extracted from the regions of measured temperature was then characterized with optical metallography. Tensile tests and conventional four point bend tests were also conducted on specimens cut from the cast plates. Scanning electron microscopy was then used to observe the microstructure on the fracture surface of the specimens. The results of this study revealed the existence of abundant oxide film defects, similar to those observed in aluminum alloys. Remnants of oxide films were detected on some pore surfaces, and folded oxides were observed in fracture surfaces indicating the presence of double oxides entrained during pouring.

Introduction

Magnesium cast alloys, such as AZ91, are gaining increasing attention in the struggle for weight saving in the automobile industry [1]. However, in many cases the consistent production of sound AZ91 castings is marred by the stubborn persistence of some defects that are difficult to remove: porosity, macrosegregation, oxide entrainment, irregularity of microstructure, etc. The formation of microporosity in particular is known to be one of the primary detrimental factors controlling fatigue lifetime and total elongation in cast light alloy components.

Many efforts have been devoted to investigate the mechanisms of porosity formation in the last 20 years. More recently, new mechanisms of pore formation based on entrainment of oxide films during the filling of aluminum alloy castings have been identified and documented [2-7]. Oxide film defects are formed when the oxidized surface of the liquid metal is folded over onto itself and entrained into the bulk liquid. A layer of air is trapped between the internal surfaces of the oxide film, which leads to the porosity formation in the solidified castings. The entrainment process due to surface turbulence is usually rapid, in the order of milliseconds; therefore the time is very limited to form new oxide film on the fresh surface, so that the entrained oxide film can be very thin, in the order of nanometers [2].

Oxide film defects may be contained in most reactive liquid metals such as Al and Mg due to surface turbulence during the melting, pouring and transfer processes in casting. These defects have been observed on the fracture surfaces of tensile test specimens and the oxides have been identified by SEM-EDX analysis [7-9]. In contrast with the efforts devoted to Al-based cast alloys, few studies have been done in Mg alloy castings. Griffiths and Lai [8] investigated the nature of the oxide film defects in pure Mg castings. They found double oxide film defects comprised of folded MgO films on the fracture surface of tensile test bars taken from the castings. Mirak et al. [9] recently studied the characteristics of oxide films in AZ91 alloys, where the formation of oxide films was induced by the impingement of bubbles.

In this study, we examined the microstructure of magnesium alloy AZ91 ingots gravity-poured in plate graphite molds. Temperature data during cooling was acquired with type K thermocouples at 60 Hz in two locations of each casting. The microstructure of samples extracted from the regions of measured temperature was then characterized using optical metallography, tensile tests, four point bend tests and Scanning Electron Microscopy (SEM) of the fracture surfaces. The nature of oxide film and porosity defects in AZ91 was investigated.

Experimental Procedure

Design of castings

The cast ingots or slab castings were produced in a graphite plate mold at the facilities of Oak Ridge National Laboratory (Oak Ridge, TN). The mold was rectangular and the thickness of the wall was 0.5 in (12.7 mm). The width, height, and thickness dimensions were 5.5×11×2.25 in (140×279×57 mm), respectively, as shown in Figure 1. Three thermocouples were placed in the empty molds at distances of approximately 2.5, 5 and 8 in (64, 127 and 203 mm) from the casting end per each casting.

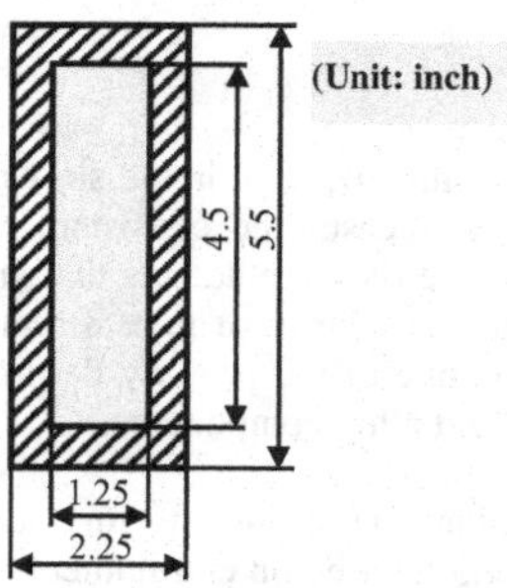

Figure 1. Cross section of plate graphite mold used for AZ91 castings. Height was 11 in (279 mm).

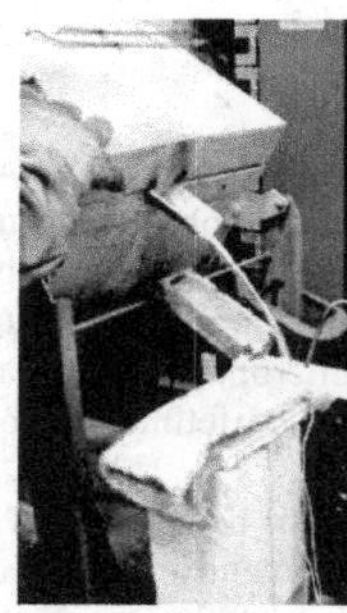
Figure 2. Pouring of casting type C.

The tested AZ91 alloy composition was Mg, 9.0%Al, 0.7%Zn, 0.2%Mn. The furnace charge was in the form of pre-alloyed ingot. The weight of the melt was 8 kg and the alloy was melted in an electrical resistance furnace. For protection, Ar and CO_2+3%SF6 were used as cover gases. The pouring temperature was approximately 700 °C. No degassing procedures were used. All castings were poured from one melt. The melt was poured directly from the crucible to minimize temperature decrease during pouring (Figure 2). The mold was not preheated and was coated with boron nitride. In order to assess the reproducibility of the results, two molds were used. Temperature data was acquired with thermocouples type K at a sampling rate of approximately 60 Hz. The measured cooling curves are shown in Figure 3. The cooling curves are labeled in the following format: xn_m, where x – is a letter, indicating the mold type, n – indicates casting number (1 or 2), and m – indicates thermocouple location (b-bottom of casting, c-center of casting). The cooling curves show an excellent reproducibility. The data measured by the thermocouple near the top of the casting was discarded because of turbulence in this region. As shown in Figure 3, the cooling rate during solidification for the AZ91 alloy castings was approximately 3.0 °C/s. In this article, results for only one mold type, denoted as type C, are reported. Analysis of castings in other graphite and ceramic molds of different dimensions will be reported elsewhere.

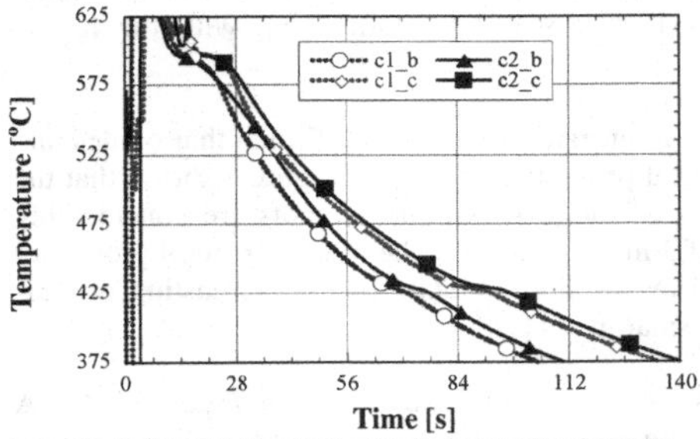

Figure 3. Cooling curves for AZ91 Mg alloy castings in mold type C.

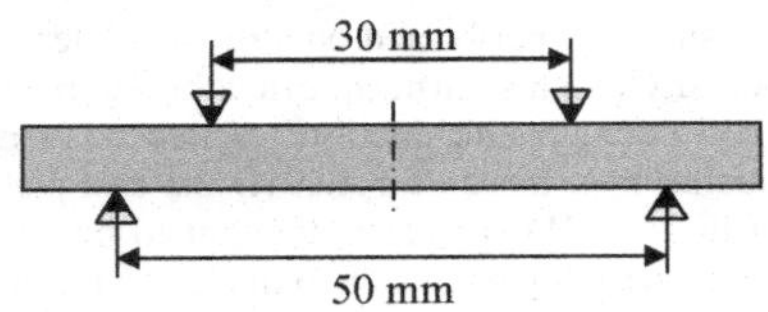

Figure 4. Sketch of four point bending test geometry.

Sample preparation for optical metallograhy

The samples were cut near the location of the thermocouple for each as-cast ingot and then hot-mounted in phenolic resin, with one side of the plate flush with the mounted surface. The samples were then polished using a machine disc grinder. The silicon carbide abrasive papers of grade 500 and 2400 μm grits were used successively. In between papers the samples were cleaned by ethanol thoroughly. The samples were then cleaned in a sonic bath before being examined by optical microscope. Approximately 20 to 30 images were taken for each sample.

Sample preparation for tensile test and Four Point Bend (FPB) test

Two tensile test strip specimens and two Four Point Bend (FPB) test specimens with dimensions of 115 mm long, 10 mm wide, and 3 mm thick, were cut from each of the cast samples to characterize the casting mechanical properties. These specimens were tested using an EM Model 5869 Instron machine at a strain rate of 0.001/s for the tensile test and a cross-head speed of 0.05 in/min for the FPB test. Figure 4 depicts the sample arrangement employed in the FPB test. The fracture surfaces of the test specimens were examined using a field-emission gun scanning electron microscope (FEG-SEM) equipped with an energy dispersive x-ray spectrometry (EDX).

Results and Discussion

Porosity was the major defect observed in the tested specimens. Pores ranging in size from 100 μm to 500 μm were found in many of the polished surfaces. Figure 5 shows typical pore morphology at a location close to the thermocouple in the AZ91 C1 sample. A magnified view (Fig. 5(b)) reveals dendrites protruding into the pore as well as pieces of oxides on the surface of the pore. EDX spectroscopy shows a three-fold increase of the oxygen content inside the pore compared with the surrounding matrix. This pore was most probably caused by interdendritic shrinkage, however, the presence of oxides might suggest also a pore formed by an entrained double oxide that was torn apart by shrinkage-induced shear forces.

Long pieces of oxide films, some longer than 1 mm, were observed in AZ91 samples through optical microscopy and in SEM images. Figure 6 shows a "dragon-shaped" oxide film found on a polished surface of the specimen. The distinct precipitation upon both sides of the film might suggest the former existence of a double oxide that was later torn open, with the higher precipitation occurring on the wetted side.

Tensile tests performed on strip specimens at a strain rate of 0.001/s confirmed that oxides and porosity had a significant effect on the mechanical properties of AZ91. Figure 7 shows that the yield strength and ductility of two C1 samples of the gravity-poured ingots are considerably smaller than those of a AZ91D die cast plate of 3 mm thickness. The relatively good properties of the AZ91D sample is thought to be caused by the high velocity of the die casting process which possibly breaks the oxide films into very small parts [10].

The details of fracture surfaces of tensile test AZ91 samples are shown in Figures 8-10. A distinct interface between the dendritic matrix and an oxide region can be observed (Figure 8). Figure 9 shows two symmetrical oxide films on either side of a fracture surface. This agrees well with the observation by Griffiths and Lai [8] for pure Mg castings. A magnified view of the oxide region (Figure 10) reveals a pleated surface, similarly as observed in double oxide films in aluminum alloys.

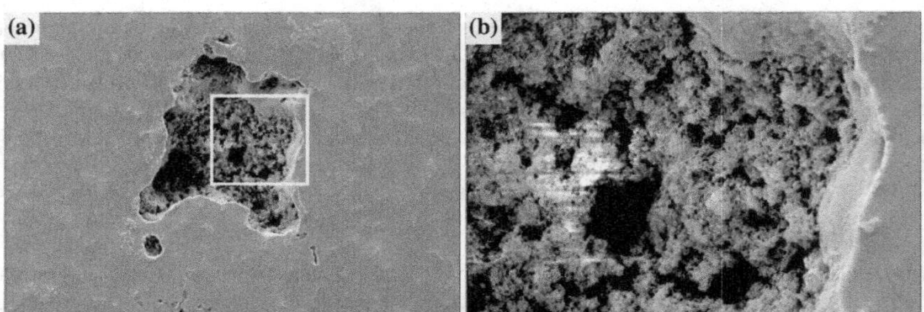

Figure 5. (a) Typical pore morphologies formed at the location close to the thermocouple in casting AZ91 C1 sample; (b) higher magnification (2000X) of image (a).

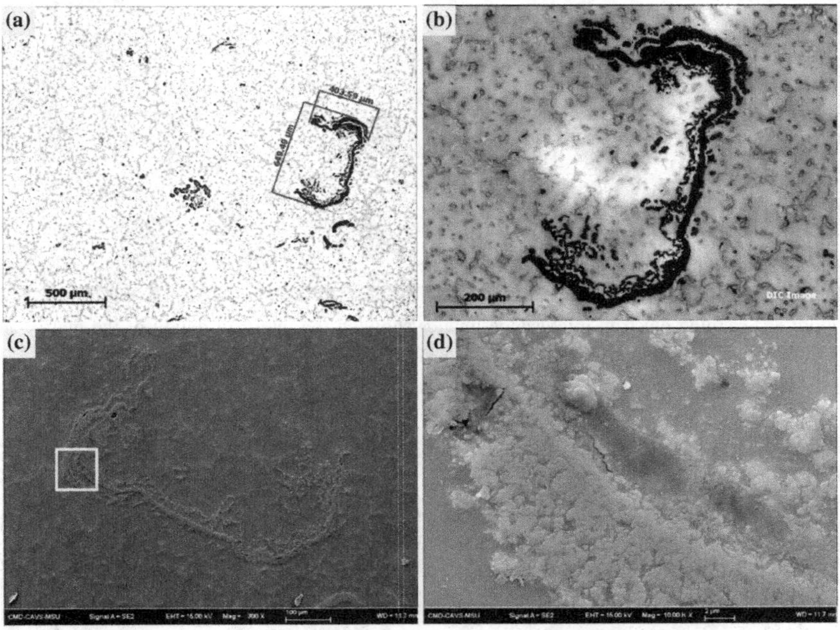

Figure 6. Oxide film in AZ91 sample C1. (a) Optical micrography; (b) Transmitted light differential interference contrast (DIC) image; (c) SEM image; (d) Higher magnification (10000X) of image (c).

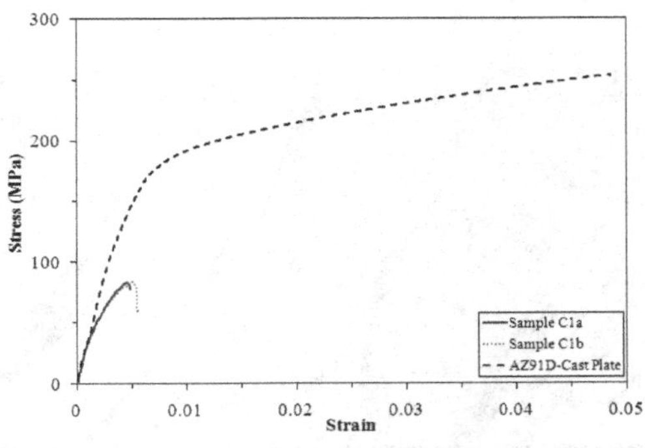

Figure 7. Tensile test results for strip specimens with a strain rate of 0.001/s; Specimen C1a and C1b were taken from different locations in AZ91 sample C1; The third specimen was taken from a AZ91D die cast plate.

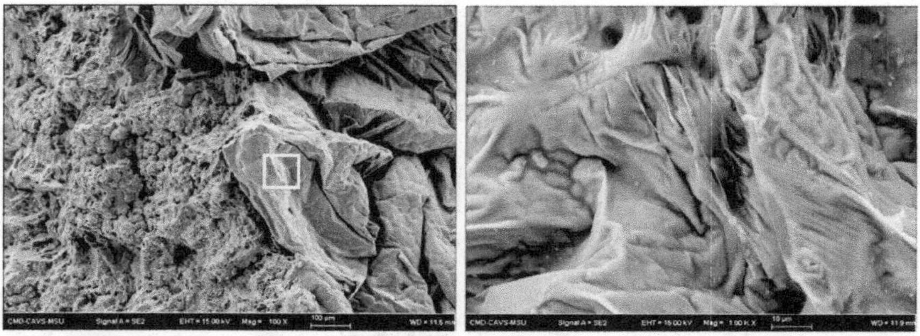

Figure 8. Scanning electron microscope images showing the interface region of the matrix and oxide film on fracture surfaces of a tensile test specimen taken from AZ91 sample C1.

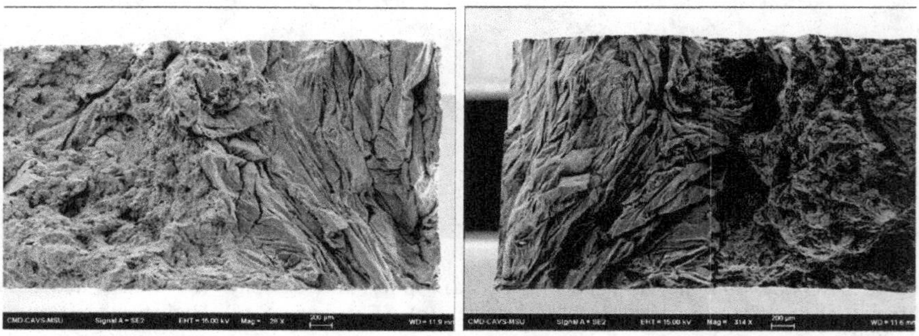

Figure 9. Scanning electron microscope images of oxide films on the two sides of the fracture surfaces of a tensile test specimen taken from AZ91 sample C1.

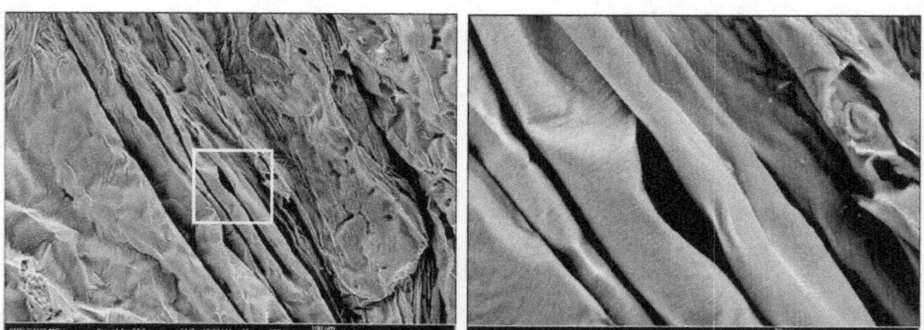

Figure 10. Higher magnification views of the oxide film found on the fracture surface shown in Figure 9.

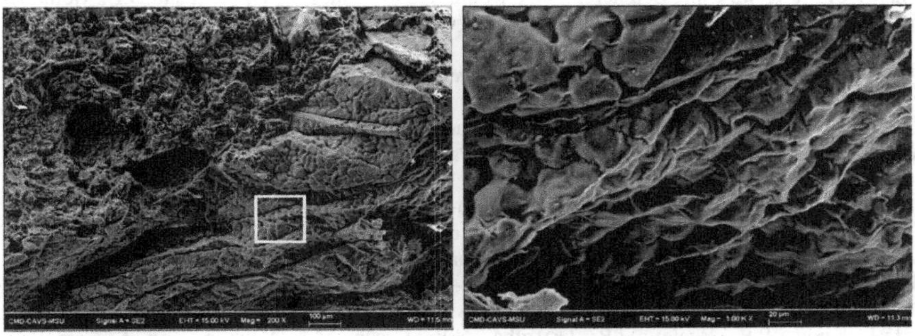

Figure 11. Scanning electron microscope images show the interface region of matrix and oxide film on fracture surfaces of a four point bending test specimen taken from AZ91.

SEM images of the fracture surfaces of a four-point bend specimen are presented in Figure 11. Again, an interface between the matrix and an oxide region can be detected. However, these surfaces look different than the fracture surfaces of the tensile test samples. They are darker and the oxide film is not pleated as the one in the tensile test fracture surfaces; instead, an apparently thinner film is folded over dendrites. Possibly, the different states of stress in comparison to the tensile test samples are in part responsible for these features.

Conclusions

The microstructure of plate-shaped ingots of magnesium alloy AZ91 obtained by gravity-pouring in a graphite mold was analyzed by EDX spectroscopy, SEM and optical metallography. Abundant porosity was found throughout the ingots, with some pores as large as 500 µm. Pieces of oxide film, some of them 1 mm or longer, were also detected in many polished cross sections. Distinct features on both sides of the films suggest that they might be remnants of torn double oxide films or bifilms entrained during the pouring. This fact is supported by the analysis of the fracture surfaces of tensile and four point bend samples, which revealed pleated layers of oxides similar to those observed in aluminum alloys. Oxide fragments were also detected inside some pores, indicating that they might have formed from oxide bifilms that were torn by shrinkage shear or dendrite action.

Acknowledgements

This work was sponsored by the National Science Foundation through grant number CTS-0553570. The authors are thankful to Prof. William Griffiths of the University of Birmingham (UK) for the helpful discussions held during his visit to MSU in Fall 2007. This work was performed in collaboration with the United States Advanced Materials Partnership (USAMP), United States Council for Automotive Research (USCAR). This research was also sponsored by the U.S. Department of Energy, Assistant Secretary for Energy Efficiency and Renewable Energy, Office of Transportation Technologies, Lightweight Vehicle Materials Program, under contract DE-AC05-00OR22725 with UT-Battelle, LLC. The authors acknowledge that this research was supported in whole by Department of Energy Cooperative Agreement No. DE-FC05-02OR22910. Such support does not constitute an endorsement by the Department of

Energy of the views expressed herein. We would like to thank E.C. Hatfield and D.C. McInturff of Oak Ridge National Laboratory for assistance with casting experiments.

References

[1] C.H. Caceres, C.J. Davidson, J.R. Griffiths, and C.L. Newton, "Effects of Solidification Rate and Ageing on the Microstructure and Mechanical Properties of AZ91 Alloy," *Materials Science and Engineering A*, 325 (2002), 344-355.
[2] John Campbell, *Castings 2nd ed.* (Butterworth-Heinemann, London, 2003), 17-69.
[3]X. Yang, X. Huang, X. Dai, J. Campbell, and R. J. Grant, "Quantitative Characterization of Correlations between Casting Defects and Mechanical Strength of Al–7Si–Mg Alloy Castings," *Materials Science & Technology*, 22 (2006), 561-570.
[4] J. Campbell, "Entrainment defects,"*Materials Science & Technology*, 22 (2) (2006), 127-145.
[5] J. Knott, P.R. Beeley, J.R. Griffiths, N.R. Green, C.J. Newton, and J. Campbell, "Commentaries on 'Entrainment Defects' by J. Campbell," *Materials Science & Technology*, 22 (2006), 999-1008.
[6] R. Raiszadeh, and W.D. Griffiths, "A Method to Study the History of a Double Oxide Film Defect in Liquid Aluminum Alloys," *Metallurgical and Materials Transactions. B*, 37 (2006), 865-871.
[7] J. Mi, R.A. Harding, M. Wickins, and J. Campbell, "Entrained Oxide Films in TiAl Castings," *Intermetallics*, 11 (2003), 377-385.
[8] W.D. Griffiths, and N.W. Lai, "Double Oxide Film Defects in Cast Magnesium Alloy," *Metallurgical and materials transactions A*, 38 (2007), 190-196.
[9] A.R. Mirak, M. Divandari, S.M.A. Boutorabi, and J. Campbell, "Oxide Film Characteristics of AZ91 Magnesium Alloy in Casting Conditions," *International Journal of Cast Metals Research*, 20 (2007), 215-220.
[10] W.D. Griffiths, "An SEM Study of High Pressure Die Cast Mg Alloy Castings" (Report #MSU.CAVS.CMD.2008-R0023, Center for Advanced Vehicular Systems, Mississippi State University, 2007)

Shape Casting: The 3rd International Symposium
Edited by: John Campbell, Paul N. Crepeau, and Murat Tiryakioğlu
TMS (The Minerals, Metals & Materials Society), 2009

ASSESSING CASTING QUALITY USING COMPUTED TOMOGRAPHY WITH ADVANCED VISUALIZATION TECHNIQUES

Georg F. Geier[1], Joerdis Rosc[1], Markus Hadwiger[2], Laura Fritz[2], Daniel Habe[1], Thomas Pabel[1], Peter Schumacher[1,3]

[1]Austrian Foundry Research Institute, Parkstraße 21, 8700 Leoben, Austria
[2]VRVis Research Center for Virtual Reality and Visualization, Ltd, Donau-City-Strasse 1, 1220 Vienna, Austria
[3]Chair of Casting Research, University of Leoben, Franz Josef Strasse 18, 8700 Leoben, Austria

Keywords: Casting quality, computed tomography, visualization, multi-dimensional transfer functions

Abstract

Increasing demand for high quality castings has increased the importance of computed tomography (CT) in the casting industry. With computed tomography it is possible to cover the whole sample-volume. Apparent differences in density can be detected and size and positions of these inhomogeneities can be determined in three dimensions. CT is a valuable tool because various casting defects can be detected and quantified. This paper highlights the possibilities and limitations of computed tomography for quality control and assessment of castings from materials to finished products. Considerable improvements can be achieved using volume rendering with novel multi-dimensional transfer functions for visualising the volume data. In particular the quantification of casting defects were addressed and compared to standard metallographic procedures and common CT analysis-tools.

Introduction

The progress in CT technology and the growth in available computing capacity during the last few years has made this technology increasingly suitable for technical applications. By enabling a non-destructive 3-dimensional view into the centre of castings this technology opens up completely new options for materials research, components development and process optimization. Computed tomography can be used with various materials, such as metallic and ceramic materials, plastics, refractories, construction materials, composite materials and reinforced components. In the foundry industry, and above all amongst users of castings, there is growing interest in using CT for 3D non-destructive testing of complex component geometries. With CT it is possible to detect classical casting defects such as gas pores, shrinkage cavities, inclusions, spongy structure, etc.. Compared to radioscopic testing, CT enables the display of smaller defects in higher contrast. Moreover it provides information about the geometry and position of the detected flaws. During CT testing an X-ray source penetrates the object, while it is rotating step-wise through 360°. The penetrating X-rays are captured by a detector. From this information clustered computers reconstruct a 3D model of the specimen for different visualization and analysis tasks.

Computed Tomography of Castings

Shrinkage cavities and pores have always been important quality criteria of castings. Non-destructive determination of the overall porosity content only used to be possible by weighing according to the Archimedes principle. This method is too inaccurate nowadays for the assessment of technologically achievable and customer-specified porosities of <1.5 % [1]. Therefore evaluation of porosities is only possible by destructive methods, as described in the VDG technical standard P201 "Volume deficits of non-ferrous metal castings" [2].

Radioscopy versus CT

Through non-superposed representations in CT scanning, even very small volume deficits can be represented in high contrast. The aluminium die casting shown in Figure 1 was subjected to X-ray testing using a high-resolution micro-focus X-ray source and a 16-bit flat panel detector (Figure 1a). The visibility of volume deficits was enhanced by contrast stretching.

A CT model of this casting that was computed employing identical imaging conditions as well as using the presented radioscopic image, shows additional volume deficits. These deficits can be characterized as gas pores, (Figure 1b). Hence in the CT model more details can be identified than in a radioscopic image which is taken using identical parameters.

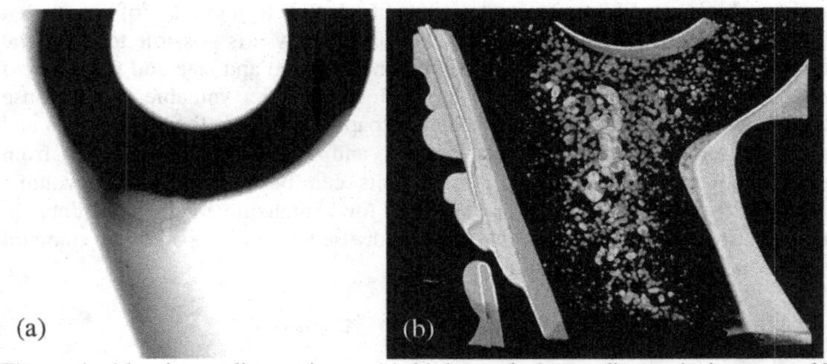

Figure 1. Aluminum die casting – (a) high-resolution radioscopic image and(b)high-resolution CT image.

Automatic Pore Detection

The quality of an aluminium melt and the success of the refining process is usually evaluated using reduced pressure testing (RPT) [1]. Samples are taken before and after the refining process for solidification in air and vacuum. From the resulting differences in density the density index is calculated as a characteristic value for the melt quality. The corresponding reduced pressure test samples contain gas pores and shrinkage cavities. A CT model of such an RPT sample (Figure 2) was used to demonstrate automatic pore detection.

The software VG Studio MAX which is generally used for the analysis of technical CT data incorporates a defect analysis module for automatic volume deficit detection. This tool captures internal deficits, their individual position, size and distribution. For this type of detection various parameters must be defined, the most important being the definition of the gray values of the material and the pores. Figure 2 shows the results of pore detections evaluated with different parameter sets. The corresponding results for the determined volume porosities and the number of detected pores are given in Table I. The detected volume porosities between 2.3 % and 5.2 % represent such a wide bandwidth that, depending on the parameters, the test result could even lie outside specification for a casting at hand. While in evaluation variant (a) certainly not all defects have been detected, variant (c) can be assumed to have included some false positives, i.e. not true defects. This shows that useful and reliable pore detection requires standardized procedures. As these are not yet available, the user must select individual and suitable parameters for each casting based on his experience, possibly supported by additional investigation methods.

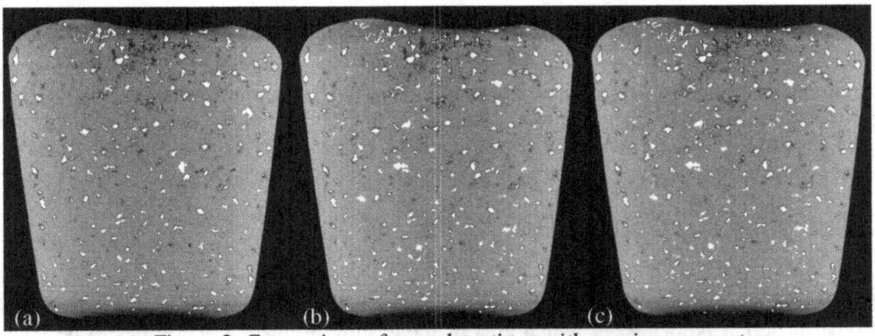

Figure 2. Comparison of pore detections with varying parameters.

Table I. Overview of the difference in volume porosity and number of pores derived from three different sets of parameters.

	Figure 1 (a)	Figure 1 (b)	Figure 1 (c)
Volume porosity	2.3 %	4.8 %	5.2 %
Number of pores	6,191	8,029	20,032

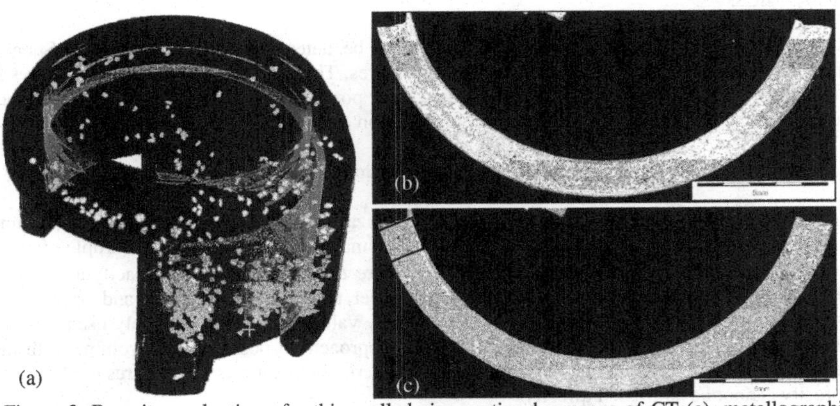

Figure 3. Porosity evaluation of a thin walled zinc casting by means of CT (a), metallographic examination of the whole microsection (b) and according to VDG technical standard P201 (c).

Evaluation of Porosity

Not only the detection but also the evaluation of size, position and distribution of pores is of great importance for each individual component. A thin-walled zinc die casting was used for the following porosity evaluations (Figure 3): volume porosity was determined by CT (0.25 %), surface porosity was determined for the whole microsection (0.9 %) and the latter also according to the VDG technical standard P201 (9 %).

It can be generally stated that there are significant differences in the porosities determined by different methods. Therefore it is particularly important to define appropriate thresholds for each method. It is recommendable not only to evaluate pore accumulations in critical areas, for example by means of the described VDG P201 method for surface porosity assessments, but also for the assessment of the volume porosity.

Differentiation between Pores and Shrinkage Cavities

Pores and shrinkage cavities are interacting volume deficits driven by supersturated gas and negativ metallostatic pressure respectively. Therefore it is important for the foundryman to know which type of deficit is located in which area of the casting. Remedial action has to be taken depending on the type and position of the defect. In principle the two defects differ in terms of morphology. However, due to the complexity in morphology there is still no commercial software tool available on the market capable of automatically differentiating between the two.

The differentiation between pore and shrinkage hole can only be performed semi-automatically by the user. The setting of suitable imaging parameters requires well-grounded know-how of the casting technology and experience in visualizing of tomographic images.

Spongy Structure

Another frequent and therefore important defect to be detected is spongy structure. Spongy structure can be interpreted as an accumulation of pores. The resolvability of individual pores is limited because of the extremely fine structure of the pores. Nevertheless spongy structures are detectable in CT because they result in a local reduction of density.

Processing and Visualization of CT-Data

Today the most widely used approach for the visualization of volume-data is "Direct Volume Rendering" (DVR) which uses the whole volume for image generation [3]. A quadruple of (r, g, b, α) is assigned to each position in the volume, where α corresponds to the opacity and r, g, b give the color of one voxel. One-dimensional transfer functions where color and opacity are assigned on the basis of the local density value (gray value) are the most widely used transfer functions. They are basically 1D tables. With this approach all positions in a volume with the same density are represented in the same way and material boundaries such as pores or shrinkage porosity cannot be addressed directly.

This leads to the standard approach of feature detection which is usually performed via some type of segmentation, which most commonly builds on region-growing and filtering operations such as morphological operators. Segmentation results in one or several static segmentation masks, which can be visualized as part of the 3D volume and also form the basis of quantification. The segmentation cannot be modified without re-computation. This decouples the detection of features from visualization and prevents working in a fully interactive manner. Most of all, it hampers interactive exploration of the volume without knowing what features are contained in the volume beforehand. Whenever the segmentation results for specified parameters are not satisfactory, the user has to modify the parameters and the entire segmentation has to be computed all over again. This is often time-consuming and tedious. Unlike this standard approach we propose a visualization-driven method for feature detection allowing features in the volume to be explored interactively without re-computing the segmentation.

Density/Gradient Magnitude Transfer Function

The most straightforward approach for a two-dimensional transfer function uses the gradient magnitude as the second dimension besides the density [4]. This type of transfer function is useful to visualize material boundaries, as the gradient magnitude is nearly zero in homogenous areas (due to intrinsic noise), whereas it reaches a maximum at the edges between two materials (e.g. metal and air). Different features of the volume can be displayed by interactively selecting the corresponding ranges of the gradient magnitude over density. This transfer function offers a quick interactive way to analyse the size and distribution of different kinds of defects in a casting [5].

Feature Size Transfer Function

This novel type of two-dimensional transfer function is defined by density and feature size [6]. In a pre-processing step, each voxel of a volume is assigned to a homogenous region (feature) to which it may belong via region-growing. In the visualization step the features can be distinguished and interactively selected according to their density and size. As this kind of transfer function is designed for the exploration of different discontinuities a separate transfer function, for instance a simple one-dimensional transer function, has to be used to visualize the the whole object. In this way the features can be shown in the context of the test volume.

Although exploration of the sample is conceptually the most important part of our concept, the basis for interactivity during exploration is a complex pre-computation stage. However, no user input is required for this stage and thus it is, although technically complex and important, decoupled from the exploration itself. The main goal for pre-computing additional information is to enable exploration of different classes of features with different parameters in such a way that, e.g. the main parameter used to control region-growing (e.g. maximum variance) can be changed interactively after the region-growing process. In order to allow this, we perform region-growing in multiple passes and track the progress of each voxel.

Two different region-growing approaches, which are outlined below, were used for the exploration and analysis of castings. One method uses a variant of seeded region-growing [7] which is also able to include a region's boundary. In this way the feature is grown by adding a voxel to the region when the difference of its density with respect to the average density of the whole region is below a given threshold $\varepsilon : |\nu - \nu_r| < \varepsilon$, where ν is a voxel's density, and ν_r is the current region's average density. After a new voxel is added, ν_r is updated accordingly. In a second step feature boundaries are added to the features after checking a gradient magnitude criterion.

The other method was suggested by Huang et al. [8]. They use a combination of region-growing based on density variance and gradient magnitude variance. They determine the variances *f* for the fixed neighborhood of a seed voxel. The main parameter is a scale factor $k > 0$:

$$f_{ca} = \frac{|\nu - \nu_s|}{k\sigma_V} \tag{1}$$

$$f_{cb} = \frac{|\nu' - \nu'_s|}{k\sigma_{V'}} \tag{2}$$

$$f_{cc}(p) = p\, f_{ca} + (1-p)\, f_{cb} \tag{3}$$

where ν is a voxel's density value, ν_s the density of the seed, ν' a voxel's gradient magnitude value, ν'_s the gradient magnitude of the seed, and σ_v and σ_g the corresponding variances in the seed neighborhood, respectively. The factor *p* can be set to a constant value but is set by default to $p = \sigma_{V'}(\sigma_{V'} + \sigma_V)^{-1}$.

The main difference between the two region-growing methods is, that the first one uses the whole feature area for the calculation of the average density, while the second method uses the average density of a given seed region. It can be shown that the first method is preferable for larger features, while the second method is best used with fine distributed features.

Multi-dimensional Transfer Functions for Casting Applications

As a reference to determine the effectiveness of the multi-dimensional transfer functions a reduced pressure test (RPT) sample with a high gas content was chosen. The gas in the melt forms pores as the metal solidifies. Furthermore, the shrinkage of the Al-Si alloy during solidification causes shrinkage cavities to be formed in the center of the upper regions of the sample. Therefore, this sample is an ideal test piece for the evaluation of feature detection, as it is virtually full of different pore sizes and shrinkage cavities.

The CT-dataset was subjected to analysis using gradient magnitude transfer functions to demonstrate the feasibility of simple differentiation between pores and shrinkage cavities with this method. One result of this analysis is given in Figure 4. The boundaries of the RPT Sample, the pores and the shrinkage cavities should be depicted, although these three feature classes have comparable densities. As a consequence, the transfer function was defined by selecting the appropriate gradient magnitude regions for each feature class. In Figure 4 pores are depicted in light gray; shrinkage cavities are shown in dark gray. While pores of different sizes can be found all over the volume of the RPT Sample, shrinkage cavities are concentrated mainly around the center of the sample. With this concept differentiation between pores and shrinkage cavities was easily accomplished. Thus in general, the gradient magnitude transfer functions offer a simple way to purposively visualize and evaluate different features of the dataset.

The dataset of the RPT sample was also analyzed with feature size transfer functions (Figure 4). The results were then compared to results obtained using standard CT-software tools as well as to metallographic examination. Since metallographic evaluation is the standard procedure today for the evaluation of the porosity of castings, it was used as a reference for the CT based measurements. Two different software tools were utilized: VG Studio MAX as the most widely distributed software tool for handling industrial CT-data, and the software package using the novel multi-dimensional transfer functions.

The pores within the RPT sample were subdivided into five classes according to their volumes, disregarding shrinkage cavities. This classification was performed via selecting the appropriate feature sizes in the transfer function (Figure 4). The result of this classification can subsequently be quantified.

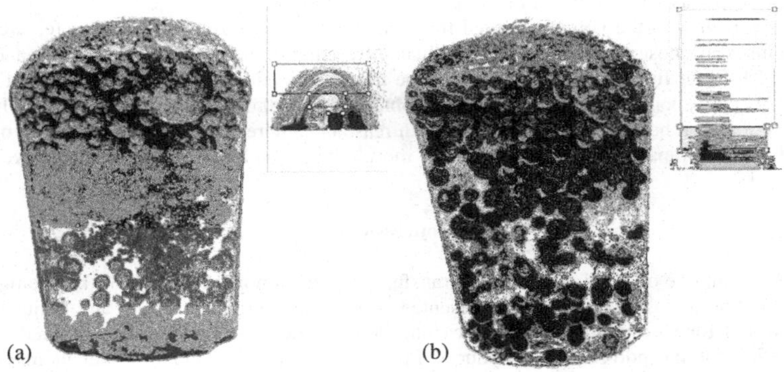

Figure 4. CT visualization of a reduced pressure test sample by using gradient magnitude (a) and feature size (b) transfer function. The applied transfer functions are given on the upper right hand side of each image.

For the comparison to metallographic examination a number of axial slices from the volume were selected and the sizes of the pores within these slices were evaluated by metallographic examination. These results were then compared with the results from the standard software tool to verify that reasonable values were achieved with the dataset used. Figure 5**Error! Reference source not found.** depicts one of the evaluated slices. The left part of Figure 5 shows the CT data as analyzed with the software package; the right part pictures the result from metallographic examination. In both images the same pore is selected to show the congruence of the two results.

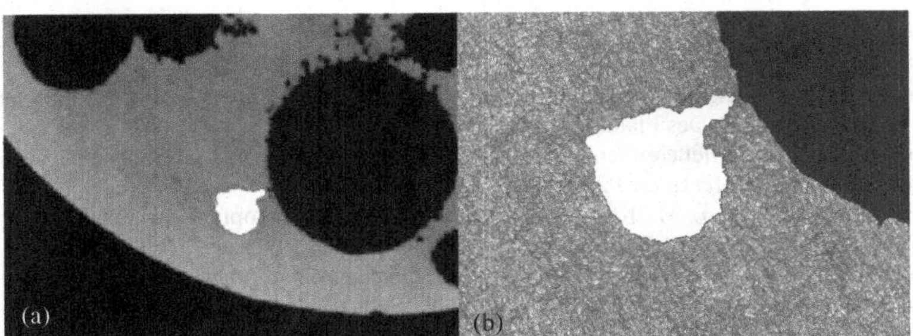

Figure 5. Comparison of a specific pore in CT representation (a) and metallographic examination (b).

Further analysis reveals that the pore sizes derived from the two methods have a mean deviation of 4.6 %, giving higher pore volumes for the CT-dataset. The deviation of the individual slices analyzed varied between -3 to 9 % depending on the morphology of the pore in that slice. From these results it can be concluded that metallography and CT data deliver comparable results where CT data slightly overestimates the size of the pores.

The next step was the evaluation of the pore volume with the appropriate transfer function and the comparison with the volume derived from the standard software. One specific feature from the volume was chosen and its size was evaluated by using both, the standard software and with the novel transfer function. Evaluation with the transfer function gives a pore volume of 8.189 mm^3. This corresponds well with the volume obtained from the standard CT software, which is 8.222 mm^3. It can be seen that the results are in reasonable agreement for the evaluation of pore sizes. Further comparisons for different pore morphologies and different pore sizes have to be conducted.

Conclusions

With the technique of multi-dimensional transfer functions a novel concept for the examination and evaluation of casting quality was presented. After discussing possibilities and limitations of CT as a tool for investigating different casting defects, the feasibility of the new method was exemplified for the porosity in a Reduced Pressure Test (RPT) sample. Gradient magnitude transfer functions allow a reliable differentiation between pores and shrinkage cavities. Furthermore, the novel feature size transfer functions allow the interactive classification of different pore sizes. Comparisons of the metallographic examination, standard CT-software tools and the new software concept correspond well for the given dataset from the RPT sample. As a consequence, it was shown that multi-dimensional transfer-functions represent a valuable new method for the analysis of CT data in assessing the quality of castings. This offers a unique visualization driven approach for the casting expert.

Acknowledgements

The authors wish to acknowledge the financial aid granted by the Austrian Research Promotion Agency (FFG).

References

1. S. Dasgupta, L. Parmenter, D. Apelian, F. Jensen:, *Proc. 5th International Molten Aluminium Processing Conference, AFS* (Des Plaines, 2002), 283-300.
2. Verein deutscher Gießereifachleute, “Volumendefizite von Gussstücken aus Nichteisenmetallen“, *VDG-Merkblatt P201* (2002).
3. K. Engel, M. Hadwiger, J. M. Kniss, C. Rezk-Salama, D. Weiskopf, *Real-Time Volume Graphics* (Wellesley, A. K. Peters, 2006).
4. J. Kniss, G. Kindlmann and C. Hansen, “Interactive volume rendering using multi-dimensional transfer functions and direct manipulation widgets” *IEEE Visualization 2001 Proceedings*, 255-262.
5. T. Höllt, “GPU-Based Direct Volume Rendering of Industrial CT Data”, student research project, University Koblenz Landau, 2007.

6. M. Hadwiger, L. Fritz, C. Rezk-Salama, T. Höllt, G. Geier, T. Pabel, "Interactive Volume exploration for Feature Detection and Quantification in Industrial CT Data", to appear in *IEEE Transactions in Visualization and Computer Graphics*, 14 (6), 2008.
7. R. Adams and L. Bischof, "Seeded region growing", *IEEE Trans. Pattern Anal. Mach. Intell.*, 16(6), 1994, 641–647.
8. R. Huang, K.-L. Ma, P. McCormick and W. Ward, "Visualizing industrial CT volume data for nondestructive testing applications", *Proceedings IEEE Visualization* 2003, 547–554.

Shape Casting: The 3rd International Symposium
Edited by: John Campbell, Paul N. Crepeau, and Murat Tiryakioğlu
TMS (The Minerals, Metals & Materials Society), 2009

RECONSTRUCTION, VISUALIZATION, AND QUANTITATIVE CHARACTERIZATION OF MULTI-PHASE THREE-DIMENSIONAL MICROSTRUCTURES OF CAST ALUMINUM ALLOYS

Harpreet Singh[1], Arun Gokhale[1], Yuxiong Mao[1], Asim Tewari[2]

[1]Georgia Institute of Technology, Atlanta, GA-30332, USA
[2]General Motors Corporation, Bangalore, India

Keywords: Visualization, Characterization, 3D microstructure, Aluminum alloys

Abstract

Serial sectioning technique is well known for reconstruction of three-dimensional microstructures of opaque materials. During the recent years, techniques have also been developed for reconstruction of high fidelity large volume segments of three-dimensional microstructures using montage serial sections; robot assisted automated acquisitions of montage serial sections are also reported. Nonetheless, the past work of three-dimensional microstructure reconstruction from serial sections is restricted to microstructures containing at the most two phases, or in the multi-phase microstructures, the three-dimensional geometry of only one or two phases is reconstructed. In this contribution, we present three-dimensional reconstruction of multi-phase microstructures of a series of cast Al-alloys containing porosity, Si particles, and numerous intermetallic inclusion phases. All the phases are segmented and separately reconstructed, rendered, and quantitatively characterized in three-dimensions, which clearly brings out the complex three-dimensional morphologies of all phases. The technique is useful for characterization of any multi-phase three-dimensional microstructure.

Introduction

Characterization and visualization of microstructure are of considerable importance in materials research owing to the fact that microstructure directly influences the properties and performance of materials. Material microstructures are usually of three-dimensional nature and, therefore, characterization and visualization of *three-dimensional* (3D) microstructures is of primary interest. Even though most of the materials are opaque, the microstructural observations are generally carried out on the *two-dimensional* (2D) metallographic sections through 3D microstructural domains of interest. The microstructure observed in such a two-dimensional metallographic section consists of intersections of the features in the 3D microstructure with the sectioning plane. Therefore, in a metallographic plane, the volumes (e.g., grains, voids, particles) in a 3D microstructure appear as areas, and the surfaces (e.g., grain boundaries, precipitate interfaces) appear as lines. It is apparent that a 2D metallographic section does not contain all the information concerning the true 3D microstructural geometry. In particular, the information concerning topological aspects of microstructure such as connectivity of particles, formation of particle clusters, percolation events, etc., cannot be obtained from independent 2D metallographic sections. Therefore, 3D microstructural reconstruction and visualization are of significant interest for understanding such aspects of 3D microstructural geometry. Numerous techniques including X-ray computed tomography, magnetic resonance imaging (MRI), and serial sectioning can be used to render a 3D microstructure depending on the material chemistry, processing, and microstructural length scales of interest. This contribution concerns visualization of 3D microstructure from a stack of montage serial sections to observe and characterize various features present in the multi-phase microstructure of a series of cast Al-alloys.

The 3D microstructure of a relatively small microstructural volume of an opaque material can be reconstructed using the classical serial sectioning technique developed in the 1970s [1]. The classical serial sectioning technique has been used in numerous investigations to study 3D microstructures of opaque materials [2], [3], [4], [5] and [6], and it is quite useful for visualization of 3D particle/feature morphologies and short-range microstructural details at sufficiently high resolution. However, classical serial sectioning is not useful for quantitative characterization of topological attributes such as coordination numbers, and important descriptors of spatial arrangement of microstructural features such as higher order nearest neighbor distributions and radial distribution function due to serious bias (systematic error) resulting from edge effects [7], [8] and [9]. Classical serial sectioning is also not useful for truly unbiased estimation of 3D grain/particle size distributions due to the same troublesome edge effects that create significant bias [10]. Further, the classical serial sectioning technique cannot be used to reconstruct a large volume of 3D microstructure (say, ~ 1 mm^3) at sufficiently high-resolution (say, ~1μm). Therefore, it is not useful for characterization of long-range correlations between features of various phases/sizes present in certain multi-phase microstructures.

An efficient montage-based serial sectioning technique is available [7], [10], [11] and [12] that permits generation of significantly large volume (few mm^3) of 3D microstructure at a high resolution (0.5 μm). For approximately the same metallographic effort, montage-based serial sectioning yields a microstructural volume containing a few thousand particles/grains, which provides sufficiently large statistical sample for efficient, reliable, unbiased, and assumption-free direct estimation of 3D microstructural properties as well as for study of topological aspects of microstructure such as feature connectivity. Recently, montage serial sectioning has been implemented in a completely automated serial sectioning set-up that utilizes a robotic arm to move the specimen back and forth between the metallographic equipment (polishing, etching, etc.) and light optical microscope to generate the montage serial sections in a completely automated manner [13].

In this contribution, montage serial sectioning have been applied to reconstruct large volumes of 3D microstructures of cast Al-alloy containing Si particles, porosity and intermetallics. Modern image processing and 3D image reconstruction techniques have been used to reconstruct and visualize the 3D microstructure using volume and surface rendering techniques. The image processing permits detailed observations of particle connectivity of different phases in a multi-phase microstructure.

Experimental Details

Material

In the present work, 3D microstructure of cast aluminum alloys have been characterized. The material contains silicon particles, porosity and numerous intermetallics and therefore serves as a perfect candidate to demonstrate the reconstruction of multi-phase three-dimensional microstructures.

Metallography

The metallographic samples were prepared by sectioning the longitudinal cross-section containing the extrusion axis and mounting them in cold mounting compound. The grinding and polishing steps involved grinding on SiC papers (320–600 grit size) followed by diamond polishing (9 μm to 1 μm) and finally using colloidal silica polishing suspension. During the process of serial sectioning only the last step (colloidal silica) was used to polish off the desired thickness of material.

Montage Serial Sectioning

To generate a large volume of 3D microstructure at high resolution, one may first reconstruct a small microstructural volume such as the one in Fig. 1a, and then reconstruct many contiguous small volumes surrounding it, perfectly match their boundaries, and paste them together to generate a large microstructural volume, as shown in Fig. 1b. A technique equivalent to such a reconstruction has been developed [7], [10], [11] and [12], and in this contribution, it is applied to reconstruct a large volume of the 3D microstructures of the cast Al-alloys. First a "montage" of 49 (or more, if necessary) contiguous microstructural fields observed at a high magnification (800× for the present microstructure) is created by using the large-area high-resolution montage procedure developed by Gokhale et al. [8], [14] and [15]. To create a montage, a field of view (FOV) is arbitrarily chosen in the region of interest in a metallographic plane, and the automated montage program is activated. The input fields consist of number of FOVs in horizontal and vertical direction and also the amount of overlap between consecutive images. Since the overlap is required to seamlessly match the FOVs, the percentage overlap of the images varies depending upon the amount of distinguishable features present in each field of view. Accordingly, a microstructure with more number of features in a single FOV would require a low value of overlap and vice versa. All successive contiguous images are grabbed individually and finally a seamless montage of a large number of contiguous microstructural fields is created. Figure 2 shows such a montage of 49 fields of view (FOV), which has been compressed for display. Each region of this montage has a high resolution of the image shown in Figure 3. Therefore, the montage is a microstructural image of a large area (~ 4 mm^2) having a high resolution. In the present work, image analysis was performed with AxioVision 4.5 and KS-400 image analysis systems. However, several other commercial image analysis systems also have the required capabilities.

Once the montage of the first serial section is created and stored in the computer memory, small thickness of the specimen is removed (about 0.5 μm) by polishing, and then a second montage is created at the region exactly below that in the first metallographic plane. In the present study, this polish–montage–polish procedure was repeated to obtain stack of 80 montage serial sections. Micro-hardness indents were used to locate the exact region of interest in successive serial sections and to measure the distance between consecutive serial sections.

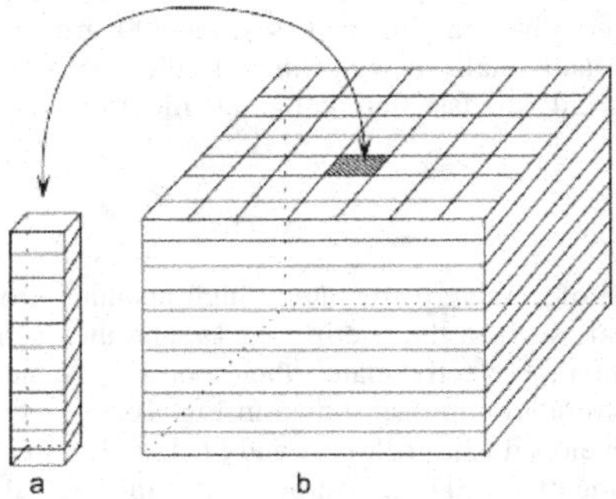

Figure 1(a). Small microstructural volume element constructed from a stack consisting of one field of view in each serial section. (b) Large volume of microstructure obtained from contiguous small volumes such as those in (a) or by using montage serial sectioning.

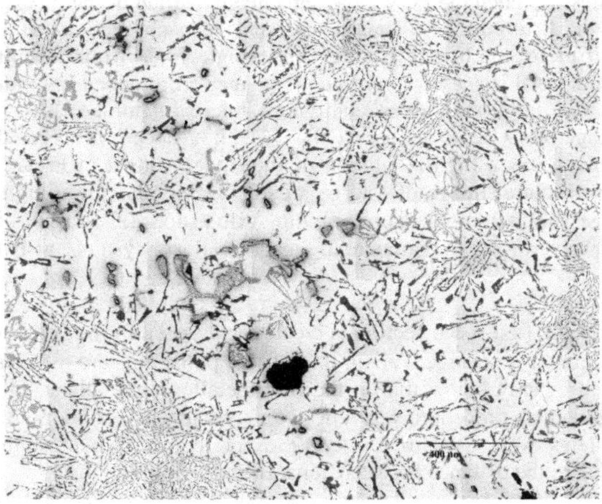

Figure 2. Montage of 49 fields of view covering an area of 4 mm^2 created by matching contiguous microstructural fields grabbed at a resolution of 0.3 μm. The montage is digitally compressed for presentation.

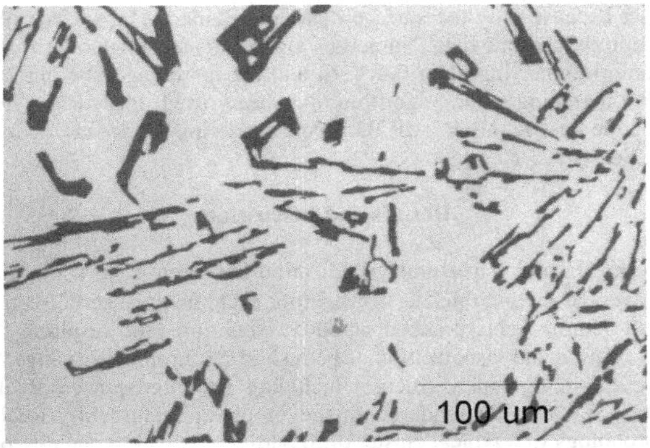

Figure 3. Single field of view showing the feature morphology at high resolution.

An important practical problem in the reconstruction of a 3D microstructure from serial sections is that the successive serial sections may not be precisely aligned; they may have some translational and rotational displacement with respect to each other. In the present study, in spite of adjusting the microscope stage, the montages of the consecutive serial sections were often displaced by about ±10 pixels and ±5°, and therefore, it was essential to precisely align successive serial sections. Alignment can be achieved by locating two common points (in the present case, micro-hardness indents were used for this purpose) in the two consecutive serial sections and translating one image until the first common point is aligned in the two images. Then the image is rotated about this point until the second common point is also aligned. In the present case, this was accomplished by using image analysis software KS400 in which the images of the montage were digitally translated and rotated until they were exactly aligned to the respective previous sections.

Reconstruction and Visualization of Three-Dimensional Microstructure

The stack of aligned serial sections essentially constitutes a volume image data set similar to those encountered in X-ray computed tomography and magnetic resonance imaging (MRI). The steps are involved in the 3D visualization of such data sets are as follows.

• data generation (in the present case, serial sections);

• pre-processing such as image alignment, grid regularization, image enhancement, and interpolation;

• rendering of 3D images.

The 3D microstructural visualization can be achieved either by surface rendering or by volume rendering. Surface rendering involves rendering of the iso-surface of the region of interest (ROI) from the volume data, whereas volume rendering is the rendering of all volume data by specifying opacity and color of each voxel (3D pixel). Surface rendering leads to reduction in the size of the data set because only the surface data are retained. The surface rendering requires fitting of a surface in the volume data. Numerous algorithms are available for surface rendering, including the contour connecting algorithm [16] and the marching cube algorithm [17]. In the present work, the marching cube algorithm has been used for surface rendering of 3D microstructures of the cast Al-alloy. All 3D image rendering work was done by using image analysis software Voxblast 3.10.

Results and Discussion

In the present work, 3D microstructural visualization has been done using 80 montage serial sections with the average section thickness of 0.5 μm; each montage serial section containing 49 contiguous microstructural fields grabbed at 500×. Therefore, the resulting 3D data sets are useful for characterization and visualization of both short range and long-range attributes of each of the phases present in the microstructure including the interconnectivity between various phases. Figure 4 shows the 3D rendered image of an individual Si particle and Figure 5 represents the morphology of a typical cluster of Si particles as seen in the microstructure. Figure 6 shows the 3D rendered image for the intermetallic phase. Figure 7 shows the 3D surface rendered image depicting the multi-phase aspect of the microstructure, the dark features represent the silicon particles and the lighter regions are the intermetallics.

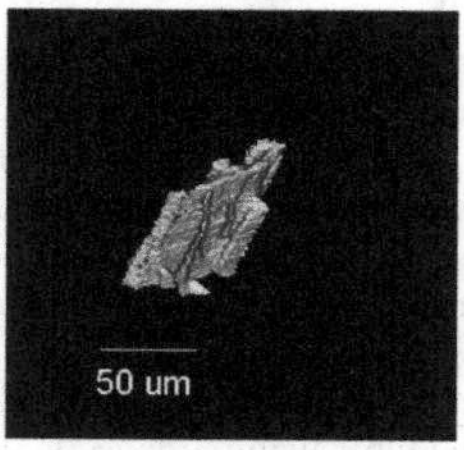

Figure 4. Surface rendered 3D microstructure showing a Si particle

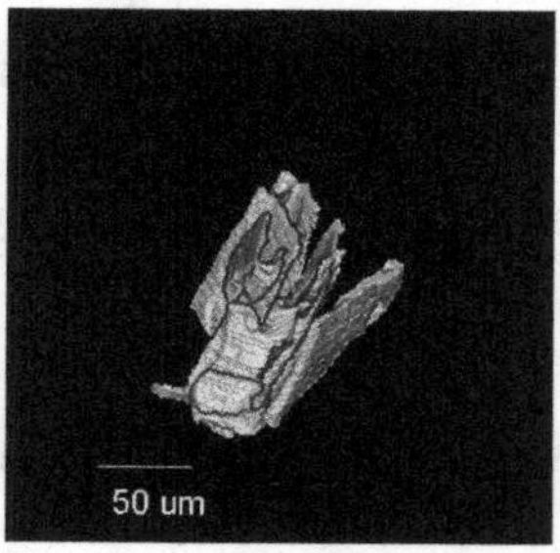

Figure 5. Surface rendered 3D microstructure showing a cluster of Si particles.

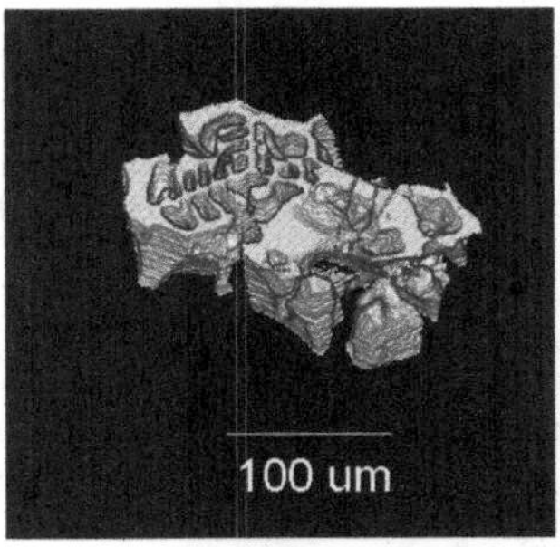

Figure 6. Surface rendered 3D microstructure showing intermetallic.

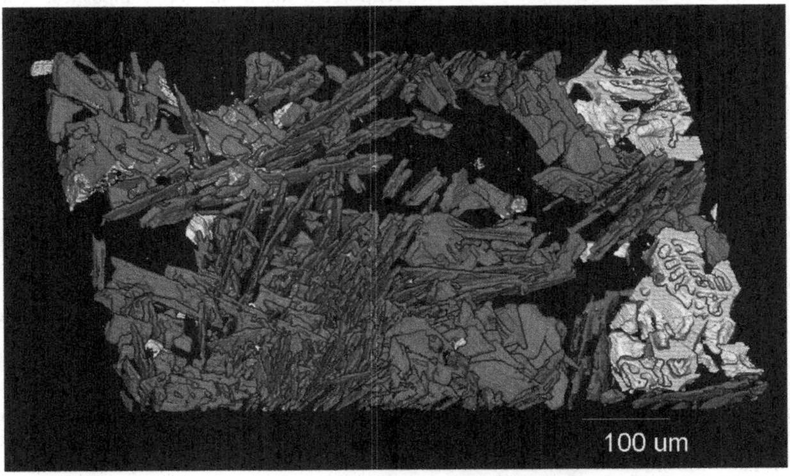

Figure 7. Surface rendered 3D microstructure showing the Si particles in dark and intermetallic in light color.

Summary and Conclusions

The montage serial-sectioning technique was used to generate high-resolution large-volume 3D microstructures of cast Al-alloys containing multiple phases. The visualization of reconstructed 3D microstructures clearly revealed the morphology of the clusters of thin plate-like silicon particles and large chunks of intermetallics present in the microstructure. 3D microstructural reconstruction and visualization are useful for characterization of individual 3D feature morphologies as well as for visualization of multiple phases as present together in the microstructure.

Acknowledgement

The research was funded by General Motors Corporation. The financial support is gratefully acknowledged. Any opinions, findings, and conclusions or recommendations expressed in this contribution are those of the authors and do not necessarily reflect the views of the funding agencies.

References

1. F.N. Rhines and K.R. Craig, "Measurement of average grain volume and topological parameters by serial sectioning analysis", *Metall. Trans., A* 7A (1976), pp. 1729–1734.
2. R.S. Sidhu and N. Chawla, "Three-dimensional microstructure characterization of Ag3Sn intermetallics in Sn-rich solder by serial sectioning", *Mater. Charact.* 52 (2004), pp. 225–230.
3. M. Li, S. Ghosh, O. Richmond, H. Weiland and T.N. Rouns, "Three-dimensional characterization and modeling of particle reinforced metal matrix composites: Part I. Quantitative description of microstructural morphology", *Mater. Sci. Eng.* A265 (1999), pp. 153–173.
4. K.M. Wu and M. Enomoto, "Three-Dimensional morphology of degenerate ferrite in an Fe–C–Mo alloy", *Scr. Mater.* 46 (2002), pp. 569–574.
5. M.V. Kral, M.A. Mangan, G. Spanos and R.O. Rosenberg, "Three-dimensional analysis of microstructures", *Mater. Charact.* 45 (2000), pp. 17–23.
6. T. Yokomizo, M. Enomoto, G. Spanos and R.O. Rosenberg, "Three-dimensional distribution, morphology, and nucleation sites of intergranular ferrite in association with inclusions", *Mater. Sci. Eng.* 344A (2003), pp. 261–267.
7. A. Tewari, A.M. Gokhale and R.M. German, "Effect of gravity on three-dimensional coordination number distribution in liquid phase sintered microstructures", *Acta Mater.* 47 (1991), pp. 3721–3734.
8. S. Yang, A. Tewari and A.M. Gokhale, "Modeling of non-uniform spatial arrangement of fibers in a ceramic matrix composite", *Acta Mater.* 45 (1997), pp. 3059–3069.
9. A.M. Gokhale and A. Tewari, "Efficient estimation of number density in opaque material microstructures: the large-area disector (LAD)", *J. Microsc.* 200 (2000) (Pt. 3), pp. 277–283.
10. A. Tewari and A.M. Gokhale, "Application of serial sectioning for estimation of three-dimensional grain size distribution in a liquid phase sintered microstructure", *Mater. Charact.* 46 (2001), pp. 329–335
11. M.D. Dighe, A. Tewari, G.R. Patel, T. Mirabelli and A.M. Gokhale, "Application of digital image processing to reconstruct three-dimensional micro-porosity in a cast A356.0 alloy", Trans. *Am. Foundry Soc.* 99 (2000), pp. 353–356.
12. Tewari A., Ph.D. Dissertation, Georgia Institute of Technology, 1999.
13. J.E. Spowart, H.M. Mullens and B.T. Puchala, "Collecting and analyzing microstructures in three-dimensions: a fully automated approach", *J. Met.* (2003 (October)), pp. 35–37.
14. P. Louis and A.M. Gokhale, "Application of image analysis for characterization of spatial arrangement of microstructural features", *Metall. Mater. Trans., A* 26A (1995), pp. 1449–1455.
15. P. Louis and A.M. Gokhale, "Computer simulation of spatial arrangement and connectivity of particles in 3d microstructure: application to model electrical conductivity of a polymer matrix composite", *Acta Mater.* 44 (1996), pp. 1519–1528.
16. E. Keppel, "Approximating complex surfaces by triangulation of contour lines", *IBM J. Res. Develop.* 19 (1975), pp. 2–11.
17. E.W. Lorensen and H.E. Cline, "Marching cubes: a high resolution 3D surface construction algorithm", *Comput. Graph.* 22 (1987), pp. 38–44.

Shape Casting: The 3rd International Symposium
Edited by: John Campbell, Paul N. Crepeau, and Murat Tiryakioğlu
TMS (The Minerals, Metals & Materials Society), 2009

CORRELATION OF THERMAL, TENSILE AND CORROSION PARAMETERS OF ZN-AL ALLOYS WITH COLUMNAR, EQUIAXED AND TRANSITION STRUCTURES

Alicia E. Ares[1,2], Liliana M. Gassa [1,3] Sergio F. Gueijman[2] Carlos E. Schvezov[1,2]

[1] CONICET (Consejo Nacional de Investigaciones Científicas y Técnicas), Rivadavia 1917, Buenos Aires, C.P. 1033, Argentina
[2] University of Misiones; 1552 Azara Street, Posadas, Misiones, 3300 Argentina.
[3] INIFTA (Instituto de Investigaciones Fisicoquímicas Teóricas y Aplicadas). Universidad Nacional de la Plata. Diagonal 113 y 64, La Plata. Argentina.

Keywords: columnar-to-equiaxed transition, Zn-Al alloys, thermal, tensile and corrosion parameters.

Abstract

The columnar to equiaxed transition (CET) has been examined in different wrought and casting alloys for many years and the metallurgical significance of CET has been treated in several articles. Experimental observations in the literature have focused on thermal parameters like cooling rate, velocity of the liquidus and solidus fronts, local solidification time, temperature gradients and recalescence. The objective of the present research consist on studying the influence of solidification thermal parameters on the type of structure (columnar, equiaxial or with the CET) and on the secondary dendritic spacing in Zn-Al alloys (Zn-1%Al and Zn-4wt%Al, weight percent). Also, correlate the thermal and structural parameters of these alloys with tensile and corrosion behavior. The results show that Zn-4wt%Al alloys are more resistant to the corrosion that Zn-1wt%Al, independently of the structure. Also, in both alloys the equiaxed structures presented a better tensile resistance than the columnar and CET zones.

Introduction

A recent alloy development is generation of a family of zinc foundry alloys suitable for sand, permanent mold, plaster mold, shell mold and investment casting. The mechanical properties of zinc alloys make them attractive substitutes for cast iron and copper alloys in many structural and pressure-tight applications [1]. Because zinc is less costly than copper, these zinc alloys have a distinct cost advantage over copper-base alloys. The ease of machining of zinc and its inherent corrosion resistance give it advantages over cast iron. Structural parameters such as type of grain (columnar, equiaxial or columnar to equiaxed transition, CET), grain size and dendritic spacing are highly influenced by thermal behavior of the metal/mould system during solidification [2,3].

The objective of the present research consist on studying the influence of solidification thermal parameters on the type of structure (columnar, equiaxial or with the CET); and on the dendritic spacing (primary and secondary) in Zn-Al (ZA) alloys (Zn-1%Al to Zn-4wt%Al, weight percent). Also, correlate the thermal and structure parameters of these alloys with tensile and corrosion behavior. The results show that the CET zone and the equiaxed structures presented a better tensile and corrosion resistance than the columnar zone.

Experimental Procedure

Directional Solidification

The casting assembly used in solidification experiments has been detailed in a previous article [4,5]. The Zn-1wt%Al (ZA1) and Zn-4wt%Al (ZA4) alloy samples were solidified directionally upwards. Initially the melt was allowed to reach the selected temperature and then, the furnace power was turned off and the melt was allowed to solidify from the bottom. During the solidification process, temperatures at different positions in the alloy sample were measured and the data were acquired automatically using a data logger. For the measurements, a set of five thermocouples type K, previously calibrated, were arranged at equal distance between thermocouples of 2cm.

The temperature profiles were determined from the measurements during solidification at the different thermocouple positions. The temperature versus time for one experiment corresponding to alloys Zn-1wt%Al is presented in Figure 1. The thermocouple T_1 is at the lowest position and the first to reach the solidification front and T_5 is at the highest position. From the temperatures versus time graphs it is possible to calculate the cooling velocity in the melt, $C.R._{LIQ.}$. The velocity associated to each experiment is the average value of the slopes determined from the graphs. The start and the end of solidification at each thermocouple determine the positions of the solidification fronts versus time, which correspond to the liquidus and the solidus temperature, respectively. Both points are detected by the changes in the slopes of the cooling curve at the start and end of solidification. This criterion was chosen in order to allow for undercooling to occur before solidification and possible recalescence during solidification of equiaxed grains, since this process is characterized by nucleation and solidification of grains in the melt rather than for what is observed in a normal solidification process where there is a dendrite tip front advancing in the melt. The local solidification time at each thermocouple location is determined by the period of time taken for the temperature to go from the liquidus to the solidus temperature. The velocity of the liquidus solidification front is calculated as the distance between the thermocouples divided by the time taken by the liquidus temperature to go from the lower to the upper thermocouple. These velocities are named as V_L for the liquidus velocity. The liquid thermal gradient, G_L, at all times are calculated straightforward, dividing the temperature difference between two thermocouples by the separation distance between them.

Microstructure Analysis

After solidification the samples were cut in the longitudinal direction, polished with emery paper and etched to reveal the structure. The reagent used was a solution of HCl acid (70%) during 120 seconds [6]. A typical resulting macrograph can be seen in Figure 2 for Zn-1wt%Al alloy. To reveal the microstructure a solution containing 5 g CrO_3, 0.5 g Na_2SO_4 and 100 ml H_2O (Palmerston´s reagent) was used. The etching time varied from 5 to 15 s, depending on the alloy solute content. After etching, the samples were rinsed in a solution of 20 g CrO_3 and 100 ml H_2O before optical microscopy examination using SEM and an optical microscope in order to measure the average secondary dendrite arm spacings (15 measurements for each selected position in the sample). The secondary arm spacing was measured in cross sections of the half of the sample. Each section was mounted, polished and etched and the spacing was determined by the number of interception in a straight line.

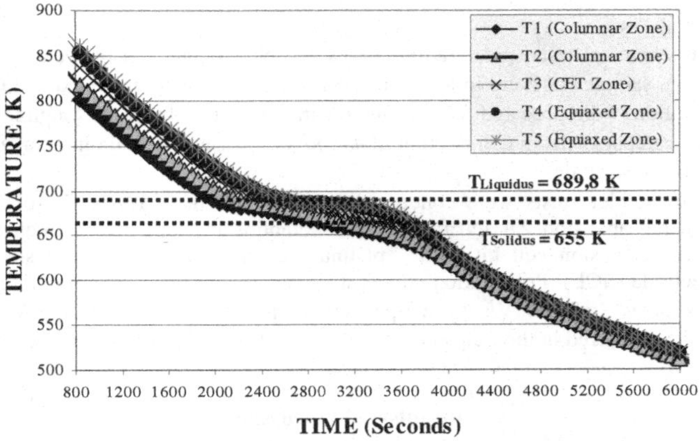

Figure 1. Temperature versus time curve. Zn-1wt%Al.

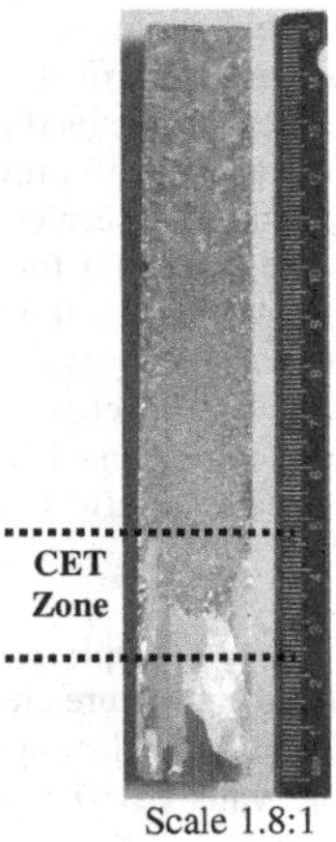

Scale 1.8:1

Figure 2. Macrograph of ZA1 alloy

Table I. Principal parameters obtained from the temperature versus time curves.

#	Alloys (wt%)	C.R. $_{LIQ.}$ ($K.s^{-1}$)	$CET_{Average}$ (mm)	V_{LC} ($mm.s^{-1}$)	G_{LC} ($K.mm^{-1}$)
1	Zn-1%Al (ZA1)	2.2	32	1.1	-0.13
2	T_L = 689.8 K	2.5	40	1.4	-0.06
3	T_S = 655.0 K	2.7	52	1.9	-0.025
4	T_E = 655.0 K	1.8	27	1.2	-0.08
5		2.1	31	1.0	-0.014
6	Zn-4%Al (ZA4)	1.7	22	1.6	-0.05
7	T_L = 663.5 K	1.9	34	0.8	-0.03
8	T_S = 655.0 K	2.2	42	1.0	0.01
9	T_E = 655.0 K	1.5	22.5	1.2	-0.16
10		2.0	38.5	1.1	0.09

<u>Tensile Tests</u>

The tensile tests were performed followings standards given by the NBR 6152 and ASTM /E-8M norms and using a SHIMADZU tensile test machine. The samples were cut, polished and etched with a solution of HCl acid (70%) during 120 seconds [6] to determine the different structure parameters. Then the tensile samples were machined to the standard shape. After the test the samples were again cut, polished and etched and the fracture zone analyzed.

Electrochemical Tests

For the electrochemical tests (polarization curves and electrochemical impedance spectroscopy technique, EIS), samples of 2 cm in length of each zone (columnar, equiaxed and CET) and for each concentration, were prepared as working electrodes cutting from the longitudinal sections, polished with sandpaper (from CSi #200 until #1200) and washed with de-mineralized water and dried by natural flow of air.

All the electrochemical tests were conducted in a 300 ml of a 3% NaCl solution at room temperature using an IM6d ZAHNER® electrik potentiostat coupled to a frequency analyzer system, a glass corrosion cell kit with a platinum counter electrode and a sutured calomel reference electrode (SCE). Polarization curves were obtained using a scanning rate in the range of 0,002 V/s ≤ v ≤ -0,250 V/s from open circuit potential until to 0,250 V. Impedance spectrums were registered in the frequency range of 10-3Hz ≤ f ≤ 105 Hz in open circuit.

Results and Discussion

They were carried out a total of 10 experiments of directional solidification, 10 of tensile test and six electrochemical test. The parameters calculated for each experiment are listed in Tables I, II and III. In the case of directional solidification the parameters are liquid cooling rates ($C.R._{LIQ.}$), average CET position($CET_{Average}$), velocity of the liquidus interface at the instant of the CET (V_{LC}), and the temperature gradient at the instant of the CET transition (critical gradients (G_{LC})). The negative value of critical gradients was analyzed before and it is an indication of a reversal in the temperatures profiles ahead of the interface, which could be associated to the recalescence due to massive nucleation of equiaxed grains, and previously reported and discussed for other alloys [4,5]. The scatter in values of the gradients is associated with the fact that the transition not always happens right at the thermocouple position and therefore, the calculated value is an average value over a region which includes the mushy zone and the melt. The error in the calculated gradients could be as large as 45 %. Thereby taking into account these considerations it can be concluded that within the error, the alloy composition does not have an effect on the temperature gradient during the transition.

The secondary dendritic spacing (λ_2) versus distance from the bottom of the sample for one sample of Zn-1wt%Al is plotted in Figure 3. It is possible to appreciate that the structure changes from columnar at the bottom to equiaxed at the upper part of the sample. The secondary spacing is increasing in size from the columnar to equiaxed zone. The average values of λ_2 on the stressed zone of each sample are listed in Table II.

With respect to the mechanical test, in Figure 4 it is shown a typical result of a tensile test corresponding to a Zn-1wt%Al alloy. From such a curve the normal data such as type of fracture, tensile strength (TS), ultimate tensile strength (UTS) and yield strength (YS), showing values of 148.8 MPa, 141.2 MPa and 85.3 MPa, respectively. The process is respected for all the samples and the results are listed in Table II and were correlated with λ_2 as a structural parameter. Figure 6 (a) show the results for Zn-1wt%Al alloys. It is observed that both strength (TS and UTS) increases with λ_2, however the effect is more pronounced in the case of TS.

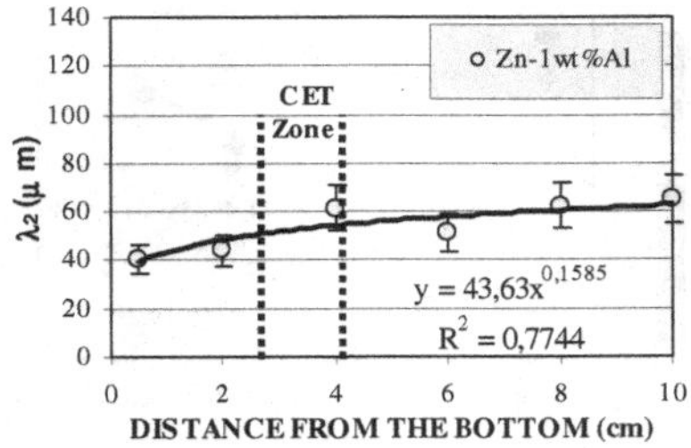

Figure 3. λ_2 versus distance.

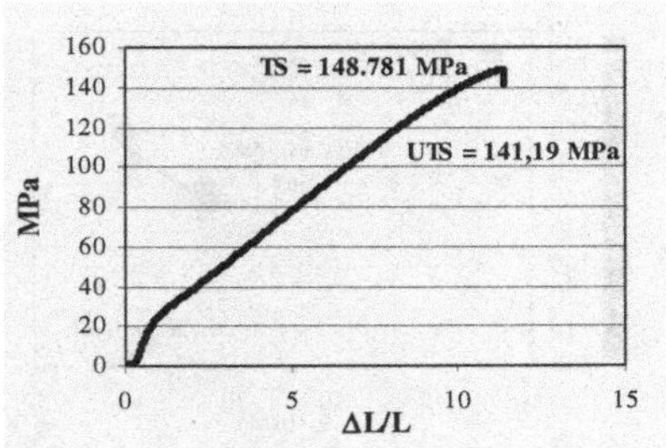

Figure 4. Tensile properties of Zn-1wt%Al alloy.

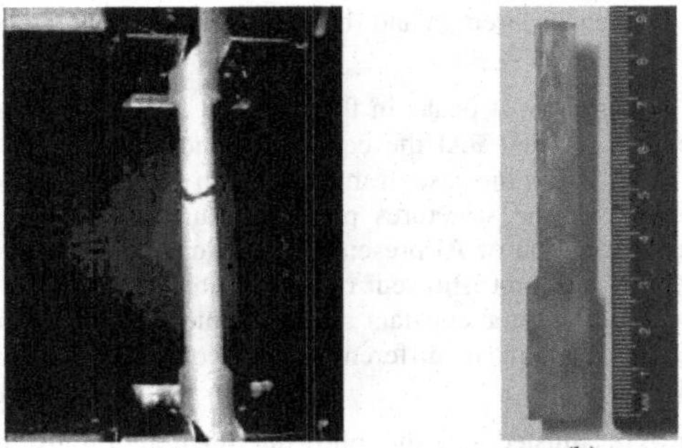

(a) (b)

Figure 5. (a) Tested sample of Zn-1wt%Al and (b) macrograph of tested sample showing the structure in the stressed zone [Scale 1.3:1].

Table II. Values of parameters for each zone and alloy concentration.

Alloy	Tensile Strength (MPa)	Yield Strength (MPa)	Ultimate Tensile Strength (MPa)	Type of Structure	Average λ_2 (μm)
Zn-1wt%Al	148.78	85.28	141.19	CET	33.51
Zn-1wt%Al	169.91	79.12	128.41	Equiaxial	41.23
Zn-1wt%Al	161.27	76.53	131.12	Equiaxial	40.52
Zn-1wt%Al	153.49	55.67	135.67	Equiaxial	32.90
Zn-1wt%Al	148.21	61.28	119.60	Equiaxial	32.50
Zn-4wt%Al	135.78	71.34	109.28	Equiaxial	33.79
Zn-4wt%Al	146.92	72.48	105.46	Columnar	42.90
Zn-4wt%Al	152.39	81.26	104.29	Equiaxial	42.50
Zn-4wt%Al	158.71	83.61	99.36	CET	51.90
Zn-4wt%Al	160.41	88.29	112.45	Equiaxial	51.76

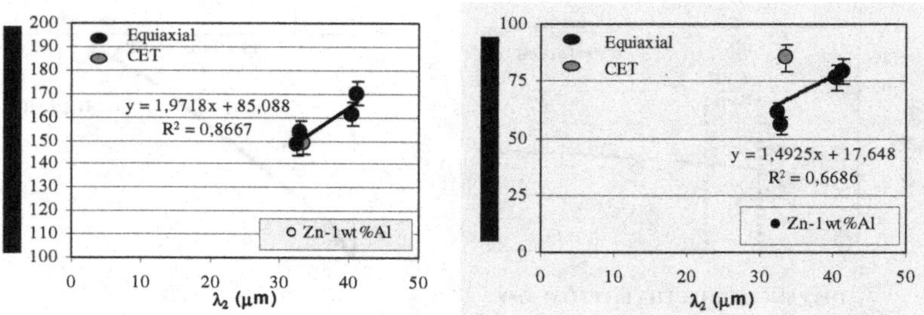

(a) (b)

Figure 6. (a) Tensile strength and (b) Yield strength as a function of λ_2.

From the analysis of the currents of peaks in the obtained polarization curves, it is possible to appreciate that in the case of Zn-1%Al the equiaxial structure is the most susceptible to the corrosion (see Figure 7(a)), and in the case of the alloy with 4wt% of Al the CET structure is the most susceptible. The rest of the structures presented currents of peaks in the same order, independently to the concentration of Al present in the alloy. The impedance diagrams showed only one capacitive time constant showed one capacitive loop at high and intermediate frequencies and non-well define time constant at low frequencies. In Figure 7 (b) are presented the experimental Nyquist diagrams in different zones (columnar, CET and equiaxed) of the samples of Zn-1wt%Al alloy.

The experimental data were adjusted with the following function of transference:

$$Z_t(j\omega) = R_\Omega + Z \qquad (1)$$

where $Z = R_{ct} + (1/j\omega C_{dl})$, $\omega = 2\pi f$; R_Ω is the ohmic solution resistance and C_{dl} is the differential capacitance of the electric double layer. In Table III are presenting the values of the adjusting parameters for each zone and alloy concentration. In both concentrations the corrosion susceptibility depends on the structure of the alloy. The alloy with only 1%Al is the less resistant to the corrosion and their susceptibility to the corrosion is independent of the structure. The alloy with 4%Al and equiaxial structure is the more resistant of all. From this analysis it is possible to appreciate that Zn-1wt%Al alloys are less resistant to the corrosion than Zn-4wt%Al, independently of the structure.

Table III. Values of the EIS adjusting parameters for each zone and alloy concentration.

Type of Alloy and Structure	R_{ct} (ohm.cm^2)	C_{dl} (F.cm^{-2})
Zn-1%Al $_{Columnar}$	22	2,91.10^{-5}
Zn-1%Al $_{CET}$	400	1,76.10^{-4}
Zn-1%Al $_{Equiaxial}$	220	4,33.10^{-4}
Zn-4%Al $_{Columnar}$	23	2,6.10^{-4}
Zn-4%Al $_{CET}$	1786	1,6.10^{-6}
Zn-4%Al $_{Equiaxial}$	3,727.10^{5}	2,1.10^{-6}

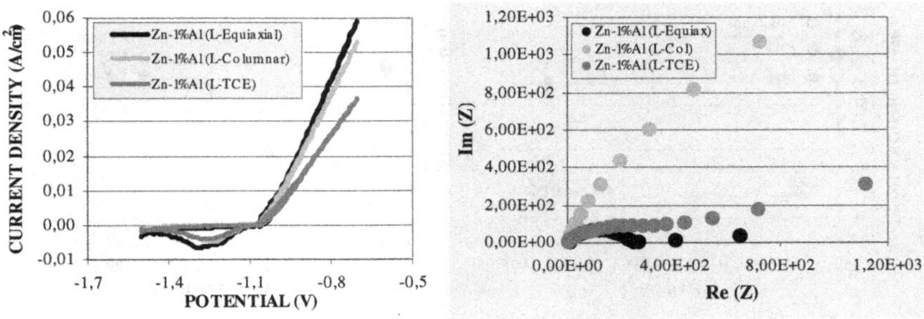

(a) (b)

Figure 7. (a) Voltammograms and (b) Nyquist of Zn-1wt%Al alloys.

Correlation between Electrochemical and Mechanical Parameters

The obtained values of the charge-transfer resistance, R_{ct}, were correlated with the values of the TS and YS for each concentration. It can be seen in Figures 8 (a) and (b), respectively.

In the case of TS versus R_{ct}, no correlation between both parameters was found, but in the case of Zn-4%Al alloys, when the YS increase from columnar to equiaxial region, the R_{ct} also increase (Figure 8 (b)).

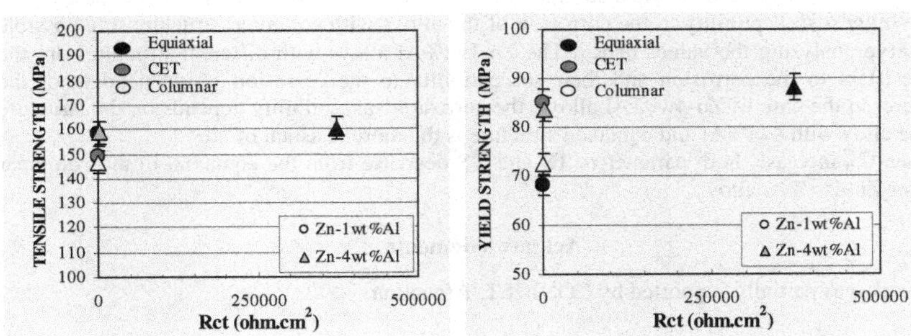

(a) (b)

Figure 8. (a) Tensile strength and (b) yield strength as a function of R_{ct}.

In the case of the differential capacitance of the electric double layer, C_{dl},, it was found a relationship between TS and YS.

In the case of equiaxial to columnar zone for the Zn-4wt%Al alloy, when C_{dl} increase, both parameters decreases. In the case of Zn-1wt%Al, no correlation between TS and YS with C_{dl} was found. The later is shown in Figures 9 (a) and (b), respectively.

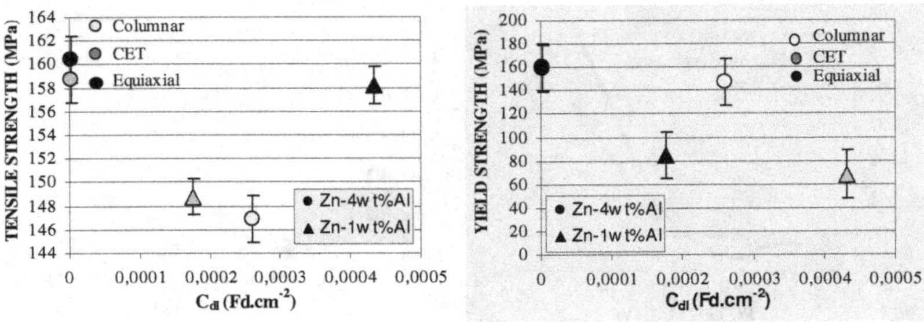

(a) (b)

Figure 9. (a) Tensile strength and (b) yield strength as a function of C_{dl}.

Conclusions

The main conclusions obtained from the present work are as follow:

1. Samples with CET were obtained and the transition occurs in a zone rather than in a sharp plane, where both columnar and equiaxed grains coexist in the melt.
2. The temperature gradient and the velocity of the liquidus front reach low critical values before the transition.
3. The average secondary spacing increases from columnar to equiaxed zones in the samples.
4. The tensile strength and the yield strength increase when the secondary spacing increases.
5. The biggest susceptibility to the corrosion of the alloys with columnar structure it is possible to observe analyzing the values of R_{ct}. The Zn-1wt%Al alloys with different structures are the less resistant to the corrosion and their susceptibility to the corrosion is independent of the structure. In the case of Zn-4wt%Al alloys, the corrosion susceptibility depends on the structure and the alloy with 4wt%Al and equiaxed structure is the more resistant of all.
6. When C_{dl} increase, both parameters TS and YS decrease from the equiaxial to the columnar zone for Zn-4wt%Al alloy.

Acknowledgments

This work was partially supported by CONICET, Argentina.

References

1. J. Birch, "New Alloys for Zinc Castings," *Material & Desing*, 11 (1990), 83-87.
2. J. Sieniawski, R. Filip, W. Ziaja, "The Effect of Microstructure on the Mechanical Properties of two-Phase Titanium Alloys," *Mater. Des.,* 18 (1997), 361–363.
3. J.M.V. Quaresma, C.A. Santos, A. Garcia, "Correlation Between Unsteady-State Solidification Conditions, Dendrite Spacings, and Mechanical Properties of Al–Cu Alloys," *Metall. Mater. Trans. A* 31 (2000), 3167–3178.
4. A.E. Ares, C.E. Schvezov, "Influence of Solidification Thermal Parameters on the Columnar-to-Equiaxed Transition of Aluminum-Zinc and Zinc-Aluminum Alloys," *Metall. Trans.A,* 38 (2000), 1485-1499.
5. A.E. Ares, L.M. Gassa, S.F. Gueijman, C.E. Schvezov, "Correlation between Thermal Parameters, Structures, Dendritic Spacing and Corrosion Behavior of Zn–Al Alloys with Columnar to Equiaxed Transition," *Journal of Crystal Growth*, 310 (2008), 1355–1361.
6. G. Kehl, *Fundamentos de la Práctica Metalográfica* (Madrid: Aguilar, 1963), 124-126.

Shape Casting: The 3rd International Symposium
Edited by: John Campbell, Paul N. Crepeau, and Murat Tiryakioğlu
TMS (The Minerals, Metals & Materials Society), 2009

SOLIDIFICATION, MACROSTRUCTURE AND MICROSTRUCTURE ANALYSIS OF AL-CU ALLOYS DIRECTIONALLY SOLIDIFIED FROM THE CHILL FACE

Alicia Esther Ares[1,2], Carlos Enrique Schvezov[1,2]

[1] CONICET (Consejo Nacional de Investigaciones Científicas y Técnicas), Rivadavia 1917, Buenos Aires, C.P. 1033, Argentina
[2] University of Misiones; 1552 Azara Street, Posadas, Misiones, 3300 Argentina.

Keywords: Directional solidification, Al-Cu alloys, thermal parameters, structure.

Abstract

The understanding of the phenomenon of the Columnar to Equiaxed Transition (CET) is very important for metallurgical applications. In the present study the CET was observed in aluminum-copper alloys of different compositions (Al-2wt%Cu, Al-20wt%Cu and Al-33.2wt%Cu, which were solidified directionally from a chill face. The main parameters analyzed include cooling rates, temperature gradients, solidification velocities of the liquidus fronts, recalescence, grain size, primary and secondary dendritic arm spacing and eutectic spacing. The temperature gradient and the velocity of the liquidus front reach low critical values before the transition. These critical values are between 0.4 to 3.8 $mm.s^{-1}$ for the velocity and -0.44 to 0.10 $K.mm^{-1}$ for the temperature gradient. In the case of the eutectic alloys, the transition from columnar to equiaxed occurs with similar characteristics as for the other concentrations, the resultant lamellar, rod-like or mixed solidification microstructure have parameters which also depend on solidification conditions, mainly velocity.

Introduction

During the last decades, much work has been done to understand the interaction between the parameters involved in the transition from the columnar to the equiaxed zone (CET), for example, Ziv et al [1] using an unidirectional heat transfer model for the Al-3wt%Cu alloy, determined that the CET occurred when the gradient fell to 0.06 $K.mm^{-1}$ and also this result is in agreement with a gradient prediction from Hunt´s model [2]. Suri et al. [3] observed the CET in Al-4.5%Cu alloys and reported that the CET occurred if $G < 0.74V^{0.64}$. Siqueira et al. [4] studied the CET in Al-2%Cu, Al-5%Cu, Al-8%Cu and Al-10%Cu alloys and the CET occur rapidly on a near horizontal plane and further from the chill with increasing heat transfer coefficient and increasing superheat, occurred at tip growth rates ranging from 0.28 to 0.88 $mm.s^{-1}$, for temperature gradients in the liquid ranging from 0.28 to 0.75 $K.mm^{-1}$ and when the cooling rate fell below the critical value of 0.2 $K. s^{-1}$ [5].

The results presented in this paper focus on the columnar to equiaxed transition studies in Al-Cu alloys at different solute concentrations (Al-2wt%Cu, Al-20wt%Cu and Al-33.2wt%Cu). In the investigation, the effect of several solidification parameters on the transition are determined and discussed. Such parameters include thermal parameters (cooling rates, temperature gradients, solidification velocities of the liquidus and solidus fronts, recalescence) and structural parameters (grain size, primary and secondary dendritic arm spacing and eutectic spacing).

Experimental Procedure

The procedure used in this investigation follows that used for Pb-Sn [6] and Zn-Al [7], in which the alloy is solidified upward vertically from water cooled copper block. The melt was prepared from aluminum (99.96 wt%) and copper (99.999 wt%) and solidified directionally upwards in an experimental setup consisting of a heat unit, a temperature control system, a temperature data acquisition system, a sample moving system and a heat extraction system. Previous to the experiments, the liquidus (T_L) and solidus (T_S) temperature was determined by differential thermal analysis [8]. In these tests, the melt was heated to the required temperature, the heat unit was turned off, and water was circulated into the heat extraction system (a cylinder of copper 60.0 mm diameter and 120.0 mm high).

Temperature measurements were made during solidification with six K-type thermocouples (1.8 mm) covered with refractory slurry. The thermocouples were previously calibrated using aluminum and copper at their fusion points. During a solidification experiment, the temperature measured by each thermocouple was recorded at regular intervals of time. Different intervals were previously tested, and as a result of this exercise, an interval of 1 second was selected. For the data processing, the readings made every 0.1 seconds during 1 second were averaged, and this value was associated to the middle of the averaged interval.

The solidified alloys were cut in the longitudinal direction. After this, the samples were polished with sand paper and the samples with less than 10wt%Cu were etched with a solution containing 15 ml HF, 4.5 ml HNO_3, 9.0 ml HCl and 271.5 ml H_2O. The samples with 20wt%Cu and 33.2wt%Cu were etched with a solution containing 320 ml HCl, 160 ml HNO_3 and 20 ml HF. Etching was performed at room temperature [9]. The position of the transition was located by visual observation and optical microscopy, and the distance from the bottom of the sample was measured with a ruler. The columnar and equiaxed grain size was measured using the ASTM E112 standard [10], at equally spaced intervals. The microstructure was analyzed using Scanning Electron Microscopy (SEM) and image processing systems Neophot 32 (Carl Zeiss, Esslingen, Germany) and Leica Quantimet 500MC (Leica Imaging Systems Ltd, Cambridge, England). The primary and secondary dendritic arm spacing and the eutectic spacing were measured in cross sections of the other half of the sample. Each section was mounted, polished and etched with 1 g NaOH and 100 ml H_2O during 5 to 15 seconds at room temperature and the spacing was determined by the number of interception in a straight line.

Results and Discussion

<u>Characterization of the Thermal and Kinetic Parameters</u>

A number of fifteen experiments in a range of alloy composition and cooling rate were performed. The list of compositions is shown in Table I. The compositions covered 2wt%Cu, 20wt%Cu and 33.2 wt% Cu (eutectic) alloys. In Table I, it is also shown an average position of the transition from the bottom of the sample. Typical results of the transition are shown in Figure 1 (a) for Al-20wt%Cu and Figure 2 (a) for Al-33.2 wt%Cu.

From the macrographs in both figures it is possible to appreciate that the transition in grain size is not sharp as it was found for Pb-Sn [6] and Zn-Al [7] alloys, occurring in a region where both kinds of grains are observed and that after the transition there is a gradual increase in equiaxed grain size which could decrease, increase or remain approximately constant in the fully equiaxed zone and it is noted that in no case the columnar growth is restored.

No effect of the set of the thermocouples in the transition was observed; either acting as nucleating sites or changing the solidification structure. However, Ziv et al [1] observed that the CET in Al-3wt%Cu occurred relatively abruptly, on a near horizontal plane and Siqueira et al [4] obtained the same results that Ziv et al [1] in Al-2wt%Cu, Al-5wt%Cu, Al-8wt%Cu and Al-10wt%Cu alloys. The different observation was attributed to smaller diameter of the samples in the present case which produce less fluid flow than in the other cases.

In Figures 1 (b) to (d) and Figures 2 (b) to (d) it is also shown the representative microstructures of solidification in the three region; columnar, transition and equiaxed for two compositions, Al-20wt%Cu and Al-33.2wt%Cu, respectively. In Figure 1, the microstructure is dendritic and in Figure 2 the microstructure is lamellae and rod-like shapes.

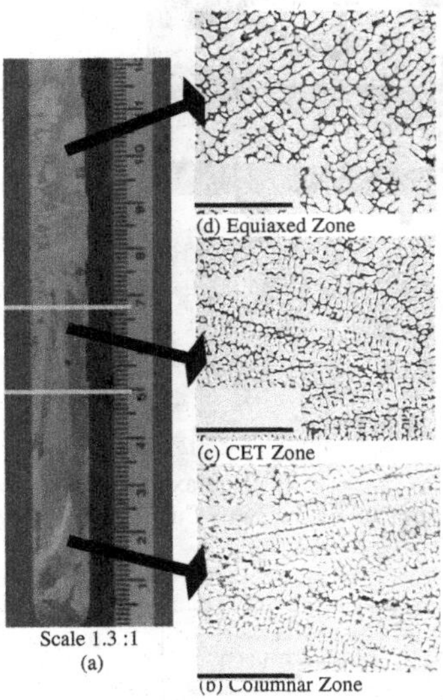

Figure 1. Macro and microstructures of Al-20wt%Cu.

Table I. Parameters of Al-Cu alloys.

#	Alloy wt%Cu	$\dot{T}$ Average ($K.s^{-1}$)	$CET_{Ave.}$ (mm)	$V_{L\ (Critical)}$ ($mm.s^{-1}$)	$G_{L(Critical)}$ ($K.mm^{-1}$)
1	Al-2wt%Cu	1.44	43.5	0.8	-0.38
2		1.12	77	0.9	-0.17
3		1.66	79.95	1	-0.06
4		1.87	71.5	0.9	0.09
5		2.15	82.4	1.3	-0.15
6	Al-20wt%Cu	2.1	85.5	0.6	-0.14
7		1.83	78.4	3.1	0.05
8		1.56	70.5	2.9	-0.12
9		1.7	68	3.8	0.10
10		1.36	87.5	0.4	-0.35
11	Al-33.2wt%Cu	2.13	74	1.6	-0.44
12		1.80	74.1	2.8	-0.17
13		1.06	43.5	0.7	-0.12
14		1.73	-----	1	0.002
15		1.13	-----	1.8	-0.05

The results of the present investigation show that at the eutectic concentration there is also a transition from columnar to equiaxed type of solidification (Figure 2 (a)); the columnar type growth is characterized by different grains along the direction of heat extraction, each presenting a mixed structure of lamellae and rod-like shapes, and both, grain and lamellae oriented in the direction of heat extraction (Figure 2 (b)). The white arrows in Figures 1 and 2 indicate the direction of heat extraction.

The temperature measured by the six thermocouples inserted in the sample during the whole process were stored and analyzed for all the experiments. A typical set of cooling curves is shown in Figure 3. The thermocouple T_1 is at the lowest position and the first to reach the solidification front and T_6 is at the highest position. In all the curves it is possible to identify a period corresponding to the cooling of the melt, a second period of solidification and the final period of cooling of the solid to ambient temperature. In some particular cases it is possible to identify a short period of recalescence when the columnar to equiaxed transition occurs at a thermocouple position.

As reported before [6,7], from the data shown in Figure 3 the following thermal parameters can be extracted; melt superheat, cooling rate of the melt, position and velocity of the solidification fronts for the solidus and the liquidus temperature, local solidification time, length of the mushy zone and temperature gradients.

Comparing the cooling rates with the distances, which correspond to the length of the columnar zone, for all alloys, it is observed that increasing the velocity increases the length of the columnar grains.

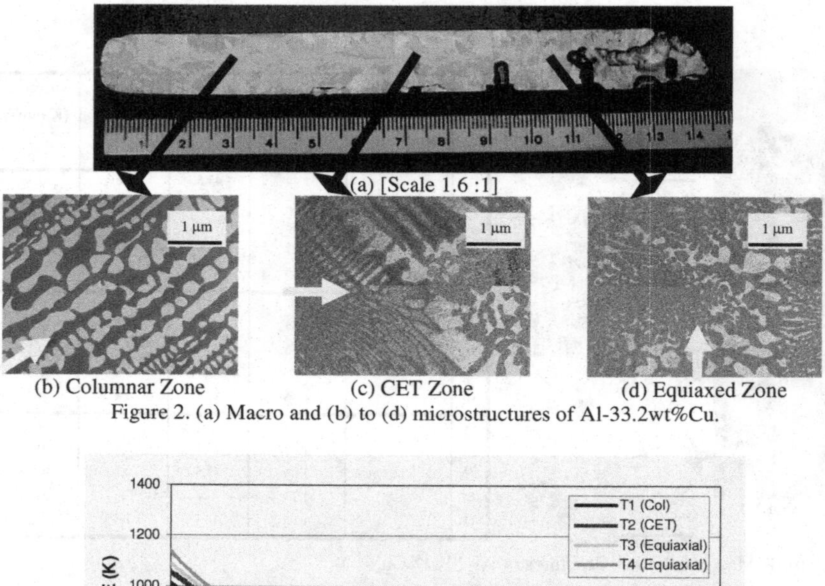

(b) Columnar Zone (c) CET Zone (d) Equiaxed Zone

Figure 2. (a) Macro and (b) to (d) microstructures of Al-33.2wt%Cu.

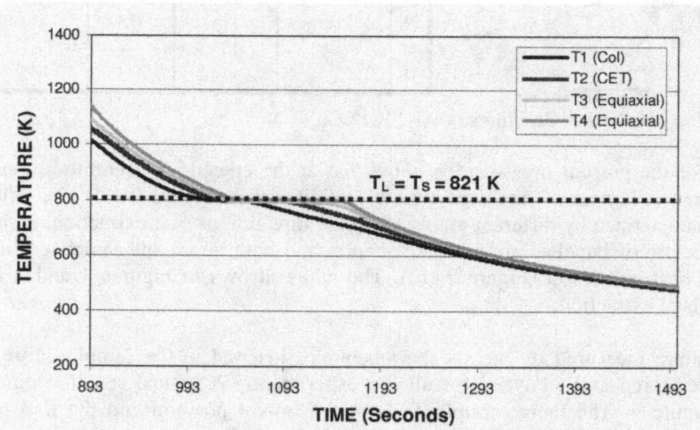

Figure 3.Temperature versus time dependence for one experiment with Al-33.2wt%Cu.

With respect to melt superheat the results show no correlation with columnar length which is in agreement with the results reported by Ziv et al [1] and by Ares et al [6,7].

It was previously demonstrated by the authors [6,7] that the temperature gradient and the liquidus interface velocity are critical parameters related to the CET. The values of temperature gradient and interface velocity at the moment of the CET are listed in Table I and they are called critical values, that is $V_{L(Critical)}$ and $G_{L(Critical)}$. These critical values are between 0.4 to 3.8 $mm.s^{-1}$ for the velocity and -0.44 to 0.10 $K.mm^{-1}$ for the temperature gradient. Other results in the literature are 0.28 to 0.88 $mm.s^{-1}$ and 0.28 to 0.75 $K.mm^{-1}$ which are in the order of values reported here [1,4]; however they did not include negative values for the temperature gradient.

<u>Characterization of the Structural Parameters</u>

The results of the measurement of the grain size show that in the transition region the equiaxed grains are small and that after the transition in the equiaxed zone, the grain size may increase, decrease or remain constant. For example, in Al-2wt%Cu the average grain size in the CET zone is 1.55 mm and in the equiaxed zone is 2.74 mm, for Al-4%Cu is about 1.32 mm in the CET zone and 1.80 mm in the equiaxed zone, for Al-20wt%Cu the average grain size in the CET zone is 1.29 mm and is 1.63 mm in the equiaxed zone and for Al-33.2wt%Cu the average grain size in the CET zone is 1.42 mm and is 3.24 mm in the equiaxed zone, see Table II.

Comparing these results with published results for similar experiments it is observed that for the same alloys Al-2%Cu in experiments evolving the transition similar equiaxed grain sizes have been obtained [26], however the columnar width reported is about 50 % wider. This could be attributed to the larger cooling velocity employed in the present investigation which affect columnar growth, and to a larger extent to the smaller size which reduce convection and as a result produce larger temperature gradients.

The average primary spacing for each alloy and composition is listed in Table II in the three regions; columnar, transition and equiaxed regions. With respect to the primary spacing, λ_1, it is observed that they are larger in the columnar and transition zone where columnar grains co-exist with equiaxed grains, than in the fully equiaxed zone. For instance, for Al-2%Cu alloy, λ_1 drops from 430 μm and 429 μm in the columnar and transition zones, respectively to 299 μm in the equiaxed zone. Similarly, for the Al-20%Cu alloy the spacing drops from 334 μm and 335 μm for the first two zones to 301 μm in the equiaxed zone. In general, the primary spacing reaches a peak value for the columnar grains in the transition zone and then decreases around 30% in the equiaxed region. In addition the primary spacing decreases as the alloy composition increases in all three regions.

With respect to the secondary spacing, λ_2, is larger for the transition and equiaxed zone than for the columnar zone. For a Al-2%Cu alloy, for instance the spacing is 49 μm and 52 μm for the transition ad equiaxed zones respectively, and 35 μm for the columnar zone. Similarly, for the Al-20wt%Cu alloys the respective values are 93 μm and 92 μm for the transition and equiaxed zones and 70 μm for the columnar zone. This pattern is repeated for all concentrations below the eutectic composition.

Table II. Alloy compositions, local solidification time, t_{SL}, average grain sizes and average λ_1 and λ_2 spacings in each zone of the samples.

#	Alloy wt%Cu	$GS_{(Average)}$ (mm) Columnar	$GS_{(Average)}$ (mm) CET	$GS_{(Average)}$ (mm) Equiaxed	λ_1 (μm) Columnar	λ_1 (μm) CET	λ_1 (μm) Equiaxed	λ_2 (μm) Columnar	λ_2 (μm) CET	λ_2 (μm) Equiaxed	$\lambda_{2\,[Equiaxial]}$ (μm)
1		4.05 ± 0.5	1.48 ± 0.3	3.21 ± 0.5	-------	-------	-------	43.76 ± 4.5	58.49 ± 5.0	57.26 ± 5.0	
2		3.14 ± 0.5	1.72 ± 0.3	2.46 ± 0.3	425.23 ± 42	443.2 ± 42	226 ± 20	38.03 ± 4.0	53.32 ± 5.0	55.11 ± 5.0	
3	Al-2wt%Cu	3.04 ± 0.5	1.23 ± 0.3	2.74 ± 0.4	-------	-------	-------	36.88 ± 4.0	47.44 ± 4.0	52.5 ± 5.0	$\lambda_2 = 11.9 * t_{SL}^{1/3}$ [11]
4	($t_{SL} \cong 34.4$ s)	2.55 ± 0.4	1.98 ± 0.4	2.14 ± 0.4	397.46 ± 42	398.9 ± 36	191.7 ± 20	36.32 ± 4.0	46.86 ± 4.5	50.26 ± 4.5	$\lambda_2 = 10 * t_{SL}^{1/2}$ [12]
5		2.25 ± 0.4	1.35 ± 0.3	3.19 ± 0.5	469.23 ± 50	446.78 ± 42	252.82 ± 20	20.72 ± 2.5	41.02 ± 4.0	45.94 ± 5.0	
6		1.50 ± 0.4	1.5 ± 0.3	1.55 ± 0. 3	270.22 ± 25	245.21 ± 25	251.23 ± 25	63.82 ± 5.0	84.12 ± 5.0	82.10 ± 5.0	
7		1.62 ± 0.4	1.35 ± 0.3	1.79 ± 0. 3	389.46 ± 36	387.53 ± 36	356.53 ± 36	65.96 ± 5.0	89.46 ± 5.0	86.11 ± 5.0	
8	Al-20wt%Cu	1.75 ± 0.4	1.21 ± 0.3	1.25 ± 0. 3	360.22 ± 36	375.69 ± 36	282.8 ± 35	73.68 ± 5.0	97.08 ± 5.0	98.87 ± 5.0	$\lambda_2 = 9.8 * t_{SL}^{1/3}$ [11]
9	($t_{SL} \cong 126.1$.s)	1.96 ± 0.4	1.13 ± 0.3	1.96 ± 0. 3	392.53 ± 42	405.13 ± 42	370.11 ± 36	70.47 ± 5.0	96.23 ± 5.0	91.07 ± 5.0	$\lambda_2 = 10 * t_{SL}^{1/2}$ [12]
10		2.05 ± 0.4	1.29 ± 0.3	1.64 ± 0. 3	260.61 ± 25	265.28 ± 25	245.68 ± 30	77.58 ± 5.0	100.14 ± 6.0	104.2 ± 6.0	

#	Alloy wt%Cu	$GS_{(Average)}$ (mm) Columnar	$GS_{(Average)}$ (mm) CET	$GS_{(Average)}$ (mm) Equiaxed	$\lambda_{(Average)}$ (μm) Columnar	$\lambda_{(Average)}$ (μm) CET	$\lambda_{(Average)}$ (μm) Equiaxed
11		1.25 ± 0.2	1.23 ± 0.3	3.92 ± 0.5	0.05 ± 0.005	0.06 ± 0.005	0.06 ± 0.005
12		1.12 ± 0.2	1.24 ± 0.3	3.26 ± 0.3	0.04 ± 0.005	0.03 ± 0.005	0.04 ± 0.005
13	Al-33.2wt%Cu	1.52 ± 0.2	1.53 ± 0.3	3.03 ± 0.4	0.08 ± 0.005	0.12 ± 0.005	0.17 ± 0.005
14		1.96 ± 0.3	1.38 ± 0.4	2.74 ± 0.4	0.1 ± 0.05	0.15 ± 0.05	0.25 ± 0.05
15		2.13 ± 0.3	1.76 ± 0.3	3.27 ± 0.5	0.09 ± 0.05	0.17 ± 0.05	0.19 ± 0.05

As it is well established, the secondary spacing is associated to the local solidification time, t_{SL}, quantitatively related to $\lambda_2 \propto t_{LS}^{1/3}$ [11], as a good approximation; or $t_{LS} \propto \lambda_2^3$. In Table II the secondary spacing is a function of $t_{LS}^{1/3}$, which is presented with a coefficient adjusted for each alloy within 5 % of scatter. It is observed that the coefficient decreases monotonically as the alloy concentration increases from 11.9 for Al-2%Cu to 8.8 for Al-20%Cu indicating a slightly lower effect of local solidification time with spacing.

The average solidification time increases with alloy concentration from 34.4 s for Al-2%Cu to 126 s for Al-20%Cu. It is therefore important to point out that there is a very good correlation between λ_2 and $t_{LS}^{1/3}$. However, the correlation proposed by Grugel [12] as $\lambda_2 \propto 10.t_{LS}^{1/2}$ consequently is not as good and the error is around 21 %.

As a conclusion, the correlation that best fit the value of λ_2 and t_{LS} is $\lambda_2 \propto t_{LS}^{1/3}$ or $t_{LS} \propto \lambda_2^3$ which can be used to establish the solidification conditions in each region using this relation.

The effect of increasing the concentration is to increase the secondary spacing in the three regions, for a Al-20%Cu alloy for the same region is 35 to 70 (200%) for the columnar zone, 40 to 93 (190%) for the transition zone and 52 to 93 (180%) for the equiaxed zone. The increase in spacing and solidification time with Cu concentration is associated with the phase diagram where it is shown an increase in the time gap between the liquidus and solidus lines which produce a much longer mushy zone at higher concentrations with similar heat extraction and cooling rates.

In the case of the eutectic alloys it has been shown that the transition from columnar to equiaxed occurs with similar characteristics as for the other concentrations, the resultant lamellar, rod-like or mixed solidification microstructure have parameters which also depend on solidification conditions, mainly velocity. The measurements show that the average spacings increases from 0.072 μm in the columnar zone to 0.106 μm for the transition region and to 0.142 μm for the equiaxed region indicating an important reduction in solidification velocity (Table II]).

Conclusions

The main conclusions of this investigation on the columnar to equiaxed transition in aluminum-copper alloys are:

1) The transition occurs in a zone rather than in a sharp plane, where both columnar and equiaxed grains in the melt co-exist, as it was previously reported for Pb-Sn and Zn-Al alloys.

2) The transition at the eutectic concentration present similar characteristic as for the lower concentrated alloys.

3) The length of the columnar zone decreases with cooling rate and there is a minimum cooling rate for columnar growth which increases with Cu composition.

4) The temperature gradient and the velocity of the liquidus front reach low critical values before the transition. These critical values are between0.4 to 3.8 mm/s for the velocity and -0.44 to 0.10 K/mm for the temperature gradient.

5) The measurements of the grain size show that in the transition region the equiaxed grains are small and that after the transition in the equiaxed zone ahead the grain size may increase, decrease or remain constant.

6) The primary spacing for all Cu concentrations is larger in the columnar and transition zone where columnar grains co-exist with equiaxed grains, than in the fully equiaxed zone. The

primary spacing reaches a peak value for the columnar grains in the transition zone and then decreases around 30% in the equiaxed region. In addition, the primary spacing decreases as the alloy composition increases in all three regions.

7) The secondary spacing is larger for the transition and equiaxed zone than for the columnar zone. This pattern is repeated for all concentrations below the eutectic composition.

8) The correlation that best fit the value of of λ_2 and t_{LS} is $\lambda_2 \propto t_{LS}^{1/3}$ or $t_{LS} \propto \lambda_2^3$ which can be use to establish the solidification conditions in each region.

9) In the case of the eutectic alloys the transition from columnar to equiaxed occurs with similar characteristics as for the other concentrations, the resultant lamellar, rod-like or mixed solidification microstructure have parameters which also depend on solidification conditions, mainly velocity.

Acknowledgments

This work was partially supported by CONICET (Consejo Nacional de Investigaciones Científicas y Técnicas, Argentina).

References

1. I. Ziv, and F. Weinberg, "The columnar-to-equiaxed transition in Al 3 Pct Cu," *Metallurgical and Materials Transactions A,* 20 (1989), 731-734.

2. J.D. Hunt, *Solidification and Casting of Metals*, (London: Metals Society, 1979), 3.

3. V.K. Suri, N. El-Kaddah and J.T. Berry, "Control of Macrostructure in Aluminum Casting, Part I: Determination of Columnar/Equiaxed Transition for Al-4.5%Cu Alloy," *AFS Transactions*, 99 (1991), 187-191.

4. C.A. Siqueira, N. Cheung and A. Garcia, "Solidification Thermal Parameters Affecting the Columnar-to-Equiaxed Transition," *Metallurgical and Materials Transactions A*, 33 (2002), 2107-2118.

5. J.A Spittle, "Columnar to Equiaxed Grain Transition in as Solidified Alloys," *International Materials Reviews,* 51 (2006), 247-269.

6. A.E. Ares and C.E. Schvezov, "Solidification Parameters during the Columnar-to-Equiaxed Transition in Lead-Tin Alloys," *Metallurgical and Materials Transactions* A, 31 (2000), 1611-1625.

7. A.E. Ares and C.E. Schvezov, "Influence of Solidification Thermal Parameters on the Columnar to Equiaxed Transition of aluminum-Zinc and Zinc-Aluminum Alloys," *Metallurgical and Materials Transactions A,* 38 (2007), 1485-1499.

8. R. F. Speyer, *Thermal Analysis of Materials*, (New York, NY: Marcel Dekker Editor, 1994) 30-109.

9. G. F. Vander Voort, *Metallographic Principles and Practice*, (Metals Park, Ohio: ASM International, 2001), 525-661.

10. H. E. Boyer and T.L. Gall, *Metals Handbook*, (Metals Park, Ohio: American Society for Metals, 1984), 35-18 -35-19.

11. A.E. Ares, C.T. Rios, R. Caram and C.E. Schvezov, "Dendrite Spacing in Al-Cu and Al-Si-Cu Alloys as Function of the Growth Parameters," (Paper presented at the 131st TMS Annual Meeting & Exhibition, Seattle, Washington, USA, 17-21 February 2002), 785-792.

12. R.N. Grugel, "Secondary and Tertiary Dendrite Arm Spacing Relationships in Directionally Solidified Al-Si Alloys," *J. Materials Science,* 28 (1993), 677-689.

Shape Casting: The 3rd International Symposium
Edited by: John Campbell, Paul N. Crepeau, and Murat Tiryakioğlu
TMS (The Minerals, Metals & Materials Society), 2009

THE MODIFICATION OF CAST AL-Mg_2Si *IN SITU* MMC BY LITHIUM

R. Hadian[1], M. Emamy[2] and J. Campbell[3]

[1]Department of Materials Science, Sharif University of Technology, Tehran, Iran
[2] School of Metallurgy and Materials, University of Tehran, Tehran, Iran
[3]Department of Metallurgy and Materials, University of Birmingham, UK

Key words: Al/Mg_2Si composite; MMC; Modification; Casting; Oxides; Bifilms

Abstract

The effects of both Lithium modification and cooling rate on the microstructure and tensile properties of an in-situ prepared Al-15% Mg_2Si composite were investigated. Adding 0.3%Li reduced the average size of Mg_2Si primary particles from ~30 μm to ~ 6 μm. The effect of cooling rate was investigated by the use of a mold with different section thickness from 3 to 9 mm. The results show a refinement of primary particle size as a result of both Li additions and increased cooling rate, and their effects were additive. Similarly, both effects increased UTS and elongation values. The refinement by Li and enhanced cooling rate is discussed in terms of an analogy with the effect of Sr and cooling rate in Al-Si alloys, and is similarly attributed to the effect of the alkali and alkaline earth metals deactivating oxide double films (bifilms) suspended in Al melts as favoured substrates for intermetallics.

Introduction

Al and Mg based composites, reinforced with Mg_2Si particles have been lately introduced as a new group of particulate metal matrix composites (PMMCs) that offer attractive advantages such as low density, good wear resistance and good castability. Al-Mg_2Si hypereutectic alloys seem to be potential candidates to replace Al-Si alloys used in aerospace and engine applications. In-situ preparation of Al-Mg_2Si composites seems to be the best way of providing such PMMCs since there are benefits of an even distribution of the reinforcing phase, good particle wetting and low costs of production. However, their coarse morphology has been thought to lead to the low ductility observed in these materials which has inhibited applications.

Thus investigations have been carried out to modify the primary and eutectic Mg_2Si structure in an effort to improve the properties of these composites. Zhang et al. [1] and Ourfali et al. [2] have both reported success in refining the coarse Mg_2Si structure by increasing cooling rate. Other efforts have been focused on the modification of the structure with addition of different alloying elements such as Sr, Ce, sodium salt (assumed to introduce Na), P, K_2TiF_6, rare earth elements and extra silicon content [3].

Assuming an analogy with the modification of Al-Si alloys by such elements as Na and Sr, a refinement of the microstructure is expected to result in an improvement of the mechanical properties. The influence of particle size on the mechanical properties of MMCs has been investigated [4] and the results confirm the expectation that the finer the particle size, the higher the strength. So far the finest size of primary Mg_2Si particle reported in the literature is about 8 μm [5].

The objective of the present study is to evaluate the tensile properties of Al-15%Mg_2Si composites and fracture characteristics of the material with the addition of up to 0.3 % Li.

Experimental Procedure

Alloys with different compositions of 0, 0.1 and 0.3 % wt Li were prepared as reported previously [3]. They were poured into cylindrical ingot molds (30 mm radius and 60 mm height) that were used for microstructural studies. Specimens were cut from a standard location 25 mm from the base of the ingots. To evaluate the effects of both cooling rate and Li modification on tensile properties, test bar castings varying in thickness from 3 to 9 mm were poured in a sand mold.

Tensile test bars were machined, according to ASTM-B577 [6] flat sub-size specimens, to 25 mm gauge length, 6 mm width and 3 mm thickness for each condition. Specimens for microstructural characterization were sectioned from the gauge length portion of the test bars. Metallographic specimens were polished and etched by HF (1%). Quantitative data on microstructure were determined using an optical microscope equipped with an image analysis system (Clemex Vision Pro. Ver. 3.5.025). The microstructure was examined by scanning electron microscopy (SEM) performed in a Vega©Tescan SEM operated at 15kV with energy dispersive x-ray analysis (EDAX).

Results and Discussion

Figure 1 shows a typical microstructure of Al-15% Mg_2Si as-cast alloy (a) before and (b) after refinement with Li. Addition of 0.3% Li has been proved to modify the microstructure more effectively. Since the alloy has a hypereutectic composition, the microstructure should contain dark faceted primary particles of Mg_2Si in a matrix of Al-Mg_2Si eutectic. The fact that also clearly seen are bright primary α-Al grains indicates that undercooling has occurred, allowing both primary phases to form together with the eutectic in the form of a matrix of well developed Al-Mg_2Si eutectic cells.

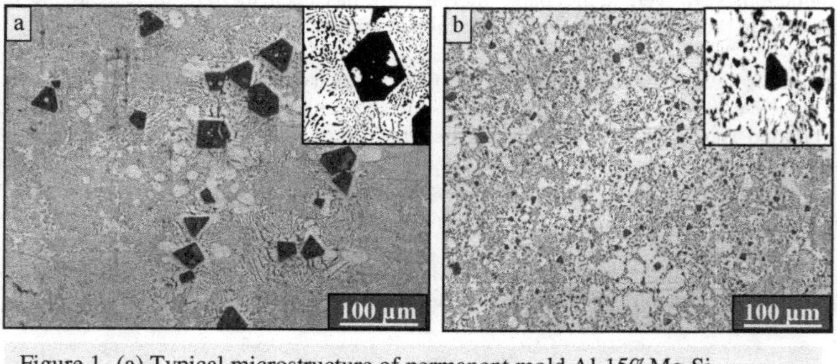

Figure 1- (a) Typical microstructure of permanent mold Al-15%Mg_2Si
(b) the microstructure after modification with 0.3%Li

Our main interest in the MMC structure centres on the mean size of Mg_2Si particles. Here in the unmodified 9 mm section it is about 30μm. The coarse Mg_2Si primary crystals in an unmodified alloy often grow as hollow shapes (Figure 1) similar to coarse primary silicon in the Al-Si system [7]. These hopper crystals are not seen in the finer Mg_2Si particles in Li-modified alloys. Addition of Li up to 0.3% reduces their size to ~6 μm and forms an even distribution of particles in the matrix. In addition to refinement of size, the morphology of the particles changes from faceted to a smoother and more rounded form (Figures 1(a) and (b)).

The refinement in size of the Mg_2Si intermetallics could be the result of two major possibilities:

(i) Conventional physical metallurgical thinking might attribute the microstructural refinement to a change in phase diagram due to an increased constitutional undercooling. As clearly seen in the magnified sections of Figures 1(a) and (b), in the unmodified structure, the pseudo-eutectic structure is mainly fibrous inside the cells, corresponding to a high rate of freezing, but a lamellar structure is observed at the cell boundaries. With the addition of Li the eutectic at the cell boundaries changes from a flake morphology to a structure in which some of the Mg_2Si eutectic particles are approximately of the same size as the refined primary Mg_2Si particles. This coarsening of the eutectic structure in the inter-cellular regions is likely to be the result of some Li segregation, and the accompanying reduction in interfacial energy of the intermetallic [8].

(ii) Turning now to the possibility of the influence of process metallurgy and extrinsic factors, in common with many other alloy systems, a large population of oxide bifilms is expected to be present in suspension in the melt as a result of turbulent melt handling. Initially these are relatively harmless because their unbonded center interface, acting as a crack in the liquid, is raveled into a compact, convoluted form by the internal turbulence in the flowing liquid. Furthermore, there is increasing evidence that the intermetallics nucleate and grow on oxides, and that during growth some intermetallics that grow in a plate-like form force the straightening of the bifilms into serious planar cracks [9].

The analogy with the precipitation of primary Si in Al-Si alloys appears almost exact: the primary Si normally precipitates on oxides in the melt, straightening the oxides from their normally convoluted and raveled forms as the crystal grows, thus straightening the crack in the centre of the oxide. The straightened and thus more effective crack thereby reduces properties [9]. When Sr is added, the Sr appears to deactivate the favoured status of the oxide, possibly by coating and blocking growth sites on the oxide. The Si is then prevented from precipitating as a primary phase, and is forced to form as a eutectic. We call this structure a 'modified' structure. The action of Li in the present MMC appears to be similar. In the absence of Li the intermetallics precipitate on and straighten bifilm cracks and so degrade properties, but in the presence of Li the oxide is prevented from acting as a substrate, forcing the growth of the intermetallic as a eutectic phase. The bifilm defects then remain in their convoluted, compact morphology in which they are relatively harmless as cracks, so that the properties of the casting are improved. Interestingly, however, in contrast with the action of Sr in Al-Si alloys, the transformation of the intermetallic is not complete in this alloy, so that some small primary phases remain after the addition of Li. It raises the question whether sufficient Li was added to complete the transformation, or whether the Li is simply somewhat less effective in this alloy. Future work will be required to clarify this.

The similar refinement of the intermetallics by the increase of freezing rate appears explicable because of the general rule that faster cooling results in higher undercoolings which subsequently leads to more prolific nucleation and faster growth rates, and of course a more limited time for growth; finally yielding a finer microstructure (Figures 2 and 3).

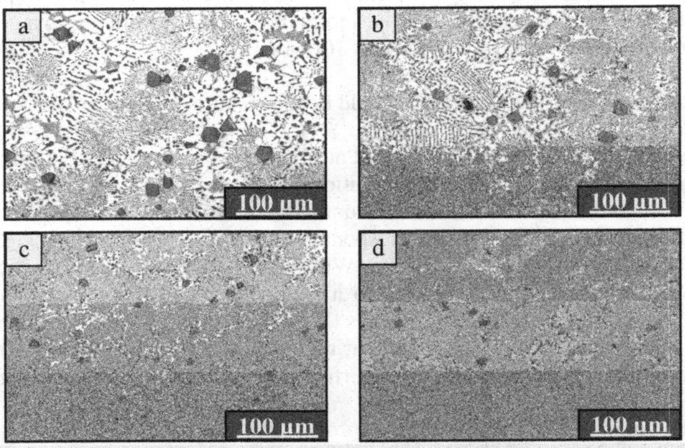

Figure 2- Effect of cooling rate on microstructure of Al-15%Mg_2Si without Li modification. (a) 9mm, (b) 7mm, (c) 5mm and (d) 3mm sections

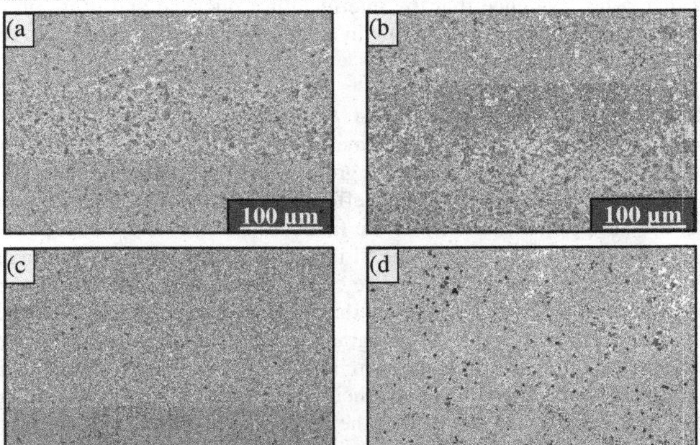

Figure 3- Effect of cooling rate on microstructure of Al-15%Mg_2Si with 0.3% Li modification. (a) 9mm, (b) 7mm, (c) 5mm and (d) 3mm sections

The ultimate tensile strength (UTS) and elongation values of the specimens are plotted in Figures 4 (a) and (b) as a function of test bar thickness. It is evident that both Li modification and faster freezing raise the tensile strength of the unmodified alloy. Over the ranges of Li and freezing rate used in this work, Li is somewhat more effective, although not so much that could not be envisaged to be reversed in more favourable freezing conditions, such as might occur in metal moulds for instance. Thus the chemical and freezing rate modifications are of a similar magnitude, and are seen to be additive.

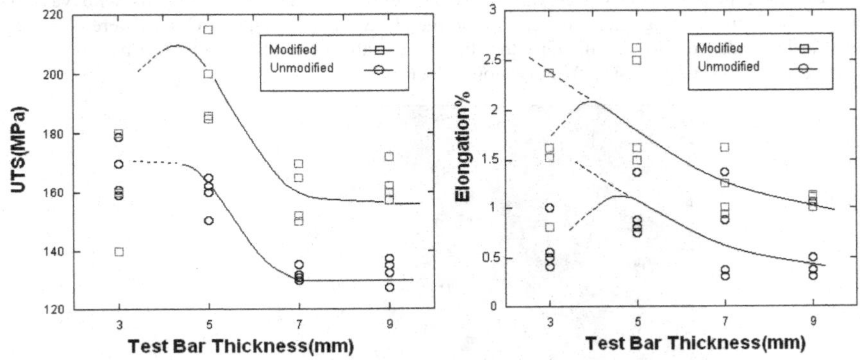

Figure 4-Tensile properties as a function of test bar thickness, (a) Ultimate tensile strength (UTS) (b) Elongation%.

The results generally confirm expectations based on the metallographic results. Some scatter is observed in the results within each section, especially in elongation values. As is well understood, ductility is highly sensitive to the presence of defects in the alloy, particularly in this case, the presence of cracks opened to varying degrees by the precipitation and growth of intermetallic crystals on bifilms. The scatter is most notably seen in 3 mm sections in both modified and unmodified samples. In addition the 3 mm results are an exception to the observed trend of improved properties with reducing casting section thicknesses. It seems reasonable to attribute the variability in thinner sections to additional casting defects which are exaggerated in thin wall castings, since any given defect will occupy a greater proportion of the thickness of the sample. The defects are almost certainly present as a result of the relatively poor casting conditions [3]. For the future, improved casting techniques will refine the preliminary results presented here. In the meantime, the tentative extrapolations noted in Figure 4 indicate truly impressive potential; the possibility to achieve elongations well above 10 per cent, which remains an incentive for future effort.

The appearance of the intermetallics on the fracture surfaces, most of which are seen to contain a single central crack (Figure 5) is directly explained by the bifilm concept. In Figure 6 an Mg_2Si particle is seen to be peeled open from what appears to be a central plane of weakness. It has not fractured into pieces, but has been deformed into a highly curved shape, clearly revealing that the Mg_2Si phase is not brittle. Brittleness has often been assumed from the presence of cracks, but the cracks are an 'extrinsic import' into the particle, becoming an integral feature unfortunately introduced as a direct consequence of the presence of a crack in their favoured substrates. We can conclude that the Mg_2Si particles are not only intrinsically hard, but also ductile if not weakened by the presence of their pre-cracked substrates.

The path of propagation of the crack was observed to generally follow the eutectic cell boundaries. This observation is consistent with the expected presence of additional oxide bifilms at these locations. The oxide films are pushed by the growing solid phase, thus finally concentrating in grain boundaries and cell boundaries, providing these regions with varying quantities of pre-existing cracks depending on the cleanness of the alloy. Cracks were observed even in Li-modified alloy, confirming that the bifilms are not totally deactivated by Li. (In this respect Sr in Al-Si alloy is significantly more effective.)

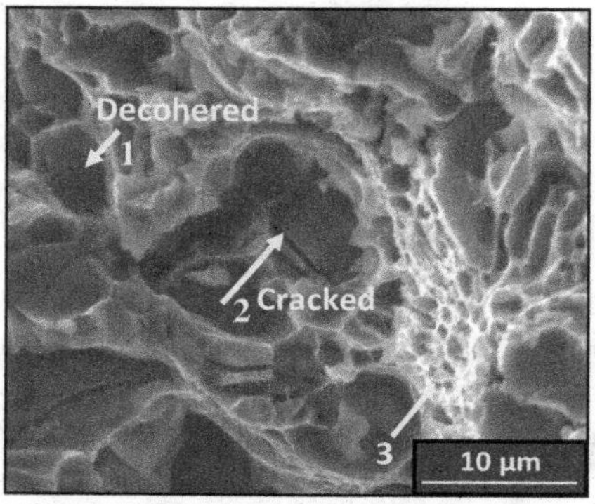

Figure 5-Cracked and decohered particles in an unmodified 3mm section

On fracture surfaces of typical MMCs two major phenomena are well known in the literature: particle decohesion and particle fracture [10]. Both phenomena are observed in this study (Figure 5). In terms of the bifilm concept, a particle will exhibit a central crack if the intermetallic forms on both surfaces of the bifilm, and so is likely to crack through its centre. If, however, the particle has formed on only one surface of the bifilm, it is likely to appear to decohere from the matrix when subjected to a tensile stress (since in a sense it was never attached to the matrix across this interface, being always separated by the residual air layer in the doubled-over oxide film). The study showed that Li-modified specimens had fewer decohered particles, in agreement with the reduction in potency of bifilms by Li, and explains the higher ductility of the MMC. In Figure 5, fine dimples which are rarely observed in these MMC fracture surfaces are shown with an arrow '3'. These features confirm the intrinsic ductility of the matrix.

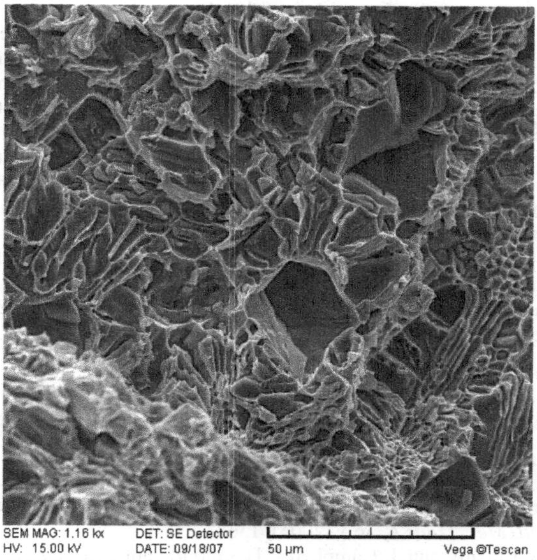

Figure 6- Typical SEM image of a fracture surface, the large Al-Mg_2Si particle at the upper right of the image appears to have peeled open from its central bifilm.

Some researchers have emphasized the importance of homogeneous particle distributions to improve ductility [11]. Adding Li has clearly homogenized the structure and changed the clustering behavior of polyhedral primary particles (Figure 1b). If precipitated early from the melt as primary phases they can be pushed into inter-cellular regions by the advance of the freezing front. Although such pushing is not expected from interfacial energy considerations, it is seen to be easily feasible if oxide bifilms are also present to separate the advancing front from phases that would otherwise be automatically incorporated. From the microstructures, it is clear that some intermetallics have been incorporated into the eutectic matrix, but others appear to have been pushed. In all cases, any bifilms remain invisible, as is usual and to be expected. After the addition of Li, the suppression of the formation of primary intermetallics (or possibly their much later arrival) reduces this unfavourable pushing action because the bifilms retain their original compact form and so take a reduced role in the creation of the microstructure.

Conclusions

1. Al-15%Mg_2Si composite was prepared by casting direct from the melt. The observed microstructure consists mainly of coarse primary Mg_2Si particles in a matrix of equiaxed eutectic cells.
2. Adding 0.3% Li refined the primary and eutectic cell structures, reducing the average size of primary Mg_2Si from 30 µm to ~ 6 µm.
3. Increasing the freezing rate refined the structure to nearly a similar extent.
4. Li and freezing rate refinement effects were additive.

5. The refined alloy was stronger and more ductile.
6. The benefits of refinement are suggested to occur principally by a suppression of the unfurling and gradual extension of pre-existing cracks in the liquid (bifilms).
7. Bifilms appear to explain (i) the apparent brittle fracture of Mg_2Si particles that display apparent cleavage facets and (ii) apparent decohesion of the particles from the matrix.
8. Large primary Mg_2Si particles and their clusters appeared to be the favored path for crack propagation but eutectic cell boundaries were a second common path. Both these paths appeared consistent with the expected locations of bifilms; thus the crack merely follows pre-existing cracks resulting not from intrinsic solidification phenomena but from extrinsic processing phenomena mainly associated with pouring.
9. Evidence was found to confirm the intrinsic strength and ductility of both the matrix and, perhaps surprisingly, the ductility of the Mg_2Si particles.
10. Significant additional potential appears to exist for increased properties by improved casting techniques to reduce macroscopic casting defects.

References

[1] J. Zhang, et al., "Effect of cooling rate on the microstructure of hypereutectic Al-Mg_2Si alloys," *J. Mat. Sci. Lett.* , 19 (2000), 1825-1828.

[2] M.F. Ourfali, I. Todd, H. Jones, "Effect of Solidification Cooling Rate on the Morphology and Number perUnit Volume of Primary Mg_2Si Particles in a Hypereutectic Al-Mg-Si Alloy,"*Metall. Mat. Trans.A,* 36A (2005), 1368.

[3] R. Hadian, M. Emamy and J. Campbell, "The modification of cast Al-Mg_2Si metal matrix composite by Li," *International Journal of Cast Metals Research*, submitted January 2008.

[4] Y. Flom, R. J. Arsenault,"Effect of particle size on fracture toughness of SiC/Al composite material,"Acta Mater., 37 (1989), 2413.

[5] J. Zhang, et al., "Microstructural development of Al–15wt. %Mg_2Si in situ composite with mischmetal addition," *Mat. Sci. Eng. A*, 281 (2000), 104–112.

[6] Annual Book of ASTM Standards, Vol.02.2, 506, B557-84, (American Society for Testing and Materials, 1986)

[7] Ru-yao Wang, Wei-hua Lu, L.M. Hogan," Growth morphology of primary silicon in cast Al–Si alloys and the mechanism of concentric growth," *J. Crystal Growth*, 207 (1999), 43-54.

[8] R. Hadian, M. Emamy," Microstructure modification of Al-15%Mg_2Si insitu composites with Ti/B and Li addition," (Proceeding of 6th conference on Materials Processing, Properties and Performance, Beijing China, Sep. 2007).

[9] J. Campbell, "Castings" Elsevier 2003 and "Entrainment Defects" *Mat. Sci. Tech*, 22 (2) (2006) 127-145 plus (8) 999-1008.

[10] L. Babout," On the competition between particle fracture and particle decohesion in metal matrix composites" *Acta Mater.* , 52 (2004), 4517–4525.

[11] J. Segurado, J. LLorca, "A computational micromechanics study of the effect of interface decohesion on the mechanical behavior of composites," *Acta Mater.* , 53 (2005), 4931–4942.

Shape Casting: The 3rd International Symposium
Edited by: John Campbell, Paul N. Crepeau, and Murat Tiryakioğlu
TMS (The Minerals, Metals & Materials Society), 2009

EFFECT OF STRONTIUM ON VISCOSITY AND LIQUID STRUCTURE OF Al-Si EUTECTIC ALLOYS

Sumanth Shankar[1], Srirangam VS Prakash[1], Minhajuddin Malik[1], Manickaraj Jeyakumar[1], Michael J Walker[2] and Mohamed Hamed[1]

[1]Light Metal Casting Research Centre (LMCRC), Department of Mechanical Engineering, McMaster University, Hamilton, ON, Canada L8S 4L7
[2]General Motors, Research and Development Center, Michigan, USA

Keywords: Al-Si alloys, Eutectic alloy, Modification of Al-Si, viscosity of metals, structure of liquid alloys

Abstract

This study aims to present conclusive evidence that trace level addition of Sr in Al-Si hypoeutectic alloys change the liquid melt characteristics and alter the nucleation environment of the eutectic phases. High temperature rheological experiments measuring viscosity of Al-Si eutectic melt with and without Sr addition show that Sr significantly alters the melt viscosities at various shear rate regimes. Further, liquid diffraction experiments have been carried out on Al-12.5wt%Si (eutectic) alloy using high-energy synchrotron X-ray beam source to determine the effect of Sr on various liquid structure parameters such as structure factor, pair distribution function and coordination numbers at various melt superheat temperatures. The analysis of the data suggests that Sr changes the nucleation environment of the eutectic Si phase. Further, the effect of Sr on the atomic arrangement of the Si atom with respect of Si and Al atoms in the liquid will be quantified and presented.

Introduction

The effect of trace levels of Sr added to the Al-Si hypoeutectic alloys have been an important and popular topic of study since the middle of last century [1-8]. Trace levels of Sr between 50 to 250 ppm is known to alter the morphology of the eutectic Si phase from a coarse plate like to a fine fibrous structure. Recently [6], it has been observed that the Sr addition not only changes the Si morphology but significantly refines the grain size of the eutectic Al phase as well. It has been hypothesized that Sr additions at trace levels alters rheological properties of the liquid Al-Si eutectic melt, which is the final composition to solidify in the inter-dendritic regions of the solidified microstructure [6]. The change in rheological properties of the liquid eutectic melt can be due to a change in the atomistic arrangement of the liquid phase. The eutectic liquid with Sr addition will not nucleate any phase (Al or Si) because of the lack of favorable interface energies. This will lead to further growth of the primary Al dendrite by the epitaxial nucleation and growth of eutectic Al on the primary phase resulting in a super-saturation of Si atoms in the inter-dendritic liquid and a suppression of the eutectic temperature. The Si phase will eventually crystallize as blocky precipitates on the primary Al dendrites and create a disconnect between the dendrite and the liquid resulting in a copious nucleation events of the eutectic Al grains between which the eutectic Si is forced to nucleate and grow as a fibrous structure. This fibrous structure in Al-Si alloys is commonly referred to as a modified structure where the plate structure found without the Sr addition is called the unmodified eutectic structure.

This proposed mechanism of modification of the eutectic phase by Sr addition need to be validated with sufficient experimental evidence to show that trace levels of Sr added to hypoeutectic and eutectic Al-Si alloys will result in a change in the atomistic arrangement in the liquid and a change in the rheological properties of the liquid. In this publication we propose to present such experimental evidence.

Rheological Properties

In this section the rheological properties, specifically the shear viscosity of the liquid Al-Si eutectic alloy with and without the addition of 0.023 wt% Sr are studied. The shear viscosity of liquid metals was measured with a high temperature rotational rheometer with a Double Concentric Cylinder measurement geometry to contain the sheared liquid [9]. Viscosities of liquid metals have been traditionally measured using the Oscillation Vessel Viscometer (OVV) and it has been convincingly shown that the OVV technique is not appropriate for the task because recently, increasing evidence has come to show most liquid metals and alloys exhibit a non-Newtonian and shear thinning flow behavior [9-11].

Materials and Procedure

An Al-12.5 wt% Si alloy was prepared from 99.99 % pure Al ingots and 99.9999% pure Si metal and the composition was verified by ICP* spectrometry. Thermal analysis during solidification (0.5 °C/sec) was obtained with a K-type thermocouple to verify the eutectic composition of the alloy. To this alloy, 0.023 wt% Sr was added using 99.9 % pure Sr metal. The alloys were cast into hollow cylindrical rods and sectioned appropriately to obtain samples for rotational rheometer.

Figure 1(a) show the picture of the rotational rheometer. The liquid sample was contained and sheared inside the measurement geometry between the rotor and the stator. Figure 1(b) shows the schematic of the measurement geometry with the rotor and stator. The measurement geometry is contained inside an environmental furnace capable of reaching temperatures in excess of 800 °C. The furnace was flushed with ultra high purity Ar gas (<2 ppm oxygen) to maintain a minimum possible oxidation environment. The maximum temperature difference between any two points in the liquid at the experiment temperatures was measured to be less than 1 °C. The liquid sample was rotated with a constant specific angular velocity and the torque required to maintain this angular velocity was measured by the torque head. The torque measured and the angular velocity was mapped to shear stress and shear rate experienced by the sheared liquid, respectively.

Angular velocity was gradually increased from 0 and 60 rad.s^{-1} and the equivalent shear rates experienced by the liquid were between 0.16 s^{-1} to 467 s^{-1}, respectively. Each angular velocity was applied for a period of five seconds and the torque was recorded at the rate of about 6 readings per second. Experiments were carried out at four melt superheat temperatures of 585 °C, 605 °C, 675 °C and 700 °C. The eutectic temperature is 578 °C. Time-dependency curves wherein the variation of measured shear stress was observed with time and if the liquid does not contain any impurity or solid phases, the shear stress should not vary with time for a specific angular velocity.

Results and Discussion of Rheology Experiments

In this section the results of shear viscosity as a function of shear rate and melt superheat temperatures are presented. Figure 2 shows the comparison of shear viscosities of eutectic Al-Si alloys with and without the addition of 0.023 wt% Sr.

---

* Inductively Coupled Plasma Spectrometry to measure chemical composition of the alloy.

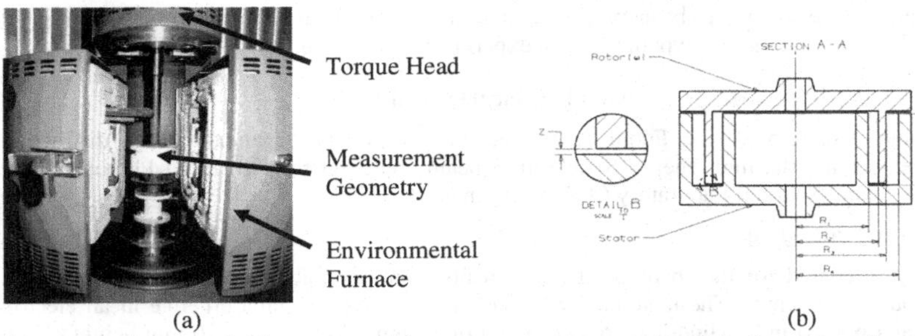

Figure 1: Rotational Rheometer with measurement geometry. (a) rheometer showing the environmental furnace, geometry and torque head, and (b) schematic of the measurement geometry showing the rotor and the stator.

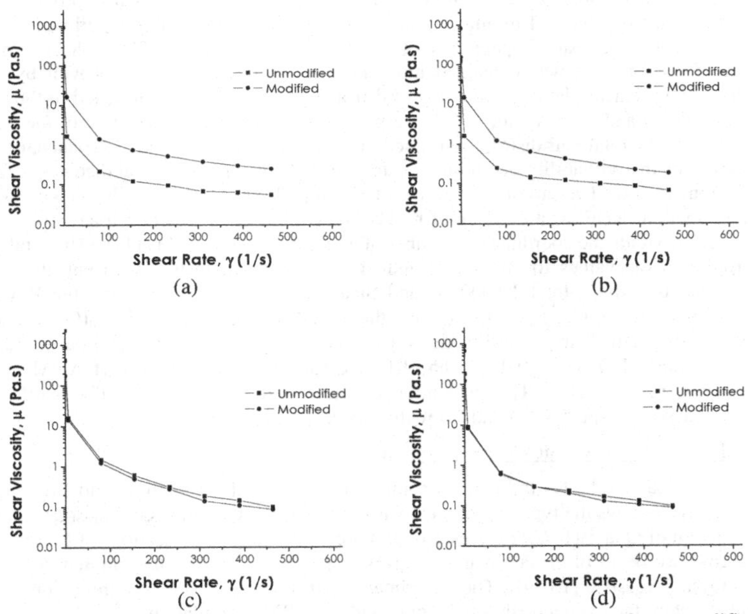

Figure 2. Comparison of shear viscosities as a function of shear rate for unmodified and Sr modified alloys. (a) 585 °C, (b) 605 °C, (c) 675 °C and (d) 700 °C.

It can be observed that Sr increases the melt viscosity significantly at lower melt superheat temperatures (585 °C and 605 °C) and the effect fades at high melt superheat temperatures (675 °C and 700 °C). It can also be observed in Figure 2 that Al-Si eutectic liquid exhibits a non-Newtonian and shear thinning flow behavior in both the unmodified and Sr modified conditions.

A hypothesis to explain why liquid metals and alloys will tend to exhibit a non-Newtonian and shear thinning behavior has been proposed in Malik et al [9] and further investigations are underway to substantiate the hypothesis with experimental evidence.

Atomic Characterization

The atomic structure of the liquid Al-Si eutectic alloys were evaluated from diffraction experiments using the high energy Synchrotron beam source at the Advanced Photon Source (APS), Argonne National Laboratory (ANL), Argonne, IL, USA.

Materials and Procedure

The alloys prepared for the rheological experiments mentioned above were also used for the diffraction experiments. The materials were melted in a clean alumina crucible in an electric furnace and poured in a cylindrical copper mould of 15 mm diameter and 70 mm height. Thin cylindrical samples of 1.5 mm diameter and 10 mm length were machined out of the casting and placed in thin transparent quartz tubes of 2 mm in diameter and 10 mm in length. These tubes were subsequently evacuated and sealed in Ar to minimize the melt interaction with atmosphere. A highly monochromatic 98.099 KeV beam (0.1264A° wavelength) was chosen to provide high penetration through the bulk and minimize multiple scattering. Diffraction experiments were carried out at melt superheat temperatures of 594°C, 616 °C and 705 °C. The chemical composition of the samples was evaluated after the diffraction experiments as well by ICP spectrometry. The intensity data as a function of diffraction angle, θ were converted to the total structure factor, S(Q) and wave vector Q [12], respectively. Fourier transformation of the curve S(Q) vs. Q yielded the total pair distribution function, g(r) as a function of radial distance, r [12]. The g(r) represents the probability of finding an atom in the structure at a radial distance, r from the central atom at r=0. The atomistic density of the liquid can used to evaluate the Radial Distribution Function (RDF) as a function of r. The integration of RDF between any two radial distances, r_1 and r_2 yields the coordination number of atoms between r_1 and r_2 [12]. The total g(r) and RDF represent the values for all the Al and Si pairs of atoms in the Al-Si eutectic melt. Atom arrangements using a total 10,000 Al and Si atoms was carried out using the Reverse Monte-Carlo Analysis (RMCA[†]) [13, 14] to yield the partial structure factors, $S_{Al-Al}(Q)$, $S_{Al-Si}(Q)$ and $S_{Si-Si}(Q)$; partial pair distribution functions, $g_{Al-Al}(r)$, $g_{Al-Si}(r)$ and $g_{Si-Si}(r)$; and partial radial distribution functions, RDF_{Al-Al}, RDF_{Al-Si}, and RDF_{Si-Si} for the individual pairs of Al-Al, Al-Si and Si-Si atoms, respectively. The evaluation of these partial quantities will quantify the distribution of each atom species (Al and Si) with respect to each other.

Results and Discussion of Atomic Characterization

In this section typical results for the various parameters in the analysis of the liquid structure is presented. Figure 3 shows the typical graphical variation of the S(Q), $S_{Al-Al}(Q)$, $S_{Al-Si}(Q)$ and $S_{Si-Si}(Q)$ as a function of Q at 594 °C. Figures 3(b) and 3(c) show magnified sections of Figure 3(a) to show the first two peaks of the curve and $S_{Si-Si}(Q)$, respectively. S(Q) is the summation of $S_{Al-Al}(Q)$, $S_{Al-Si}(Q)$ and $S_{Si-Si}(Q)$ [13,14]. The experimental S(Q) was fitted by the summation of the three partial structure factors denoted as S(Q) obtained from RMC simulation.

Figure 4 shows the total and partial pair distribution functions at 594 °C obtained from Fourier transformation of the experiment g(r) and the partial structure factors obtained from RMCA. Figures 4(b) and 4(c) show magnified sections of Figure 4(a) to show the first two peaks of the curve and $S_{Si-Si}(Q)$, respectively. From theory, g(r) is the summation of $g_{Al-Al}(r)$, $g_{Al-Si}(r)$ and g_{Si-}

[†] http://www.isis.rl.ac.uk/RMC/

$_{Si}$(r). The experimental g(r) (Fourier transformation of experimental S(Q)) was fitted by the summation of the three partial pair distribution functions denoted as the g(r) from RMC simulation (obtained from the Fourier transformation of S(Q) from RMC simulation).

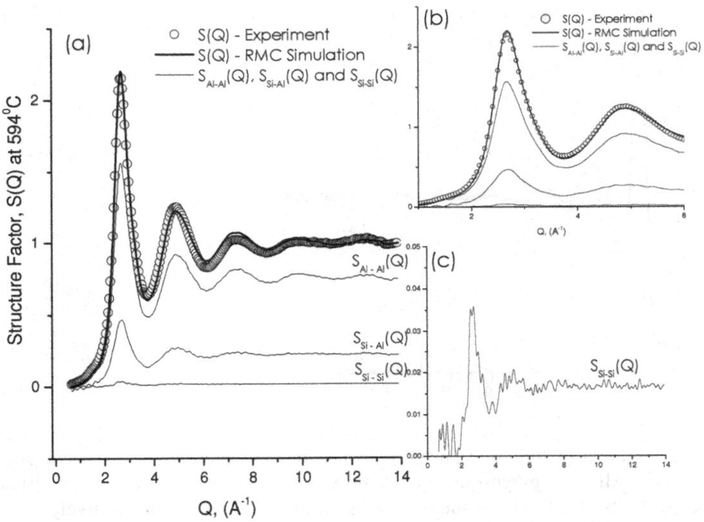

Figure 3. The variation of total and partial structure factors for Al-Si eutectic alloy with Q at 594 °C. All three graphs have the same variables in the respective axes. (b) and (c) are magnified sections of (a) to show the first two peaks and the $S_{Si-Si}(Q)$, respectively.

Figures 3 and 4 demonstrate that the RMCA simulation carried out with 10,000 atoms were accurate. Figure 5 shows the comparison of the first peaks of $g_{Al-Al}(r)$, $g_{Al-Si}(r)$ and $g_{Si-Si}(r)$ between unmodified and Sr-modified at the three experiment temperatures. Figure 5 shows that the Sr addition to the melt significantly alters the atom arrangement of the Si-Si pairs. At 594 °C, the first peak of $g_{Si-Si}(r)$ for unmodified alloy is sharper and has a smaller half peak width than the modified alloy. The atoms are more ordered as the peak sharpness is higher and the half peak width is smaller. The $g_{Si-Si}(r)$ for the unmodified alloy is sharper at 594 °C and blunt at 700 °C showing that the Si-Si atom pair becomes more ordered as temperature of melt superheat decreases. The Sr addition to the melt has only a minor effect on the ordering of Si-Si atoms at 700 °C. At lower melt superheat temperatures the Sr prevents the ordering of the Si-Si atoms thus impeding the formation of higher order Si clusters which are essential for nucleation at the eutectic temperature. Sr addition uniquely affects the distribution of the Si and not Al atoms. In Figure 5, at 700 °C and 616 °C, the Sr addition has negligible effect on the Si-Al atom pair structure. However, at 594 °C, the effect of Sr on the distribution of the Si atoms has resulted in reduced ordering of the Si-Al atom pair clusters. The effect of Sr addition to the melt causes negligible change to the Al-Al atom pair distribution as is also seen in Figure 5.

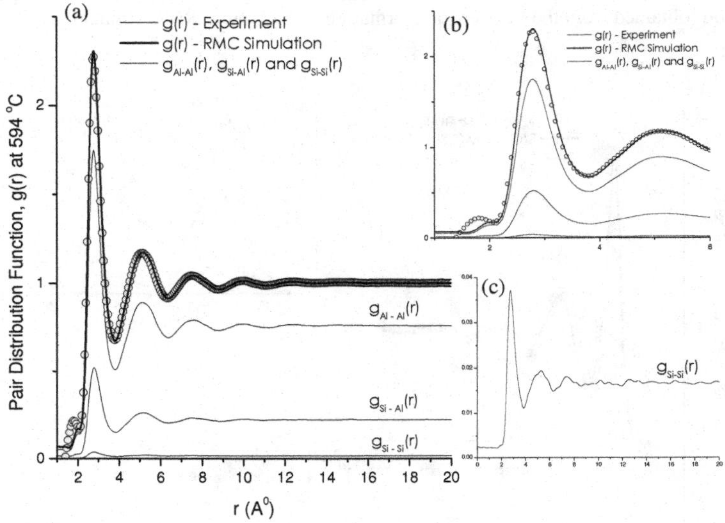

Figure 4. The variation of total and partial pair distribution function for Al-Si eutectic alloy with Q at 594 °C. All three graphs have the same variables in the respective axes. (b) and (c) are magnified sections of (a) to show the first two peaks and the $g_{Si-Si}(r)$, respectively.

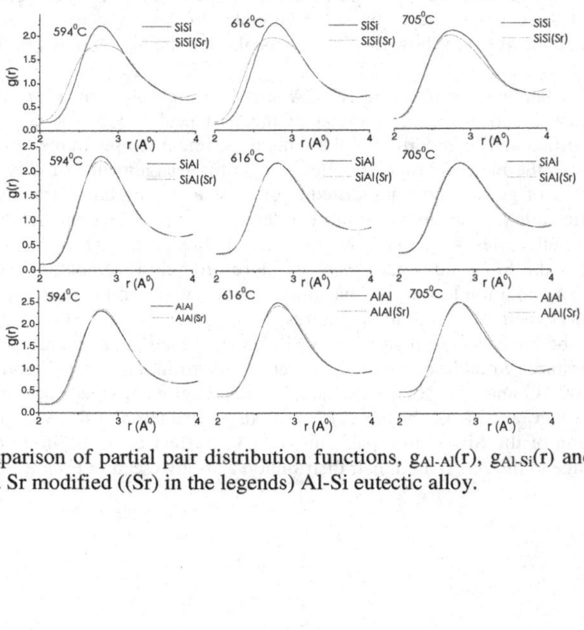

Figure 5. Comparison of partial pair distribution functions, $g_{Al-Al}(r)$, $g_{Al-Si}(r)$ and $g_{Si-Si}(r)$ for unmodified and Sr modified ((Sr) in the legends) Al-Si eutectic alloy.

Figure 6 shows the first peak of the RDF for Si-Si atom pairs at the three experiment temperatures. It can be observed that the height of the RDF_{Si-Si} peaks at all temperatures is reduced with the Sr addition to the melt. The area under the first peak in Figure 6 represents the coordination number of atoms in the first shell or cluster. Addition of Sr significantly reduces the coordination number of atoms in the first shell or cluster of Si-Si atom pairs. Further, Figure 6 shows that at 700 °C the position of the first peak of RDF_{Si-Si} is relatively the same for both unmodified and Sr modified alloys. The position of the first peak of the RDF_{Si-Si} denotes the maximum Si atom concentration from a reference Si atom at r=0 and typically the position of the first peak will reduce as the melt superheat temperature decreases as the atoms tend towards more clustering and eventual nucleation. Figure 6 shows that the position of the first peak of RDF_{Si-Si} decreases with decreasing temperature (typical behavior) in unmodified alloys whereas the position increases with decreasing temperature (anomaly) in Sr modified alloys. Sr addition not only prevents the clustering of Si-Si atoms but also reverses the clustering by creating more disorder in the Si-Si atom arrangement as melt superheat temperature decreases. Figures 7(a) and 7(b) show the quantified data for the position of the first peak in RDF_{Si-Si} and the coordination number of Si atoms in the first shell or cluster (area under first peak in Figure 6). Figure 7(a) shows that the Sr addition causes a reversal in the clustering of the Si atoms as melt superheat temperature decreases. The mechanism of such a reversal is being investigated.

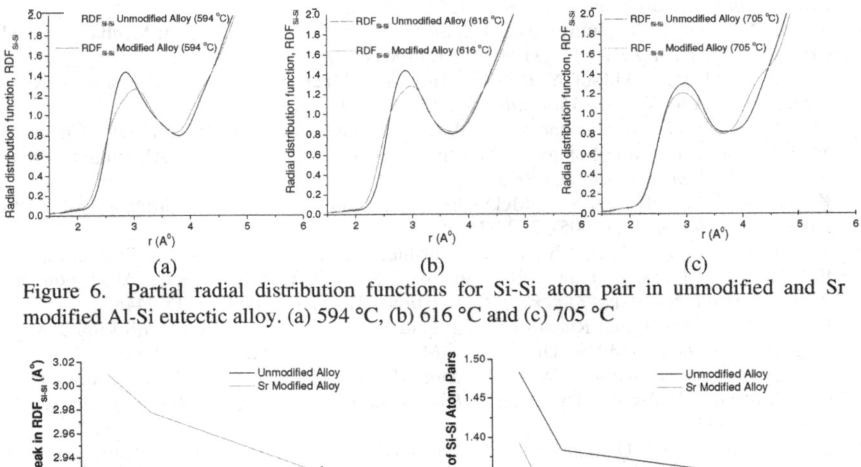

(a) (b) (c)

Figure 6. Partial radial distribution functions for Si-Si atom pair in unmodified and Sr modified Al-Si eutectic alloy. (a) 594 °C, (b) 616 °C and (c) 705 °C

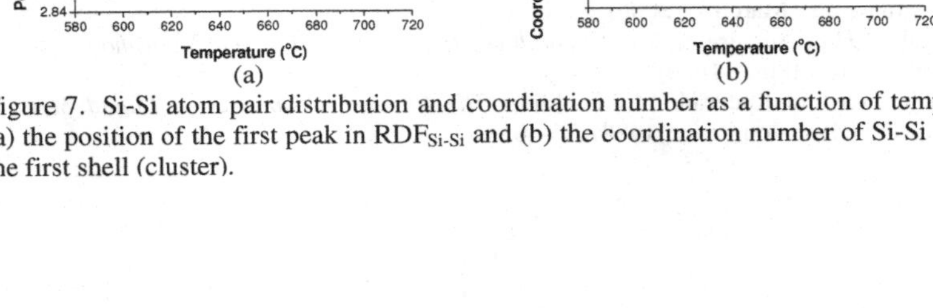

(a) (b)

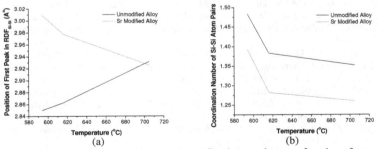

Figure 7. Si-Si atom pair distribution and coordination number as a function of temperature. (a) the position of the first peak in RDF_{Si-Si} and (b) the coordination number of Si-Si atoms in the first shell (cluster).

Figure 7(b) shows that the overall coordination number of Si atoms in the first shell of atoms is lower in the Sr modified alloy. Sr addition will delay or prevent the clustering of Si atoms which will result in the absence of nucleation of eutectic Si phase at the eutectic temperature.

Conclusions

This study has presented conclusive experimental evidence that 0.023 wt% Sr addition to the Al-Si eutectic alloy melt significantly changes the atom arrangement and hence the rheological properties of the alloy. This will result in the absence of any nucleation event at the eutectic temperature leading to an undercooling of the alloy and a supersaturation of Si atoms in the inter-dendritic liquid. This is the first stage in the mechanism that results in the modification of the morphology of the eutectic Al and Si phases [6].Further, it has been observed that the Al-Si eutectic alloy melt exhibits a non-Newtonian and shear thinning behavior wherein the shear viscosity decreases with increasing shear rate.

References

1. Pacz, *US patent 1387900*, 1920
2. J. R. Davis, "*Aluminum and Aluminum Alloys*", (Ohio: ASM International, 1993), 627.
3. A. Hellawell, "The Growth and Structure of Eutectics with Silicon and Germanium", *Progress in Materials Science*, 15, (1970), 3-78.
4. M. M. Makhlouf, H. V. Guthy, "The Aluminum–Silicon Eutectic Reaction: Mechanisms and Crystallography", *Journal of Light Metals*, 1, (2001) 199-218.
5. S. Shankar, M. M. Makhlouf, Y. Riddle, "Nucleation Mechanism of Eutectic Phases in Al-Si Hypoeutectic Alloys", *Acta Materialia*, 52, (2004), 4447-4460.
6. M. M. Makhlouf, S. Shankar, Y. Riddle, "Mechanisms of Formation and Chemical Modification of the Morphology of the Eutectic Phases in Hypoeutectic Aluminum-Silicon Alloys", *AFS Transactions*, 05, (2005), 088
7. K. Nogita, P.L. Schaffer, S.D. McDonld, L. Lu and A. Dahle, "Modification of Al-Si Alloys", *Aluminium*, 81, (2005), 330-335.
8. Y. H Cho, H-C. Lee, K.H. Oh and A. K. Dahle, "Effect of Strontium and Phosphorus on Eutectic Al-Si Nucleation and Formation of β-Al5FeSi in Hypoeutectic Al-Si Foundry Alloys", *Metallurgical and Materials Transactions A*, 3,(10), (2008), 2435-2448.
9. M. M. Malik, "Rotational Rheometry of Liquid Metal systems: A Study with Liquid Al-Si Hypoeutectic Alloys" (MASc. Dissertation, McMaster University, 2008), 19-53.
10. Y. Qi, T. Cagin, Y. Kimura, W.A. Goddard III. "Viscosities of Liquid metal alloys from nonequilibrium Molecular Dynamics", *Journal of Computer-Aided Materials Design*, 8, (2001) 233-243.
11. C. Desgranges and J. Delhommelle, "Viscosity of liquid iron under high pressure and high temperature: Equilibrium and nonequilibrium molecular dynamics simulation studies", *Physical Review B*, 76, 172102(1)-172102(4), (2007).
12. Srirangam.V.S. Prakash, M. J. Kramer and S. Shankar, "Structural Characterization of Liquid Al-Si Hypoeutectic Alloys", *136th Annual meeting and Exhibition, TMS 2006*, Orlando, FL-USA, February 25 - March 1, (2006), 33-41.
13. Y. Waseda, "*The structure of Non-Crystalline Materials, Liquids and Amorphous Solids*", (USA: McGraw-Hill Inc., 1980), 41-76.
14. Takamichi Iida and Roderick I.L. Guthrie, "*The Physical Properties of Liquid Metals*", (Oxford, NY: Clarendon Press., 1988), 33-40.

Shape Casting: The 3rd International Symposium
Edited by: John Campbell, Paul N. Crepeau, and Murat Tiryakioğlu
TMS (The Minerals, Metals & Materials Society), 2009

CHARACTERIZATION OF THE MELT QUALITY AND IMPURITY CONTENT OF AN LM25 ALLOY

Katharina Haberl[1], Peter Schumacher[1,2], Georg Geier[2], Bernhard Stauder[3]

[1]Chair of Casting Research, Metallurgy Department, University of Leoben, Franz-Josef-Str. 18, 8700 Leoben, Austria
[2]Austrian Foundry Research Institute, Parkstr. 21, 8700 Leoben, Austria
[3]NEMAK Linz GmbH, Zeppelinstr. 24, 4030 Linz, Austria

Keywords: melt quality, oxides, bifilms, reduced pressure test

Abstract

The melt quality of an LM25 aluminium casting alloy has been examined using reduced pressure test (RPT) measurements, porous disc filtration analysis (PoDFA), and fatigue and tensile tests. The aim of this study was to determine existing melt quality and thus evaluate methods used with respect to monitoring and improving melt cleanliness. Special emphasis was given to the influence of oxides. It was found that the melt quality has varying degrees of effect on the tests used. Results in particular indicate, that it was necessary to distinguish between "new" oxides and "hard" inclusions in the melt, as new oxides impact porosity whilst hard inclusions impact on ductility. Based on the results of this study, suggestions for the measurement of the melt quality have been proposed.

Introduction

Aluminium has a high affinity for oxygen and as a consequence readily forms an oxide. If a molten aluminium surface reacts with atmospheric oxygen, an oxide film is created immediately. This film can take different morphologies, depending on the alloying elements of the melt and the contact time with the oxygen [1]. A film created on the melt surface cannot be dissolved in the liquid solution; it remains solid. The formation of bifilms has been extensively described by Campbell and co-workers [1]. Essentially surface turbulences are the only possibility to entrain folded films (bifilms) into the melt [2], which subsequently lead to casting defects, e.g. cracks or pores [2,3].

Two forms of oxides can be distinguished [4]: old oxides and new oxides (bifilms). Old oxides are created before or during the melting. Their shape can be described as 3-dimensional. In contrast new oxides are created during filling and casting, and their shape is 2-dimensional. Furthermore it has been proposed that trapped atmosphere in the form of N_2, O_2, can react with the bifilm, so that the reaction products fill the bifilm and a hard inclusion is formed in the end [5,6]. For the nucleation of hydrogen pores bifilms (new oxides) are particularly relevant. Moreover the influence of other hard inclusions (carbides, nitrides, etc.) in the melt has to be considered.

Porosity has a negative influence on the casting part. It decreases mechanical properties (fracture elongation, tensile strength, fatigue life) and hinders potential applications for the castings. There are two extreme forms of porosity: gas and shrinkage porosity. The formation of each one cannot be separated from the other; usually porosity is a mixture of both these features [1,7]. However, in the Reduced Pressure Test (RPT) method, samples are well fed during solidification. Thus the influence of shrinkage on porosity formation can be viewed as negligible (although this depends also on solidification morphology). For this reason the RPT is an ideal measurement method for the determination of gas pores and a good but qualitative measurement for the gas content.

The measurement of the melt quality is not trivial, because new oxides are in at least one dimension very thin. These new oxides or bifilms are assumed to have a big influence on the formation of gas pores and thus quality. The majority of melt cleanliness assessments cannot measure bifilms. The difficulty is to find a suitable method for detecting bifilms. Different impurities are measured by different measuring methods, such as PoDFA, RPT, Prefil, etc. Thus melt quality cannot be described by one parameter alone.

An intensive study on the melt quality and the impurity content was performed. Special emphasis was given to bifilms and their effect on the melt quality. Hundreds of melt samples were collected [8]. For this publication the secondary alloy LM25 (AlSi7Mg0.5 (Fe,Cu)) was chosen because of the many potential impurities contained within it. The samples were taken in the foundry from the melt in transport ladle (after melting in the induction furnace), casting furnace before rotary degassing and casting furnace after rotary degassing with N_2. The melt quality was determined using advanced RPT, PoDFA, tensile testing, and fatigue testing techniques.

Experimental

Reduced Pressure Test (RPT)

This test is used in conventional foundry handling for controlling the gas content of liquid metals. A sample is tested by solidifying a melt in a vacuum chamber under a standardized reduced pressure (80 mbar). Figure 1(a) shows the RPT-apparatus (BOC Edwards, CG 16K) and Figure 1(b) RPT-samples. The pores in the sample cause a change in the density, which can be measured by the density index DI = $(\rho_{Atm}-\rho_{RPT})*100/\rho_{Atm}$; ρ_{Atm} is the density of the sample solidified under atmospherical pressure and ρ_{RPT} is the density of the sample solidified under reduced pressure.

Advanced RPT

The RPT-samples were examined by advanced RPT to detect pore distribution and geometry. The two methods of advanced RPT are picture analytical examination of pores and porosity evaluation with computed tomography (CT).

Picture Analytical Examination Of Pores

Pores in the cross sections were examined with a light microscope Nikon MM40; the number of pores per mm² was determined in a rectangular section (see Figure 1(c)) with computer software NIS Elements Br 2.30, Nikon. For each condition at least 4 RPT-samples were taken and their cross sections were examined. The investigation of pores was performed in accordance with the VDG-data sheet P201 [9].

In the present analysis only the number density of pores was detected, not the influence of shrinkage cavities. On this account only pores smaller than 7 mm² and bigger than the DAS with a circularity of at least 0.5 were counted. The results of measurements are given in pores per unit area. The concentration of pores in the cross section and the size distribution can thus be determined. Grinding and polishing may affect the porosity observed in the sample by stereography; as a consequence 3-dimensional CT was also employed.

(a) (b) (c)

Figure 1. (a) RPT apparatus, (b) RPT-samples (left: low gas content, right: high gas content), (c) Picture analytical examination of the RPT-sample's cross section.

<u>Computed Tomography CT</u>

Because of the geometrical and stereographical limitation of cross sections it is not possible to measure the exact diameter of pores by metallography. It is necessary to conduct 3-dimensional examination methods. With CT the pore diameter, size and volume can be determined.

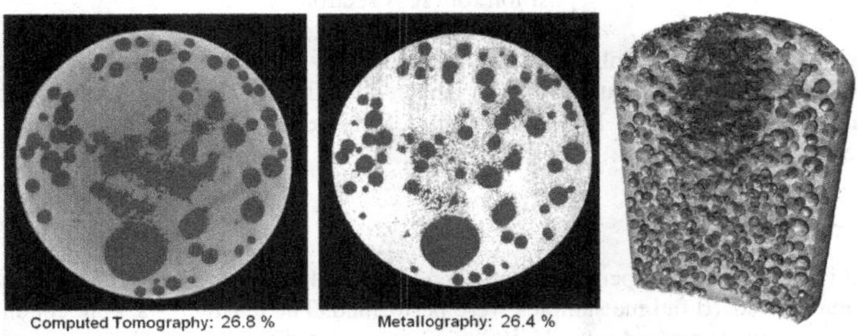

(a) (b) (c)

Figure 2. (a)&(b)Porosity evaluation at corresponding positions: (a) CT analysis, (b) metallographic section, (c) 3-dimensional view of the CT sample (separating pores (light) and shrinkage cavities (dark)).

CT delivers 3-dimensional datasets that represent the local x-ray attenuation coefficient. To interpret these datasets according to pores, the interfaces between the pores and the surrounding material have to be defined exactly. Furthermore, pores and other inhomogeneities, especially shrinkage cavities, have to be distinguished. Advanced visualization techniques can give a near exact interface of the pores [10].

Investigations of the RPT-samples have been made with a Phoenix x-ray v|tome|x-C CT system. An example of a comparison between the CT and metallographic evaluation is given in Figure 2. It shows a good correlation between the two investigative methods.

PoDFA (Porous Disc Filtration Analysis)

Undissolved hard or voluminous inclusions in the melt can be measured by filtration of the melt in PoDFA (ABB). The area fraction inclusions are measured at a standard position above the filter (Figure 3). For each condition at least 5 PoDFA-samples were taken and their cross sections were examined. During filtration a filter cake is built above the filter. This filter cake consists of particles from the melt, which can even be smaller than the pore diameter of the filter. However, participation of bifilms in the filter cake is limited because of their low stiffness. Furthermore, their detection is difficult because of their thin 2-dimensional morphology. Within PoDFA there is no specific category for bifilms. The typical spectrum of a PoDFA contains: Al-oxide, Al-carbide, Ti-boride, Al-nitride, Mg-oxide and spinel. However, coarse oxide films are usually detected with the PoDFA method.

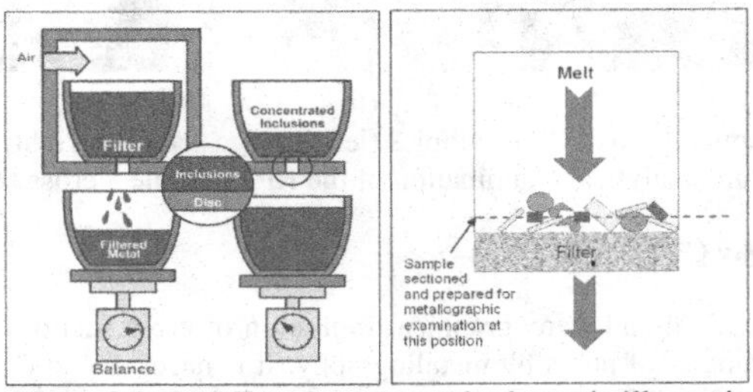

Figure 3. Schematic view of the PoDFA apparatus and schematic illustration of filter cake with position for cross section.

The unit of measurement in PoDFA for melt impurity is in mm²/kg. However, results of the PoDFA give no information about the size of detected inclusions. Size is an important parameter for mechanical properties since large inclusions readily act as cracks. Inclusions are basically hard notches.

Mechanical Testing

Tensile and fatigue tests were performed on T6 heat treated samples. For each sampling location at least 10 tensile and 10 fatigue samples were performed. The tensile tests were conducted with a Zwick/Roell Z050 (test standard: EN 10002). For the interpretation of the tensile test special emphasis is given on the fracture elongation A as this is a very sensitive parameter with high significance to melt purity. The fatigue tests were performed in the laboratory in accredited testing procedures (DIN 50100) on a Russenberger Pruefmaschinen AG Mikrotron 9201/129 resonant frequency machine. For the fatigue testing the stress ratio R was +0.05. A Log-Normal-distribution was performed for the interpretation of the measurements. The life cycles before failure for a 90% survival probability were calculated using the program Visual-XSel 9.0.

Results and Discussion

Figure 4 shows the results at different stages in the manufacturing process of the alloy LM25. The values were standardized, so that it is possible to show all measurement values on two axes: on the left side are the purity values (DI, pores/area, PoDFA) and on the right side are the mechanical values (A, life cycles). Values at the top of the diagram represent good melt quality, values at the bottom of the diagram confer to a poor melt quality.

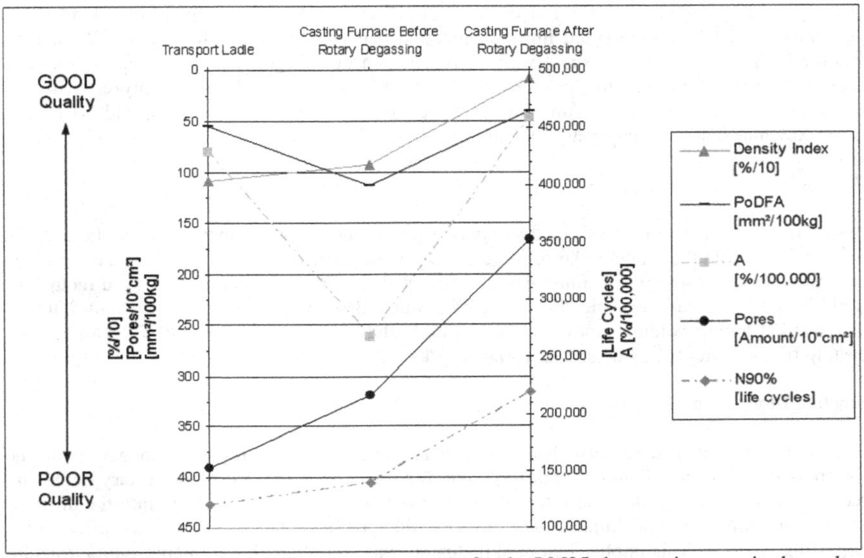

Figure 4. Trend lines of the various measurements for the LM25 show an increase in the melt quality from the transport ladle and the casting furnace before rotary degassing to the casting furnace after rotary degassing, with the exception of the fracture elongation and PoDFA; here the melt quality decreases in the casting furnace before rotary degassing during stirring of the sedimented sump by decanting of the transport ladle.

The quality of the melt in the transport ladle and the casting furnace before rotary degassing is poor as expected: low life cycles, high amounts of pores in cross sections of the RPT-sample, high DI and low fracture elongation (at the casting furnace before rotary degassing).

The quality of the melt in the casting furnace after rotary degassing is improved: high life cycles, low amount of pores in the cross section of the RPT-sample, low DI, high fracture elongation and a low value at PoDFA. A reason for this is the low gas content and the resultant lower pore diameter and better pore distribution. Special emphasis must be given to the casting furnace sedimented sump. The LM25 is a secondary alloy, for this reason the influence of the sedimented sump is more important. An accumulation and redistribution of particles from the sedimented sump occurs in the melt after each decanting of the transport ladle.

Density Index DI

The DI decreases during the melt treatment process. At the transport ladle the DI is 10.8 and at the casting furnace before rotary degassing it is 9.2. Natural degassing in an atmosphere with lower H_2 pressure facilitates this observation which should be more pronounced with longer times. In the casting furnace after rotary degassing the DI decreases to a value of 0.8, as a result of the rotary degassing with N_2.

Usually the DI is about 1. A DI of 1 corresponds close to a hydrogen content of 0.1 ml/100g. Dasgupta et al. [11] has reported linear connection between measured density of RPT-sample and the hydrogen content measured with an Alscan sensor at high hydrogen contents (see also Campbell and Dispinar). Independent measurements with a hydrogen analyzer with an electrochemical sensor confirm this [8]. However, measured hydrogen content and DI do not accurately match at low hydrogen concentrations [11].

PoDFA

A significant influence of sedimented sump after decanting of the transport ladle can be seen. In the transport ladle the value is 0.6 mm²/kg. After the decanting in the casting furnace the value before rotary degassing is 1.1 mm²/kg as a result of the melt accumulation with hard inclusions (oxides and other carbides, nitrides, etc.). The rotary degassing floats the inclusions into the dross and at longer holding times impurities sink to the bottom of the furnace. The value in the casting furnace after rotary degassing is 0.4 mm²/kg.

Fracture Elongation

Here the sedimented sump also has a significant influence. The fracture elongation of the transport ladle samples is 4.3 %, but the value for the casting furnace before rotary degassing reduces to only 2.7 %. After the rotary degassing impurities are either washed into the dross or sink to the bottom of the furnace. The fracture elongation in the casting furnace after rotary degassing is 4.6 %. This implies that hard inclusions have a dominating influence on the fracture elongation.

There is a qualitative correlation between PoDFA and fracture elongation as a general trend: at low PoDFA values high fracture elongation values are observed. For these two parameters the best melt quality is in the casting furnace after rotary degassing, good values are also achieved in the transport ladle, while the worst values are achieved in the casting furnace before rotary degassing.

Pores/Area

The amount of pores per area of the RPT-sample shows an explicit increase of the melt quality during the treatment process. The amount of pores in the transport ladle is 39 pores/cm². At the casting furnace prior to rotary degassing, 32 pores/cm² and after the rotary degassing, only 17 pores/cm² are observed. The reason for this is the reduced amount of nucleation sites in the melt as a result of the rotary degassing.

The gas pores in the melt evolve from nuclei. Nucleation sites for pores in the RPT-sample are assumed to be bifilms, which are activated by reduced pressure to form a pore. Thus, the number of bifilms or other inclusions that act as nuclei can be found by counting gas pores [8,12]. Interestingly in holding experiments the number of pores diminishes over time suggesting bifilms become inactive over time or are trapped as dross (Figure 5).

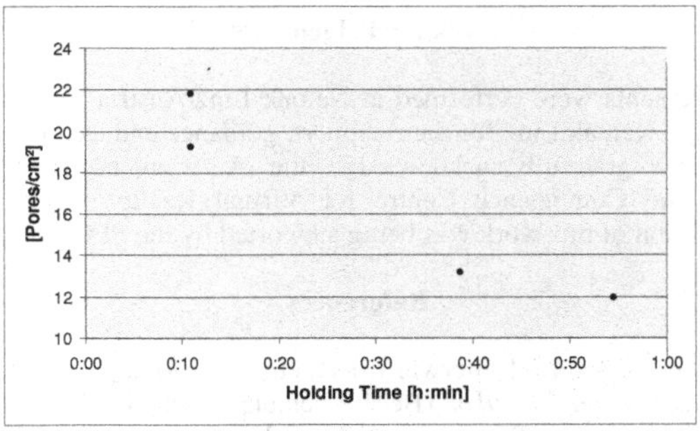

Figure 5. The number of pores per area decrease at longer holding times in the casting furnace after rotary degassing.

Life Cycles

The number of life cycles possible before failure increases during the melt treatment process. At the transport ladle the life cycles with a survival probability of 90 % are 120,000, at the casting furnace before rotary degassing it increases to 140,000 and at the casting furnace after rotary degassing the life cycles are 220,000. The reason for this is the decreasing amount of bifilms as oxides in the melt. Oxides as bifilms cause pores, and pores cause low cycles to fatigue [13,14]. Additionally, the gas content has been simultaneously lowered so that fewer and smaller pores were present.

There is a clear trend between the findings of the RPT and the fatigue tests (see Figure 4): at low values for pores/area there are high values for life cycles. The best melt quality is achieved at the casting furnace after rotary degassing, the worst quality at the transport ladle.

Conclusion

The melt quality and impurity content of the aluminium casting alloy LM25 (secondary alloy) was analyzed by various methods. Special emphasis was given to the influence of bifilms as nucleation sites for gas pores. The advanced RPT gives information about the number of nuclei for the formation of gas pores.

It is important to distinguish between old oxides and other hard inclusions or new oxides such as bifilms [4]. New oxides influence pore formation as bifilms, while the old oxides tend to act as hard inclusions and affect fracture elongation [4,8,15].

The life cycles are dominated by pores; therefore they correlate well with the amount of pores and the DI of the RPT. Hard inclusions cannot be determined with the RPT, therefore PoDFA or a similar method must be used. At low hydrogen contents, similar to that obtained in industry, the fracture elongation is ruled by hard inclusions. The fracture elongation correlates well with the PoDFA results.

For the overall determination of the melt purity the advanced RPT must be combined with PoDFA or similar methods.

Acknowledgements

The practical measurements were performed at Nemak Linz/Austria, the authors would like to gratefully acknowledge Nemak Linz for their support, guidance and excellent collaboration. The authors would like to gratefully acknowledge the Austrian Foundry Research Institute Leoben/Austria and the Competence Centre for Virtual Reality and Visualization VRVis Vienna/Austria. A portion of this work was being supported by the FFG.

References

1. J. Campbell, *Castings* (Oxford: Butterworth-Heinemann, 2003), pages 12, 17, 65, 225.
2. E. Brunnhuber, ed., *Giesserei Lexikon* (Berlin: Schiele & Schön, 1994), 864.
3. W. Hufnagel ed., *Aluminium-Taschenbuch* (Düsseldorf: Aluminium-Verlag GmbH, 1983), 377.
4. Q.G. Wang, P.N. Crepeau, J.R. Griffiths, C.J. Davidson, "The Effect of Oxide Films and Porosity on Fatigue of Cast Aluminum Alloys," *Shape Casting: The John Campbell Symposium*, ed. P. Crepeau, M. Tiryakioglu (Warrendale, PA: TMS, 2005), 205-213.
5. C. Nyahumwa, N.R. Green, J. Campbell, "Influence of Casting Technique and HIP on the Fatigue of an Al7SiMg Alloy," *Met. and Mat. Trans. A, 32A* (2001), 349-357.
6. W. Griffiths, R. Raiszadeh, A. Omotunde, "The Effect of Holding Time on Double Oxide Film Defects in Aluminium Alloy Castings," *Shape Casting: 2nd International Symposium*, ed. P. Crepeau, M. Tiryakioglu, J. Campbell, (Warendale, PA: TMS, 2007), 35-42.
7. E.J. Whittenberger, R.N. Rhines, *J. Metals, 4(4), Trans. AIME 194* (1952), 409-420.
8. K. Haberl, "Schmelzereinheit einer Al-Gusslegierung am Beispiel AlSi7MgCu0,5 und LM25" (MSc Thesis, University of Leoben 2007).
9. VDG-data sheet P 201, "Volumendefizite von Gussstuecken aus Nichteisenmetallen", VDG Verein deutscher Gießereifachleute, (Duesseldorf, May 2002), 1-16.
10. G. Geier et al, "Assessing Casting Quality using Computed Tomography with Advanced Visualization Techniques", *Shape Casting: 3rd International Symposium*, ed. P. Crepeau et al (Warendale, PA: TMS, 2009)
11. S. Dasgupta, L. Parmenter, D. Apelian, F. Jensen: *Proc. 5th International Molten Aluminum Processing Conference, AFS* (Des Plaines, 1998), 283-300.
12. D. Dispinar, "Determination of Metal Quality of Aluminium and Its Alloys" (PhD-Thesis, Birmingham 2005), 15.
13. R. Minichmayr, W. Eichlseder, "Lebensdauerberechnung von Gussbauteilen unter Berücksichtigung des lokalen Dendritenarmabstandes und der Porosität," *Gießerei, 5* (B) (2003), 70-75.
14. H. Leitner, W. Eichlseder, Ch. Fagschlunger, "Lebensdauerberechung von Aluminiumkomponenten: Von der Probe zum komplexen Bauteil," *Gießerei Praxis 3* (2006), 70-76.
15. G. E. Byczynski, J. Campbell, "A Study of Crack Initiation Sites in High Cycle Fatigue of 319 Aluminium Alloy Castings", *Shape Casting: The John Campbell Symposium*, ed. P. Crepeau, M. Tiryakioglu (Warrendale, PA: TMS, 2005), 235-244.

SHAPE CASTING:
3rd International Symposium
2009

Novel Methods and Applications

Session Chair:
Mahi Sahoo

Shape Casting: The 3rd International Symposium
Edited by: John Campbell, Paul N. Crepeau, and Murat Tiryakioğlu
TMS (The Minerals, Metals & Materials Society), 2009

ABLATION CASTING UPDATE

J. Grassi[1], J. Campbell[2], M. Hartlieb[3] and F. Major[4]
[1]Alotech Ltd., 1448 Hedgewood LN NW, Kennesaw, GA 30152, USA
[2]Alotech Ltd., Park Rd., West Malvern, WR14 4BJ, United Kingdom
[3]RioTintoAlcan Inc., 1188 Sherbrooke St. W., Montréal, Québec, H3A 3G2, Canada
[4]Rio Tinto Alcan, 945 Princess St., Kingston, Ontario, K7L 5T4, Canada

Keywords: Ablation, Ablative Casting

Abstract

Light alloy castings made in an aggregate mold are sprayed with coolant to ablate (erode) away the mold through dissolution of the mold binder. The coolant thereby gains access to the surface of the casting prior to extensive solidification, avoiding normal limitations to heat transfer via the 'air gap' commonly present between casting and conventional mold, conferring otherwise unattainable rates of cooling. The process was first announced at the 2008 TMS Annual Meeting [1], and is here updated.

Introduction

As described previously [1], filling of sand (aggregate mold) castings and investment castings can be reasonably controlled (compared to high pressure die-casting), but the freezing is normally slow or involves complicated chilling, giving relatively poor and inhomogeneous mechanical properties. Aggregate mold casting technologies allow higher complexity than permanent mold based technologies, but in general are rather limited to castings of modest accuracy. Investment casting allows significant improvements but at greater cost. In aggregate mold casting technologies the core assembly processes allow significantly improved accuracy and potential for complexity, but the resin-type binder systems and aggregate reclamation are major cost penalties. Additionally, during freezing of the casting, the casting contracts away from the mold, and the mold expands, opening the so-called 'air gap' between the casting and the mold. This air gap controls the rate of cooling, and thus the fineness of the microstructure and the mechanical properties of the casting.

The first paper described our new approach to the making of a casting in which this fundamental limitation to heat flow is eliminated by removal of the mold and the application of coolant directly onto the surface of the casting. The Ablation Casting Technology removes the aggregate mold (held together with an environmentally friendly water soluble inorganic binder) with water, which is sprayed so as to ablate away the mold, allowing the water to impinge directly on the casting. The technique achieves an easy removal of both the mold itself as well as the most complex internal cores, while providing a very high chill rate to the casting. Directional solidification is greatly enhanced, reducing typical problems commonly associated with casting alloys outside the typical Al-Si system (for instance the AA2xx series).

Results and Discussion

An interesting and remarkable outcome is not merely the enhanced mechanical properties but also microstructural homogeneity throughout thin and thick sections. For design strength targets of mean minus three standard deviations high mechanical properties, and also their variance, are of utmost importance when deciding what values to design to for a given casting technology and alloy. Table 1 describes the properties achieved in a single steering knuckle sectioned in different locations.

For comparison the data for four permanent mold cast A356-T6 parts are shown in Figure 1 from the USCAR project [2]. In that project 15 production castings were sectioned for mechanical properties in several different places; typically a designated critical area of heaviest loading and three other regions of lesser importance. In Figure 1, the statistics for the designated area alone are shown on the left while those for all regions averaged together are shown on the right. The whiskers display the ± 3σ range while the bars range from the minimum to maximum values observed. The variation over different regions is particularly interesting to consider in comparison to the results of Table 1.

Table 1 – Mechanical Properties – Ablation Cast knuckle in A356.0-T61

Sample Location	0.2% Offset Yield MPa	Tensile Strength MPa	Elongation to failure %
Arm-1	234	321	13.5
Arm-2	223	310	13.0
Edge-1	230	322	14.3
Under-1	243	332	12.1
Horn-1	237	328	11.3
Horn-2	234	322	12.1
Hub-1	231	315	10.6
Mean ± ($\sigma/\sqrt{n}$) MPa	233 ± 2.7	321 ± 2.7	12.4 ± 0.5
Mean - 3σ	214	300	–

In Table 1 the mean shown for the ablatively cast knuckle was based on 7 samples distributed over various areas of the part, giving the mean 0.2% proof stress as 233.0 ± 2.65 MPa (33.8 ± 0.4 ksi) (Note that the engineering error presented in Table 1 was based on the standard deviation / $n^{1/2}$ where n = 7 in this case.)

The 'mean minus three sigma' 0.2% proof stress here is 214.2 MPa. Comparing this with the results shown in Figure 1, shows that even a first prototype ablative casting was found to be among the "best in class" of competing casting technologies. The UTS compares just as favorably.

Although the minus 3σ values are shown in Figure 1 for the tensile elongation, these are generally espoused in favor of the minimum for design purposes [3]. The 10.6 % minimum observed in Table 1 for the ablatively cast part once again compares favorably to the results shown in Figure 1.

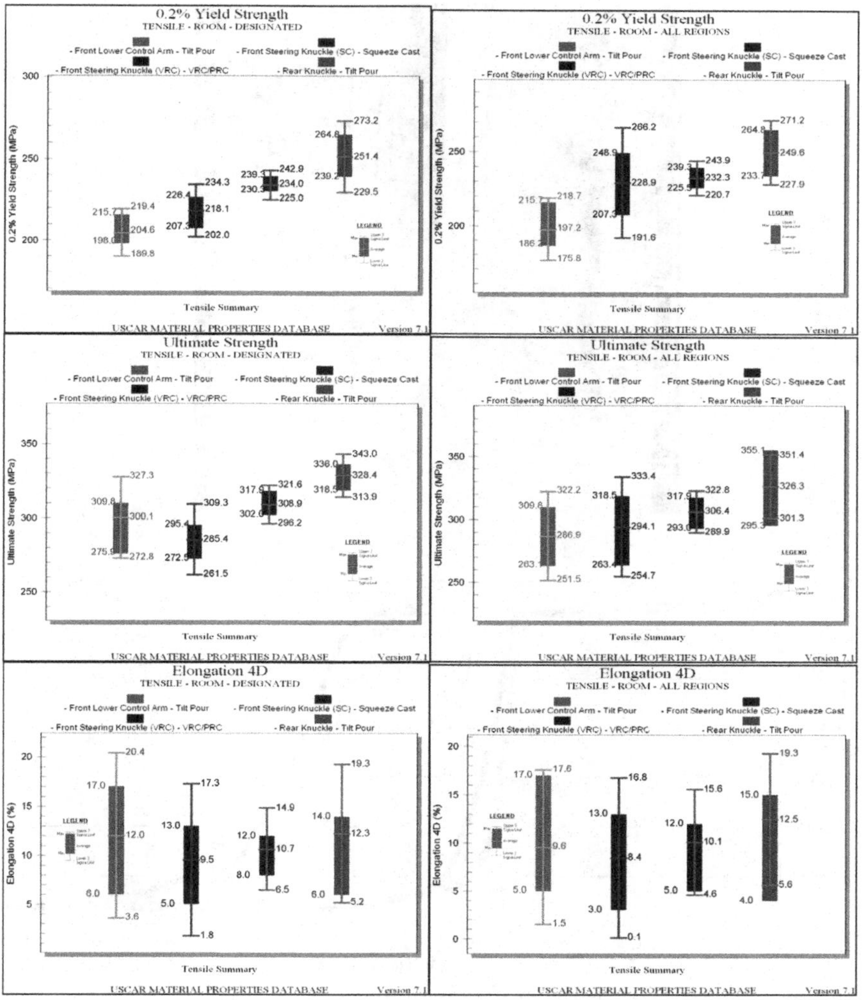

Figure 1. The plots in this figure were generated from the USCAR database [2]. The left hand plots show data from a single designated area of the casting while the right hand plots show data averaged over four different regions of the casting.

We are, in a sense, comparing apples to oranges in that a single ablative casting is being compared to data populations in the case of the USCAR study. The results are nevertheless interesting and significant as obtaining uniform properties throughout a part is more difficult than achieving low variation in a single region of a casting from part to part.

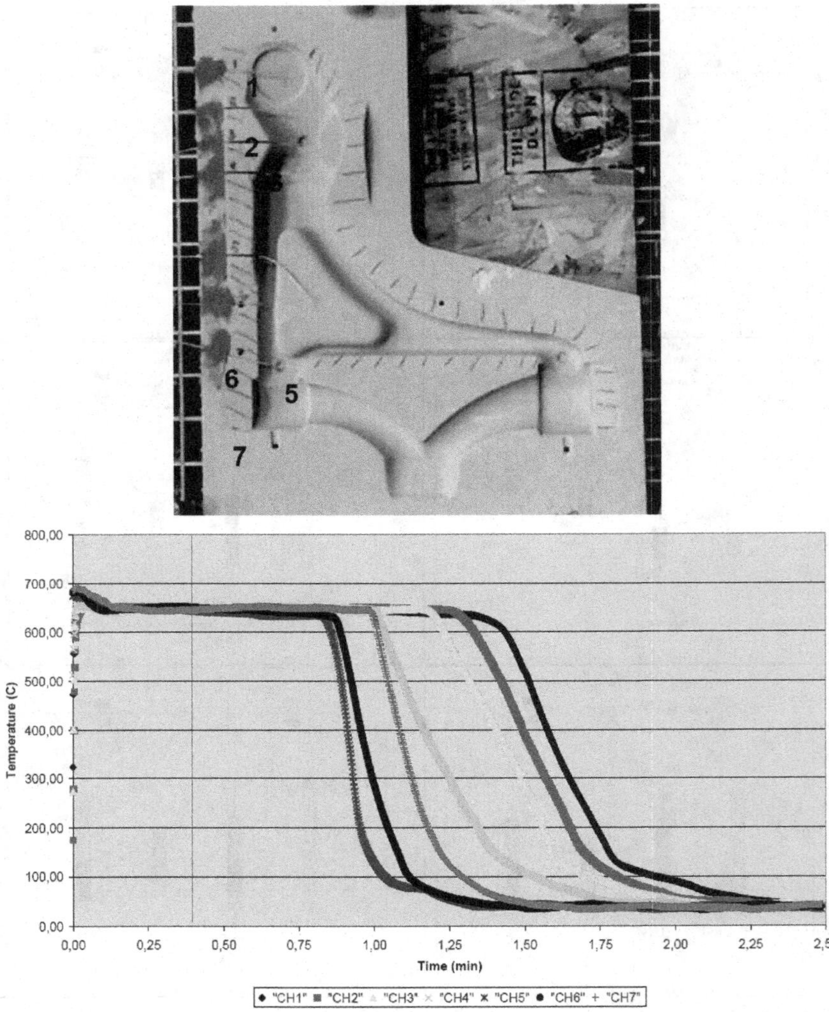

Figure 2. Example cooling curves obtained from an instrumented ablative casting. Note the large temperature differences between adjacent thermocouples at any given time.

The unique feature of ablative casting that allows for uniformity in properties like this is the strong temperature gradient established through the parts being cast. An example set of thermocouple traces is included in Figure 2.

Current Projects

Cored Castings

The properties of cores are proving to be a significant enabler with the new technology. The fact that the inorganic binder has minimal outgassing is a major advantage for complex cores that cannot be vented. Furthermore, the ability to simply 'wash out' even the most complex internal cores avoids the standard problems of thermal decoring where oxygen cannot easily reach some parts of the core, making binder burn-out problematic. An example is a complex valve plate which has only previously been made in steel as no aluminum process technology was available to make it at reasonable cost.

Ablative casting may be an ideal process for engine cylinder heads which contain complex cores and require high properties. The main advantages that can be seen in applying Ablation Casting for cylinder heads can be summarized as follows:

- Improved properties – very fine microstructure due to rapid solidification
- Homogeneous properties without use of chills
- Alloy flexibility (355, 356 and 319. High performance 2xx series alloys and semi-wrought 6XXX or 7XXX compositions)
- Controlled solidification to aid the feeding
- Conventional sand mold making equipment — faster cycle times for cost productivity
- Automatic, fast and low cost mold/core removal (in contrast with costly thermal treatment to demold)
- Possibility of complex cores with easy removal (a cost driver in current semi-permanent mold processes)

Thin and thick walled components

Foundry technologies have their limitations when it comes to wall thicknesses. High pressure die-casting is ideal when walls are between 2 and 5 mm but thicker sections risk shrinkage porosity and poor mechanical properties. Most sand and permanent mold casting technologies are typically not suitable for wall thicknesses below 3 or 4 mm over a large surface without low pressure filling. In castings to date both thin walls (2 mm) and combinations of thin and thick sections have been successfully cast. This is because the mold is not intended to extract heat from the metal (the cooling fluid achieves this in a separate and controlled action). Thus the mold can either be pre-heated or be constituted from an aggregate of low thermal diffusivity selected to extract a minimal amount of heat, thus facilitating the filling of thin sections. Additionally techniques like vacuum or low pressure assisted filling of the mold are easily applied. The directional and progressive solidification towards the feeder at temperature gradients that can reach 60 Kmm^{-1} drives effective feeding of thick sections.

The technology could lend itself to production of all sorts of structural components e.g. automobile inner door panels, engine cradles, body structure parts, etc. The additional possibility to produce many of these components with hollow sections would allow the OEMs to simplify and strengthen structures currently made by the welding together of several separate aluminium components.

Surface finish

Surface finish was initially expected to be a limitation of the aggregate mold technology although improvements to the binder are addressing this issue. Furthermore, an investment casting made by ablation (Figure 3) has also been demonstrated.

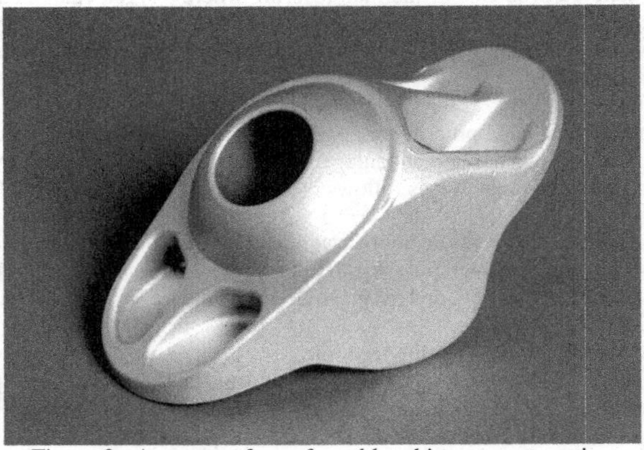

Figure 3. As cast surface of an ablated investment casting

In order to be able to offer good surface finish (at least comparable with permanent mold or possibly even pressure die-casting) a mold coating technique was developed, which promises to be an economical route to success.

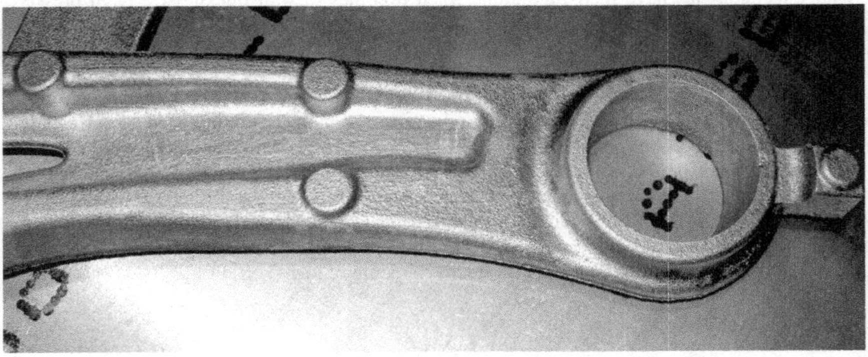

Figure 4. As cast surface of a coated aggregate mold Ablation casting.

Possibilities in Magnesium

Magnesium is not as easily cast as aluminium due to lower heat capacity, causing rapid temperature loss during casting. The use of mold aggregates of low thermal diffusivity is expected to largely counter this problem so that the benefits of ablation should directly extend to Mg-based alloys. Perhaps surprisingly, magnesium alloys have been found to be docile with respect to ablation.

Improvements of the equipment technology

The equipment shown in the first paper was based on a single 'water curtain' from the top which limited mold removal as well as the achievement of high cooling rates in the lower areas and undersides of castings. It also limited the available options for gating and risering the casting as the riser always needed to be the last thing ablated as it as it has an active feeding role during ablation. The latest equipment in development has a large number of individually controlled spray nozzles from top, bottom and both sides which allow ablation to be programmed from any direction (or even from several directions at the same time).

Figure 5. Latest Ablation Casting cell installed at Alumalloy Metalcasting Co., Avon Lake, OH

Conclusions and Future Outlook

The Ablation Casting Technology appears to offer potential for a variety of components and applications in a number of different markets. The main advantages of the technology are the ability to cast complex parts combining both thin and thick sections, together with internal cores. It also appears possible to cast any aluminium or magnesium alloy, and achieve good properties. The technology might also be suitable to cast other metals such as steels. The priority right now is to characterize more aluminium and magnesium alloys and optimize these for the process. For the equipment the priority is to work on improvements in cycle times and therefore overall competitiveness of the process as well as recycling of the sand. Developments are in hand to enhance economy and environmental performance by the recycling of the consumables, the mold aggregate and the water.

References

1. Grassi, J., Campbell, J., Hartlieb, M., & Major, J., "Ablation Casting", in Aluminum Alloys: Fabrication, Characterization, & Applications, Weimin Yin & Subodh K. Das, eds., (TMS, the Minerals, Metals, and Materials Society, 2008), pp. 73-77.
2. USAMP-LMD 110 Project, "Design and Product Optimization for Cast Light Metals", 2001. Available from the American Foundry Society.
3. GMN7152, Specification and Verification of Tensile and Fatigue Properties in Cast Components," GM Engineering Standards (2001).

Shape Casting: The 3rd International Symposium
Edited by: John Campbell, Paul N. Crepeau, and Murat Tiryakioğlu
TMS (The Minerals, Metals & Materials Society), 2009

THE NEMAK COSWORTH CASTING PROCESS - INNOVATION

Glenn Byczynski[1] and Robert Mackay[2]
[1]Nemak Europe GmbH; [2]Nemak Canada;

Keywords: Cosworth, Precision Sand, Mechanical Properties

Abstract

The Cosworth Process is well recognized for its ability to produce high quality, dimensionally accurate aluminum castings. The process was designed from first principles with casting quality as the main focus. Ford transformed this process into a high volume production system in the early 1990's at its Windsor Aluminum Plant and this plant continues to manufacture world-class aluminum cylinder block castings today as Nemak Canada. The one drawback (if any) of the original process is the resultant microstructure due to the relatively slow solidification rate in heavy sections in the sand mould. The secondary dendrite arm spacing combined with typical automotive grade alloys limits the mechanical properties in certain areas of the casting. The innovation discussed in this paper is an augmentation of the original Cosworth Process to include an integral chill that increases local solidification rates and drives casting performance to new levels.

Background

Today cast aluminum alloys are the materials of choice for automotive cylinder block and cylinder heads. They have proven themselves to be a lightweight alternative to the much heavier ferrous based versions of the past. The resultant reduction in vehicle weight remains as one of the largest single improvements to overall vehicle fuel economy. Despite having firmly established themselves in this niche, aluminum castings are poised for yet further growth, notably in the diesel engine cylinder block area where many designs today are still based on iron. The stage has been set therefore for a process/material combination that results in a high performance aluminum solution. The fact is that there are already players on the stage, notable examples include the Daimler V6 diesel[1] and the castings that make up the 2008 international engine of the year: the 2.0L inline diesel from BMW[2]. Both of these cylinder blocks are cast aluminum and both are produced by Nemak's Precision Sand casting plant in Dillingen, Germany. The property challenge for cylinder blocks in particular is met by combining the Precision Sand Process with carefully engineered chilling components contained in the mould. Recently (2006) these same concepts have been applied to the Cosworth process used in Nemak Canada's Windsor Aluminum Plant and significant improvements to mechanical and fatigue strength were achieved.

Traditional Cosworth Process

In the late 1980's Ford Motor Company set out to find a casting process for its new generation of lightweight cylinder head and block applications. A specific task force was assembled to perform a worldwide search for this optimum casting process, on which Ford would build the future of its lightweight casting business. The Cosworth casting process had established its quality and consistency in the realm of high performance racing, but at this time had yet to be proven to meet the demands of a high volume production environment. A unique prototype plant was constructed to validate the Cosworth casting process. Designed to use full-scale production equipment for moulding and casting, Ford's Cast Aluminum Research and Development (CARD) plant was built in 1988 (today this exists as the Nemak Engineering Centre) in Windsor, Canada. After successfully proving the capability of the process, Ford

Motor Company purchased an exclusive worldwide license to use the Cosworth casting process for high volume production of castings. In 1992 construction of the Ford Windsor Aluminum Plant (WAP) was completed. Today the plant has a capacity of approximately 1.3 million castings per year. There are several distinctive aspects of the Cosworth casting process (Campbell 1981). The three fundamental ideas are discussed briefly below.

Minimization of Liquid Metal Damage

Minimization of liquid metal damage is definitively the most important aspect of the Cosworth casting process. The entire liquid metal handling system is designed as not to entrain surface oxides. This starts with the treatment of the metal, whether melted 'in house' or received in the liquid form. The liquid metal is transferred by a simple launder system to the casting furnaces. The launder system is designed to be of the same elevation at all points along the delivery system so that the metal does not 'run' downhill or experience any turbulence during transfer. The size of the casting furnaces that receive the metal are sufficiently large for the metal to remain resident for upwards of ten hours under full production rates. These holding times allow for impurities to settle to the bottom of the furnace or rise to top of the melt depending on their density. The metal used for casting is pumped from the cleanest middle layer of the melt. This is performed using an electromagnetic pump that is partially submerged in the liquid metal bath. The metal is pumped in a controlled manner into a bottom-gated mould, once again in order to minimize metal surface turbulence. A feedback sensor/control system compares the actual level of the aluminum versus a preprogrammed value. The pump voltage (which varies directly with metal flow rate) is then adjusted, to ensure that the mould is filled at the correct rate by the liquid metal.

Mould Rollover

The use of a bottom-gated mould, while advantageous from a filling point of view has one major drawback. The hottest metal (or last metal into the mould) is located lower than the rest of the metal in the mould cavity. This is undesirable for two reasons: first, proper feeding will not take place, as the temperature gradient is unfavourable. Second, convective currents may occur, as the hot metal is inclined to exchange positions with the colder more dense metal. Both of these conditions can lead to a plethora of casting defects. To counter this situation, pump pressure is maintained after the mould has been completely filled and the mould is rotated 180° about the mould/pump interface point. This now allows for the correct temperature gradient and gives the added benefit of being able to separate the mould from the metal supply. This allows for the now full mould to be indexed away and the next casting cycle to begin. Production rates of 72 castings per hour for one four station casting machine are typical. Finally, the casting yield (ratio of final casting weight vs. poured weight) is high compared to other processes because the gating system essentially transforms into the feeding (risering) system after rollover thus making the most of the liquid metal introduced to the mould.

Precision Zircon Sand Mould and Cores

One of the features of the original process is the utilization of zircon sand as a mould material. Unlike silica sand, zircon sand remains dimensionally stable, having a low and linear thermal expansion throughout the temperature range of molten aluminum. An additional benefit is that cores made with zircon sand and phenolic urethane binders have a very similar density to that of molten aluminum. This eliminates any buoyant force that cores of less density experience. These forces can lead to core distortion or even floatation resulting in poor dimensional repeatability of the final casting. Other features include the fact that cores are made at room temperature from dimensionally stable core boxes, and that zircon sand is chemically neutral to most binder systems and to liquid aluminum.

Nemak Cosworth Innovation

The drive for increased casting performance led to the process innovations discussed in this paper. Specifically for a new product the main bearing saddle mechanical and fatigue properties requirements were significantly higher than what was normally achievable in the standard Cosworth process. With fixed alloy parameters the best opportunity involved increasing the solidification rate in the areas of the casting that needed the property "enhancement". Consequently, the main area of development focused on the feasibility of reliably integrating a main bearing saddle chill (crankcase area).

Despite the described advantages of zircon sand, it currently has one overwhelming disadvantage: its purchase price. Since the time of the original Cosworth process there have been several examples of precision sand castings using high quality silica sand as mould material. These designs employed careful attention to design of the smaller cores and printing of these cores in order to minimize any potential distortion issues due to buoyant forces. Considering all of these factors silica sand was deemed to be the best overall material and it was chosen for the upgraded process.

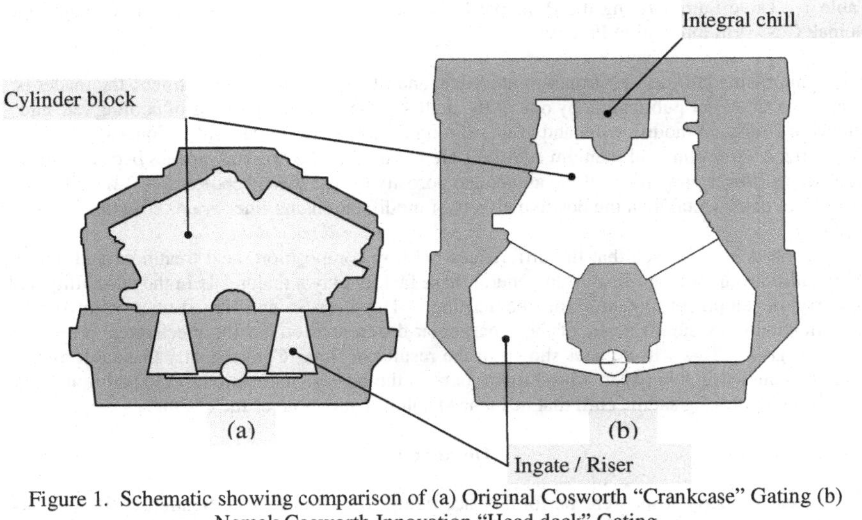

Figure 1. Schematic showing comparison of (a) Original Cosworth "Crankcase" Gating (b) Nemak Cosworth Innovation "Head deck" Gating

The filling orientation of the casting is always dependant on the casting geometry, local wall thicknesses and availability of metal feed-paths. As described earlier one of the fundamentals of the Cosworth process is the unidirectional temperature gradient. In order to maintain this when a chill is applied to the "bottom end" of the cylinder block mould the only alternative is design a gating / risering system that is attached to the head deck of the casting (Figure 1). In a "vee" engine design two banks of risers are necessary for the correct feeding. One clear disadvantage is that this negatively impacts the casting yield when compared with crankcase gated alternative.

Metal Chills	None	Integral crankcase chill	Property Increase
Mould / Core Material	Zircon sand	Silica sand	Cost Benefit
Filling Orientation	"car position" crank case gating	"inverse-car position" head deck gating	Design necessity
Casting Yield	73%	55%	Design necessity
Heat treatment	T5	T7	Property Increase
Alloy	AlSi7Cu4 (0.4 Fe max)	AlSi8Cu3 (0.6 Fe max)	Customer Requirement
Modifier / Grain Refiner	None	Sr / TiB	Improved Castability

Table 1. Table Summarizing the Principal Differences between the Cosworth process and the Nemak Cosworth Innovation Process

Regarding melt additions i.e. strontium modifiers and titanium boron grain refiners, the reader is referred to an earlier publication by one of the authors where the interactions of cooling rate and effects of strontium modification and grain refining are discussed in detail[3]. Concisely summarized: strontium and titanium additions are not used in the original process because the side effects (lower properties) of the associated porosity (in the slowly cooled areas) were found to be more detrimental than the positive effects of modification and finer grains structure.

From Table 1 one can see that the differences in alloy composition, heat treatment and master alloy additions are all significant. In general these factors play a major role in the castability and property development potential of the casting. However it was noted that with similar solidification rates the influence of these particular differences on casting mechanical properties were not great. This will be later shown in the results section. Consequently these differences are not discussed in this paper, whose main focus is the process innovation of the addition of the integral main bearing saddle chill that influenced solidification rate of the casting.

Discussion

Mechanical property tests were performed according to customer specifications. Though not identical, the sample locations and testing procedures are certainly comparable. A summary of the test results from series of mechanical test bars taken from main bearing saddles (30 samples from each process) are shown in Figure 2. These histogram charts show not only the range of values produced from the two processes but also demonstrates a clear separation in the population groups. The samples taken from the chilled process have superior mechanical properties in all cases when considering the main bearing saddle area. In Figure 2(d) the Weibull probability curves for each population are shown. Weibull analysis is useful to demonstrate the reliability or conformity of the data to the Weibull distribution[4]. The value of the slope of a line of best fit is called the Weibull modulus, the greater the slope the smaller the scatter in the data. In these cases correlation coefficients of R^2 >.90 indicate acceptable fit and a Weibull modulus difference of approximately 4-fold indicates a large improvement in the consistency of the values. Secondary dendrite arm spacing (SDAS) and porosity measurements were also carried out.

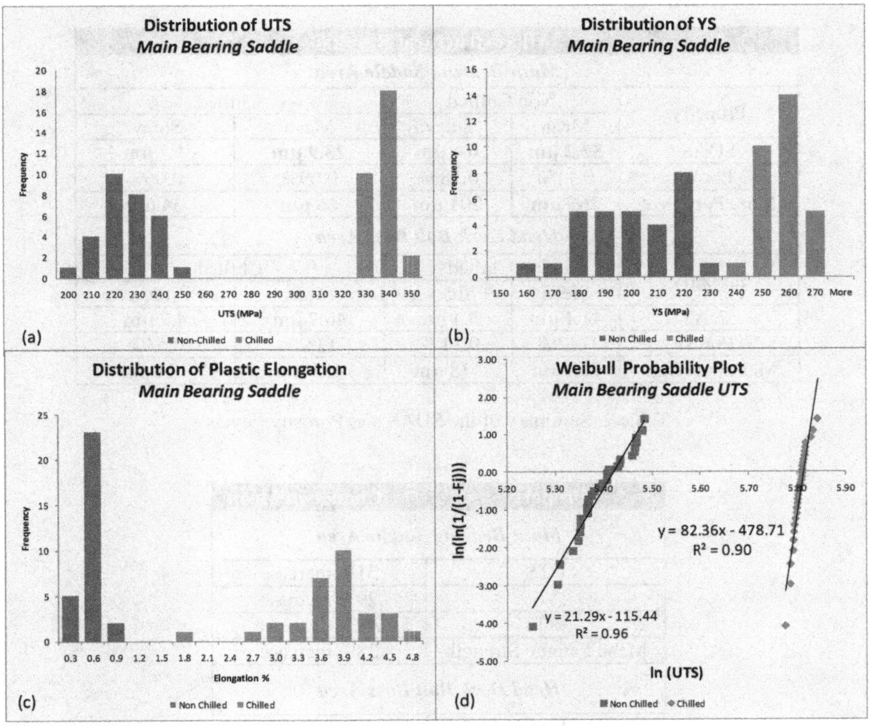

Figure 2. Charts showing the Distribution of Main Bearing Saddle Values for (a) Ultimate Tensile Strength, (b) Yield strength and (c) Plastic Elongation (d) Weibull Probability Plot of UTS

Table 2 summarizes the effects of the increased cooling rate on SDAS and % porosity and maximum pore size. The main influence was of course seen in the main bearing saddle samples that were closest to the chill. SDAS, % porosity and maximum pore size were all significantly reduced in the main bearing saddle samples. Conversely the bolt boss samples on the other hand suffered slightly in the SDAS and significantly when considering porosity. This is due to the proximity of the bolt boss samples to the risers in the head deck gated design. As these values are significantly higher than those seen in the non chilled main bearing saddle (whose SDAS is higher) the authors believe that the strontium and grain refiner additions had a negative influence in accordance with previous observations already discussed [3].

Microstructural Results				
Main Bearing Saddle Area				
Property	Non Chilled		Chilled	
	Mean	Stdev.	Mean	Stdev.
SDAS	**52.2 μm**	6.8 μm	**23.9 μm**	1.7 μm
% Porosity	0.11%	0.18 %	0.09%	0.06%
Max. Pore Size	**265 μm**	201 μm	**66 μm**	34 μm
Head Deck Bolt Boss Area				
Property	Non Chilled		Chilled	
	Mean	Stdev.	Mean	Stdev.
SDAS	41.1 μm	3.4 μm	46.7 μm	4.7 μm
% Porosity	0.12%	0.11 %	1.09%	0.54%
Max. Pore Size	250 μm	118 μm	482 μm	124 μm

Table 2. Summary of the SDAS and Porosity Levels

Property	% Change in Mean
Main Bearing Saddle Area	
UTS	51% increase
YS	29% increase
% El	705% increase
Mean Fatigue Strength	38% increase
Head Deck Bolt Boss Area	
UTS	1% decrease
YS	1% increase
% El	74% decrease
Mean Fatigue Strength	Not tested

Table 3. Mechanical Property Comparison between Casting Processes (Chilled compared with Non Chilled)

Summaries of the percent differences of the mean values of these properties are shown in Table 3. In this comparison the values are compared as referenced to the original process. It is also

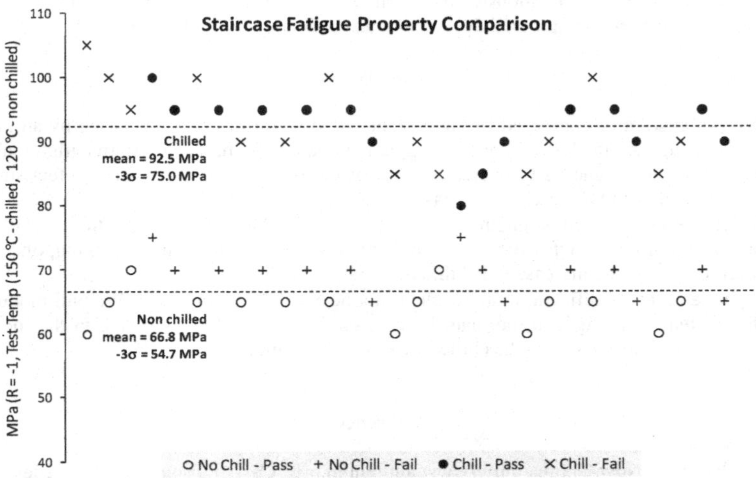

clear here that there are no significant strength differences in the non chilled areas of the casting. The only significant difference is that there is a significant decrease in the plastic elongation in samples from the bolt bosses of the head deck gated process when compared to that of the crankcase gated block.Figure 3. Staircase fatigue performance of Chilled and Non-Chilled Main Bearing Saddle Specimens

In Figure 3. we consider the high cycle fatigue performance of the main bearing saddle samples. This is a plot of the staircase fatigue tests performed on again 30 samples of each process. The staircase or "up and down" method of fatigue testing is a method of assessing the mean fatigue life of an alloy[5]. In brief the technique involves identifying a life cycle goal and testing samples sequentially at a stress level higher or lower (by a predetermined increment) than the previous test depending on whether the previous test specimen reached the life cycle goal or failed short of the goal. The fatigue strength estimate is calculated by taking the mean stress of either the failed or passed samples (whichever is lower in number). The logic behind choosing the outcome with the fewest results to calculate the mean is to minimize the effect of a poor estimate of the starting stress. A poor estimate of starting stress would lead to several tests of the same outcome before the incurring the opposite result. The merits of this technique over standard probit analysis have been summarized elsewhere[6]. The main advantages are that it uses fewer samples than traditional methods and analysis of the data is straightforward. Disadvantages are that the tests need to be sequential and a reasonable amount of insight is needed to determine the initial stress and increment levels. Provided the estimates are reasonable and the samples are homogeneous, convergence on the mean fatigue strength is rather quickly achieved.

Regarding mean fatigue strength we see a significant performance advantage of the chilled version over the non chilled variant. The performance enhancement is nearly entirely due to the reduction in the size of pores and oxide bifilm defects due to the solidification rate increase in the chilled version. The relationship between the size of the defect that forms the fatigue initiation site and fatigue life is well documented in the literature[7,8,9,10,11].

Conclusions

- The combination of controlled filling and increased solidification rate caused by an integral main bearing saddle chill is thought to greatly enhance ultimate tensile strength, elongation and high cycle fatigue performance because of the reduction in size of the defect structure (porosity and bifilms) present in the casting.
- Improvements in yield strength are thought to be related to the traditional Hall-Petch relationship of finer microstructure (grain size) to increased dislocation blocking, once again achieved here by the increase in solidification rate.
- The increase in solidification rate resulted in a nearly 4-fold increase in Weibull modulus of ultimate tensile strength in the main bearing saddle samples. This is due to the increased consistency of defect size in the chilled cast microstructure.

References

1. Aluminum Now Online, http://www.aluminum.org "Cast of Thousands, Mercedes May Make Aluminum diesels the Next Big Thing", July/August 2005.
2. http://www.ukipme.com/engineoftheyear/winners_08/bestnew.html
3. G. E. Byczynski and D. A. Cusinato, "The effects of strontium and grain refiner additions on the fatigue and tensile properties of industrial Al-Si-Cu-Mg alloy castings produced using the Ford Motor Company - Cosworth precision sand process", International Journal of Cast Metals Research, 14 (5): 315-324, 2002.
4. Weibull, W. (1951). "A statistical distribution of wide applicability." Journal of Applied Mechanics **18**: 293-297.
5. Dixon, W. J. and A. M. Mood (1948). "A method for obtaining and analyzing sensitivity data." Journal of the American Statistical Association **43**(241): 109-126.
6. Brownlee, K. A., J. L. Hodges, et al. (1953). "The up-and-down method with small samples." Journal of the American Statistical Association **48**(262): 262-277.
7. G.Byczynski and J.Campbell, "A study of crack initiation sites in high cycle fatigue of 319 aluminum alloy castings", Shape Casting: The John Campbell Symposium, TMS (The Minerals, Metals and Materials Society), 2005, p. 235-244.
8. Couper, M. J., A. E. Neeson, et al. (1990). "Casting defects and the fatigue behaviour of an aluminium casting alloy." Fatigue and Fracture of Engineering Materials and Structures **13**(3): 213-227.
9. Skallerud, B., T. Iveland, et al. (1993). "Fatigue life assessment of aluminum alloys with casting defects." Engineering Fracture Mechanics **44**(6): 857.
10. Nyahumwa, C., N. R. Green, et al. (1998). "The Concept of the Fatigue Potential of Cast Alloys." Journal of the Mechanical Behavior of Materials **9**(4): 227-236.
11. Wang, Q. G., D. Apelian, et al. (2001b). "Fatigue behavior of A356-T6 aluminum cast alloys. Part I. Effect of casting defects." Journal of Light Metals **1**(1): 73-84.

Shape Casting: The 3rd International Symposium
Edited by: John Campbell, Paul N. Crepeau, and Murat Tiryakioğlu
TMS (The Minerals, Metals & Materials Society), 2009

DEVELOPMENT OF AN ALUMINUM ALLOY FOR ELEVATED TEMPERATURE APPLICATIONS

Kumar Sadayappan [1], Gerald Gegel[2], David Weiss [3] and Mahi Sahoo [1]

1 CANMET-MTL, NRCAN, 568 Booth Street, Ottawa, Canada
2 – Material and Process Technologies, Morton, IL, USA.
3 –Eck Industries, Manitowoc, WI, USA

Keywords: Al-Cu alloy, high temperature, heat treatment

Abstract

Non-ferrous light alloys including aluminum and magnesium have maximum effective operating temperatures of about 200°C which is equal to their aging temperatures. While this temperature capability is adequate for most traditional applications, demands from important industrial sectors as well as the military require lightweight alloys that can operate in temperature ranges of 250-300°C. For example, the need to reduce the exhaust emissions of medium and heavy-duty diesel engines has led to the use of two-stage series turbocharger designs for the air system. Single stage compressors run at an outlet air temperature of approximately 175°C at sea level. This is the approximate temperature limit for the currently used cast 354-T61 aluminum alloy impellers. Second stage outlet air temperatures are predicted to reach 260°C or higher at sea level conditions and this temperature will increase with operating altitude. Efforts were made to develop an Al-Cu alloy with scandium addition which is able to retain its strength at 250°C. Details of the alloy development along with results are presented and discussed in this publication.

Introduction

Aluminum alloys have been successfully used in many automotive applications to reduce the weight of the vehicle. Some major uses for aluminum alloys include the engine block and cylinder head castings for gasoline engines and turbine-compressor wheels for diesel engines. The approximate temperature limit for the currently used 354-T61 aluminum alloy impellers is around 175°C. The maximum operating temperatures of most structural aluminum alloys, both cast and forged, is about 200°C which is limited to their aging temperatures. There is little or no history of aluminum alloys use above 200°C. The need for aluminum alloys with improved creep resistance at ever increasing temperatures has intensified the search for alloying elements that produce stable precipitates.

State of Current Technology

The high temperature performance of aluminum alloys could be enhanced by any one or combination of the following three methods; precipitation hardening, ceramic reinforcements and solid solution strengthening. The first approach is the in-situ formation of thermally stable very fine and thermally stable intermetallic phases in the aluminum matrix that will not coarsen at elevated temperatures. The second approach is the addition of small ceramic particulates or fibers, designed to work in concert with each other (using appropriate techniques) to an Al matrix alloy to produce a composite. In the third approach, elements which show complete solubility in the solid state are added to increase the yield and tensile strengths of the solvent, in this case pure aluminum. In the current study we will focus on precipitation hardened aluminum alloys.

Of those precipitation strengthened Al-Cu and Al-Si casting alloys listed in the ASM Metals Handbook few could be used for high temperature applications [1]. However, at temperatures above 230ºC the precipitates present in these alloys coarsen or dissolve rapidly, and transform into more stable phases. It has been shown that these transformations reduce coherency with the matrix resulting in a dramatic reduction of mechanical properties, most specifically ultimate tensile strength and high cycle fatigue strength.

The major task in the development of a new alloy system is to ascertain the thermal stability of candidate strengthening precipitates and then devise an alloy that contains the stable precipitates that increase ultimate tensile strength and fatigue strength at elevated temperatures. The addition of Ni, Co, and Zr to essentially an A206 composition resulted in the Rolls Royce alloy 350 that has increased temperature properties but exhibits poor castability [2].

The second group of aluminum alloys used for high temperature applications are Al-Si alloys [3-8]. Several aluminum-silicon-copper hypereutectic and eutectic alloys used to manufacture cast aluminum engine pistons contain additions of Ti, Zr, V, Cr, etc. to improve both intermediate and elevated temperature properties. These alloys provide adequate tensile and yield strengths at temperatures up to 350°C but do not meet the fatigue and fracture toughness requirements for structural applications. The addition of even higher quantities of V and Zr, in NASA alloy, has shown that the operating temperature can be increased by another 75°C to 100°C over those already available [3]. These alloys appear to be promising but there is some question as to whether the alloys will have sufficient fracture toughness to be used for applications which demand excellent fatigue performance at elevated temperatures.

It is known that many trialuminide particles including Al_3Ti, Al_3Y, Al_3Zr and Al_3Hf are stable at high temperatures [9]. However, most of these binary trialuminide particles are brittle and lack coherency in many aluminum alloys making them unsuitable as strengthening elements. Scandium is reported to form a ductile, coherent Al_3Sc particle in various aluminum alloys including Al-Mg, Al-Zn-Mg and Al-Mg-Li systems. These coherent particles contributed to the high temperature stability by increasing the recrystallization temperature.

Scandium cannot be used with all alloying elements. It is mentioned in the literature that Sc can form W- phase particles when the Cu-Sc ratio exceeds a certain limits [10-13]. An estimate of the Cu and Sc contents that may preclude the formation of W-phase particles was plotted using the data from literature. This plot (Figure 1) predicts the allowable Sc content of aluminum-copper alloys if the formation of 'W' phase is to be prevented. These reactions cause particles to grow in the melt that consume copper and scandium making these solutes unavailable to form stable Al_3Sc precipitates during heat treatment. Similarly, it has been reported that silicon can react with scandium to form primary precipitates making hardening impossible.

Work done by Yu et al., [12] showed that the addition of 0.3% Sc and 0.3% Zr to an aluminum alloy containing 2.2% Cu (the other elements were similar to aluminum alloy 2618) had a positive response to aging. Also elevated temperature mechanical properties of this alloy were improved. In the current work effect of scandium on the mechanical properties of an Al-Cu alloy designed for casting was evaluated.

Experimental Work

A 100 kW push up type induction furnace was used for melting. Either A206 ingots or pure aluminum was used as the starting material for preparing the alloys. Alloying additions were made as master alloys or pure metals. Master alloys used in this work were Al-33% Cu, Al-2% Sc, Al-25% Ni, and Al-33% Mn. Magnesium and silicon were added as pure metals. In each experiment 20kg of selected alloy was prepared. The ingots and selected master alloys were charged in a clay graphite crucible and melted in the induction furnace. The melt was degassed using C_2Cl_6 tablets due to the small melt sizes. Magnesium and scandium (as Al-Sc master alloy) were added after degassing. The composition of the alloy was tested using an optical emission spectrograph before casting operation. Most of the melts were designed as split melts where extra alloy additions were carried out after the first set of experiments. The final composition was evaluated by wet chemical analysis using the ICP method. Permanent mold cast plates (150 x 100 x 12.5 mm) or rods (19 mm diameter) were produced to obtain test coupons.

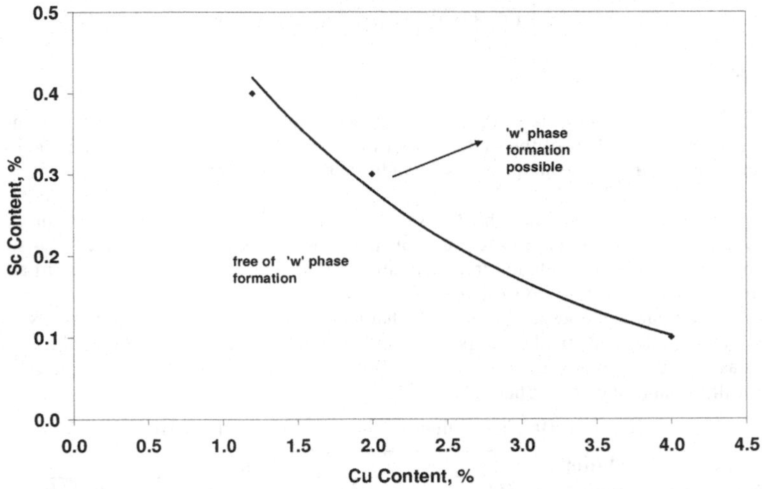

Figure 1. Relation between copper and scandium content on the formation of W-phase precipitates [data from ref 10-12]

The test coupons were subjected to T6 treatment. The solution treatment temperature was varied from 525C to 590C. The ageing temperature was 300°C for different times. Hardness was measured on test coupons to assess the strengthening during the optimization of alloy chemistry and heat treatment. After optimization, some of the selected alloys were soaked at 250°C up to 1000 hrs and their response was measured using hardness testing.

Results

Composition

The compositions of five alloys tested in this investigation are presented in Table I. Alloys 10 and 11 are magnesium free versions while Alloy 15 is silicon free. All the alloys contain titanium which is usually added for grain refinement. These alloys were not refined with Ti-B additions during melting.

Table I. Composition of the alloys prepared

Alloy #	Cu	Mg	Si	Ni	Mn	Other
10	2.2		0.43	0.47	0.48	Sc, Ti
11	2.2		0.78	0.46	0.47	Sc, Ti
12	2.0	0.84	0.80	0.47	0.46	Sc, Ti
13	2.0	0.98	0.47	0.55	0.23	Sc, Ti
15	1.9	0.92		0.52	0.48	Sc, Ti

Heat Treatment

The first three alloys, numbered 10, 11 and 12, were solution treated at 525°C and aged at 300°C. The hardness values of the samples after different ageing times are presented in Table II and graphed in Figure 2. The hardness results indicate the following:

- Alloys 10 and 11 have very low hardness in as-cast condition. On the other hand alloy 12 has high hardness. This is due to the effect of magnesium which provides some solid solution strengthening. After the solution treatment all three alloys exhibit softening but still alloy 12 is stronger than the other two variations.
- The ageing treatment does not improve the hardness of alloys 10 and 11 significantly. On the other hand, alloy 12 exhibits a peak hardness value at shorter duration which rapidly decreases as the ageing time is increased. The final hardness of all the alloys is similar or lower than that of the as-cast base alloy.

Table II. Hardness values of alloys 10, 11 and 12 (BHN)

Alloy #	As cast	Solution treated	Aged										
			15 min	1 hr	2 hr	3hr	4 hr	5 hr	6 hr	7 hr	8 hr	9 hr	10 hr
10	62	48	57	61	62	63	63	64	63	65	66	64	65
11	59	53	56	57	58	57	57	58	60	62	57	54	57
12	81	68	101	77	73	68	68	71	69	71	67	68	71

As it can be observed the best performance is that of Al-Cu-Mg-Sc alloy. However the final hardness observed is very low. One possibility for the very low hardness after ageing is unavailability of scandium for precipitation during ageing treatment. The solubility of Sc in aluminum increases with solution temperature and reaches 0.2% at about 600C. So far the heat treatment in this investigation was restricted to 525°C due to the higher copper content. Hence it was decided to increase the solution temperature. This decision makes it important to keep the copper content to less than 2%. All other elements were maintained at the same level.

Alloys 13 and 15 contain 2% copper but were solution treated at 560 and 590°C instead of 525°C. Alloys treated at 560°C were designated as 13 and 15 and those solution treated at 590°C were designated as 13a and 15a. After this high temperature solution treatment these alloys were aged at 300°C as before. The hardness values are reported in Table III.

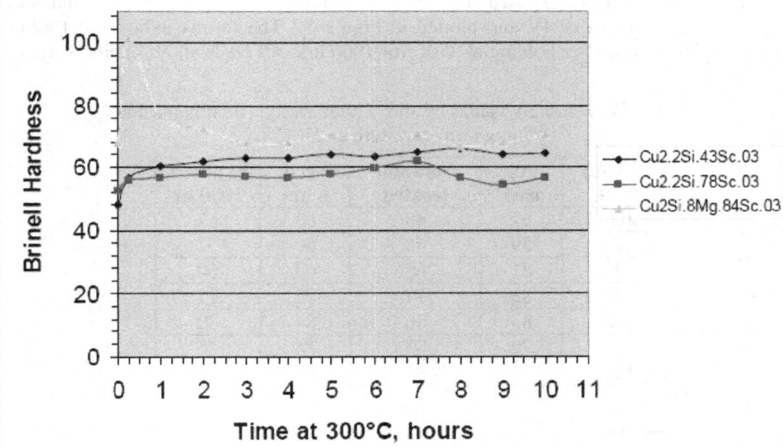

Figure 2. Ageing response of alloys 10, 11 and 12 with 2.2% copper solution treated at 525C

Table III. Hardness values of alloys 13 and 15 solution treated at 560°C and 590°C

Alloy #	As cast	Solution treated	Aged 15 min	Aged 2 hr	Aged 6 hr
13	83	77	113	90	77
13a	83	56	96	90	84
15	79	83	76	79	77
15a	79	81	82	79	80

The hardness test results indicate the following:

- Solution treatment at 590°C results in alloys with softer matrix compared to the alloys solution treated at 560°C.
- Hardness of alloy 13 increases after ageing treatment but the increase is not very significant for alloy 15.

- o Hardness reaches a peak just after 15 minutes of ageing treatment for alloy 13 but longer holding times resulted in reduced hardness.
- o Alloy 13a is much more stable than other alloys. Even though the hardness after solution treatment of alloy 13a is low compared to other alloys, the hardness is higher after 6 hours of ageing and it stabilizes much rapidly.

Long Term Exposure

After optimization of heat treatment cycle, all the alloys were subjected to long term exposure test at 250C. Samples were exposed for 100, 250 and 1000 hrs. The hardness values of the samples are reported in Table IV and plotted in Figure 3. The results indicate that alloy 13a retains the hardness even after holding at 250C for 1000 hrs. All other alloys exhibit softening to various degrees.

Table IV. Hardness values of alloys after various treatments and long term exposure at 250C

Alloy #	As cast	Solution treated	Aged, 6 hrs	Exposed, 1000 hr
10	62	48	63	62
11	59	53	60	57
12	81	68	69	60
13	83	77	77	71
13a	83	56	84	80
15	79	83	76	75
15a	79	81	80	75

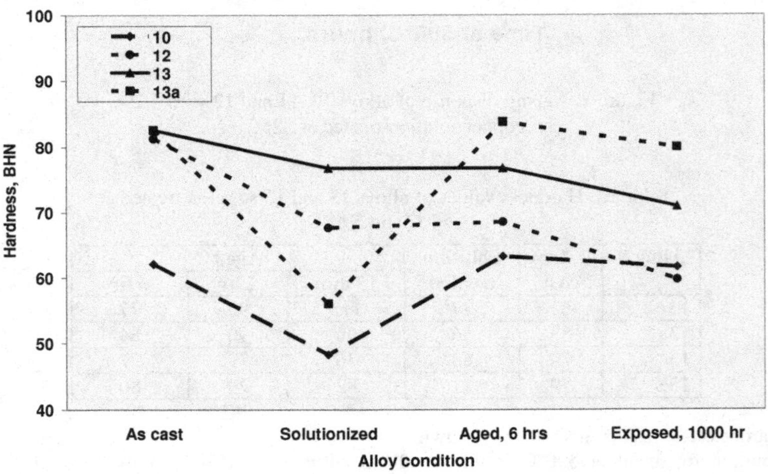

Figure 3. Hardness of alloys after various treatments and long term exposure

Tensile Properties

Tensile testing was conducted for alloy 13a only in the heat treated condition. Some of them were soaked at 250°C for 100, 500 and 1000 hrs before testing. Tensile testing was carried out at room temperature and 250°C. The test specimens contained considerable amount of porosity. Hence only yield strength values are reported in Figure 4.

The results indicate that the strength of the alloys decreases after exposure to high temperatures. The drop in room temperature strength is nearly 25% after 100 hours of exposure. The reduction is less rapid beyond this and after 1000 hours of exposure the reduction in strength is only 35%. The high temperature strength of the alloy suffers more when exposed to high temperatures. Even before exposure the high temperature strength of the alloy was reduced by 20%. Nearly 50% of the strength is lost after an exposure time of 500 hours. After this the reduction is not significant.

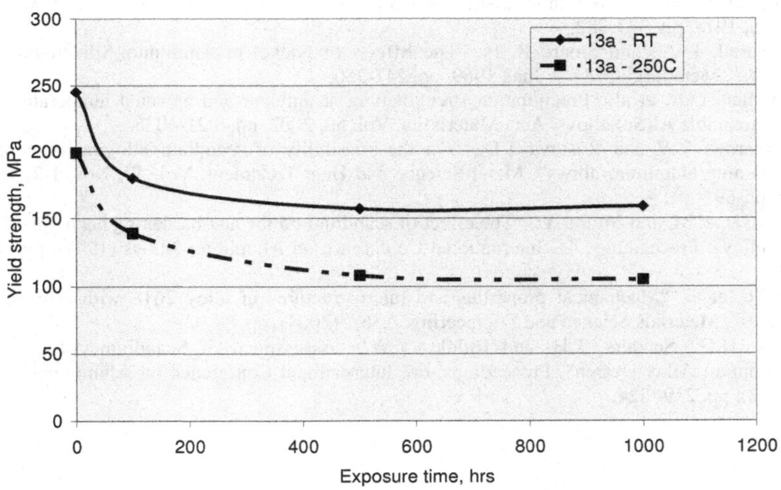

Figure 4.Tensile properties of alloy 13a at room temperature and 250C.

Summary

1. An aluminum alloy containing copper, magnesium, silicon, nickel, manganese and scandium as alloying additions was developed.
2. A new heat treatment schedule was developed as the part of the project. The alloys were solution treated at 590C and aged at 300C.
3. The yield strength of this alloy is 3.6 times higher than those of wrought aluminum alloy 2618-T6. When compared to a cast aluminum alloy (354-T6) generally used for turbine compressor wheels, the current alloy property is 2 times higher.

References

1. ASM Handbook Vol. 2 Properties and Selection: Nonferrous Alloys and Special-Purpose Materials, Published by ASM International, 1990, pp 125, 173-175.
2. Aluminum alloy RR350, Private Communication.
3. NASA 388 Aluminum alloy, Material Properties Data Sheet, Published by NASA-Marshall Space Flight Center, Metallic Materials & Processes Group (ED33), Huntsville, AL 35812.
4. Granger, D. A., Truckner, W. G. and Rooy, E. L., "Aluminum Alloys for Elevated Temperature Application", AFS Transactions, Vol. 94, 1986, pp. 777-784.
5. Crepeau, P. N., Antolovich, S. D., and Worden, J. A., "Structure-Property Relationships in Aluminum Alloy 339-T5: Tensile Behavior at Room and Elevated Temperature", AFS Transactions, Vol. 98, 1990, pp. 813-822.
6. Gundlach, R. B., et al., "Thermal Fatigue Resistance of Hypoeutectic Aluminum Silicon Casting Alloys", AFS Transactions, Vol. 102, 1994, pp. 205-223
7. Komiyama, Y., Uchida, K., and Gunshi, M., "Effects of Fe, Ni, Mn,, Cr, and Mg on Properties of Ni free aluminum casting alloy for piston", J. Japan Inst. Light Metals, Vol. 28, No. 8, 1978, pp. 377-382.
8. Catherall, J. A., and Smart, R. F., "The Effects of Nickel in Aluminum-Silicon Eutectic Alloys", Metallurgia, 78-79, June 1969, pp. 247-250.
9. Seidman, D.N. et al, "Precipitation strengthening at ambient and elevated temperatures of heat treatable Al(Sc) alloys, Acta Materiallia, Vol. 50, 2002, pp.4021-4035.
10. Zakharova V.V and Rostova T.D., "On the possibility of scandium alloying of copper-containing aluminum alloys", Metal Science and Heat Treatment, Vol. 37, Nos. 1-2, 1995, pp. 65-69.
11. Nakayama, M. and Miura, Y., "The effect of scandium on the age-hardening behavior of Al-Cu alloy", Proceedings, 4th. International Conference on Aluminum Alloys (1994), pp. 538-545.
12. Li Yu, et al, "Mechanical properties and microstructures of alloy 2618 with $Al_3(Sc, Zr)$ phases", Materials Science and Engineering, A368, (2004), pp. 88-93.
13. Paris, H.G., Sanders, T.H. and Riddle, Y.W., "Assessment of Scandium Additions in Aluminum Alloy Design", Proceedings, 6th. International Conference on Aluminum Alloys (1998), pp. 219- 224.

Shape Casting: The 3rd International Symposium
Edited by: John Campbell, Paul N. Crepeau, and Murat Tiryakioğlu
TMS (The Minerals, Metals & Materials Society), 2009

Controlled Diffusion Solidification (CDS): Conditions for Non-Dendritic Primary Aluminum Phase in Al-Cu Hypo-Eutectic Alloys

Abbas A. Khalaf, Peyman Ashtari and Sumanth Shankar

Light Metal Casting Research Centre (LMCRC)
McMaster University, Hamilton, ON Canada L8S 4L7

Keywords: CDS, SSM, Rheocasting, Al Alloy

Abstract

Controlled Diffusion Solidification (CDS) is a novel process wherein specific Al alloys can be cast by mixing two alloys of specific compositions and temperature and subsequently casting the resultant mixture. Contrary to conventional casting, the microstructure of the resultant casting in CDS is non-dendritic. In this study, a hypothesis is proposed to describe the critical details of the mixing two pre-cursor alloys and subsequent solidification process. Al – 4.5 wt% Cu is used as an example alloy system to explain the mechanisms.

Introduction

Controlled Diffusion Solidification (CDS) is a novel process wherein a non-dendritic microstructure is obtained by controlled mixing and solidification of two precursor liquid alloys into a cast component [1-2]. The process enables the casting of Al wrought alloys into near net shaped components. The process aims to improve the mechanical properties and performance of the castings while reducing the cost of the casting process.

The primary Al phase in castings made by conventional processes such as with sand and metal moulds has a dendritic morphology. Al based wrought alloys could not be easily cast into shaped components by conventional processes because they are prone to hot tearing wherein the inter-dendritic liquid at the end of solidification cannot effectively feed the shrinkage cavities created by the solidification of the primary dendrites due to the large and complex dendritic network that is created. Al based wrought alloys can be cast with the CDS process because of the absence of a large and complex dendritic network and thus alleviating the hot tearing tendencies.

CDS is a method of Semi Solid Metal Processing (SSM), specifically rheocasting. The advantages of the SSM process are low gas porosity, low pouring temperature, longer metal mould life and improved properties and performance of the cast part. There have been significant advances in the field of SSM in the last thirty years. There are typically two processing routes in SSM: *Rheocasting* and *Thixoforming* [3-9]. In rheocasting, the liquid metal is cooled to a specific temperature in the semi-solid region and subsequently held for a while to homogenize the temperature and the microstructure of the primary non-dendritic phase before casting into a shaped component by a pressure induced process. At the liquidus temperature during cooling of the liquid, it is ensured that there is copious nucleation of the primary phase and forced convection to spread the nuclei evenly in the melt to enable a non-dendritic microstructure.

Recently, there have been various commercial rheocasting processes developed for Al alloys [3]. MIT, UBE rheocasting, THT, CRP and SLC are a few examples of commercial rheocasting processes [3, 9-10]. The most time and energy consuming step in all the currently available commercial processes for rheocasting is the holding period at the semi-solid stage to create a billet for casting. This step in the processes can take a few minutes for the temperature and the microstructure to homogenize. The CDS process circumvents this problem by creating a thermally homogeneous alloy at around the liquidus temperature with copious nucleation of the primary Al phase and subsequently casting the alloy into shaped components [1, 11-12]. Moreover, the CDS process enables the casting of the Al based wrought alloys into shaped components. The 2XXX, 3XXX, 4XXX, 5XXX and 7XXX family of Al wrought alloys have already been shown to produce a non-dendritic microstructure in the CDS process [1, 11]. The 2XXX alloy has also been used to commercially cast an ABS brake housing component for automotive application [12]. Apelian et al proposed a theoretical framework for evolution of the non-dendritic microstructure during CDS [13]. The framework is based on the free energy evaluation from thermodynamic modeling of the alloy phase diagram. One critical assumption in the theory is that the nucleation of the primary phase occurs from the resultant mixed alloy.

In this study, a hypothesis is presented to describe the favorable thermal and solute fields during mixing of the two pre-cursor alloy and the favorable solute and thermal profiles ahead of the growing primary phase during subsequent solidification of the resultant alloy mixture.

Mechanism of CDS: A Hypothesis

The CDS process is described below:

STEP (1) Two pre-cursor alloys are chosen at specific concentrations of the solutes such that they yield the desired composition of the final alloy when mixed in a specific mass ratio.

STEP (2) The temperatures of the two pre-cursor alloys are chosen such that their melt superheat temperatures are close to the respective liquidus temperatures and upon mixing, the temperature of the resultant alloy mixture does not have a high superheat temperature from its liquidus temperature.

STEP (3) The pre-cursor alloy with the lower thermal mass (lower temperature and lower mass) is taken in a container.

STEP (4) The pre-cursor alloy with the higher thermal mass is poured into the same container on top of the lower thermal mass.

STEP (5) The resultant alloy mixture is cast into the shaped component with or without the assistance of pressure.

The nomenclature used in this publication to define the various alloys and parameters are given below:

Alloy 1	Pre-cursor alloy with higher thermal mass (higher temperature and higher mass).
Alloy 2	Pre-cursor alloy with lower thermal mass.
Alloy 3	Resultant mixed alloy.
T_{L1} T_{L2} and T_{L3}	Liquidus temperature of Alloy 1, Alloy 2 and Alloy 3, respectively.
T_1, T_2 and T_3	Melt Temperature of Alloy 1, Alloy 2, and Alloy 3, respectively.
T_L and T_U	Temperature of lower and upper thermocouple in experiments, respectively
m_1, m_2 and m_3	Mass of Alloy 1, Alloy 2 and Alloy 3, respectively.
m_r	Mass ratio of Alloy 1 and Alloy 2 (Alloy 1 : Alloy2).

t_{mix}	Time taken for mixing Alloy 1 and Alloy 2.
C_1, C_2 and C_3	Average solute composition in Alloy 1, Alloy 2 and Alloy 3, respectively.
C^*	Solute composition in the liquid at the end of solute field equalization.
S/L	Solid – Liquid interface.
G	Temperature gradient in the liquid.
ξ	Position of S/L interface

The following are the stages in the CDS process starting from the mixing of the two pre-cursor alloys and ending with the solidification of the resultant mixture alloy.

Figure 1 shows a schematic of the temperature and time graph that is observed during the CDS process. Figure 1 is drawn for a thermocouple that is immersed in the Alloy 2 at the beginning of the process.

There are three main stages in CDS process as shown in Figure 1 by line segments AB, BCD and DE. Point A and E denote the beginning of the mixing process and the end of solidification, respectively. The events in the three stages are presented below and the events are all described in reference to Figure 1.

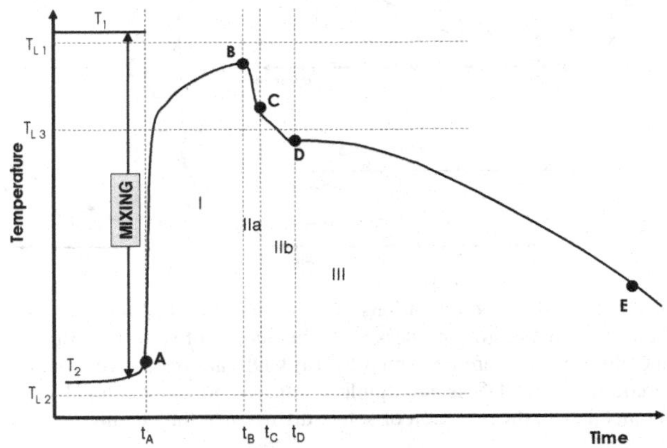

Figure 1: Schematic of the typical thermal profile observed in the CDS process. There are three stages starting from the mixing of the alloys at point A.

<u>Stage I ◊ Segment AB</u>

In this stage, the Alloy 1 is mixed into Alloy 2. The following are the events occurring in Stage I.

1. The mechanical mixing starts at A and ends at point B.
2. Nucleation of primary Al phase begins from Alloy 1. Nucleation is higher at bottom of the crucible and decreases towards the top.
3. Heat added to the system until a maximum temperature of mixing is reached at B is as follows:-
    - Enthalpy of Alloy 1.
    - Enthalpy of Mixing Alloy 1 and Alloy 2.
    - Enthalpy of fusion from nucleation.
4. The temperature and solute fields in the mixture do not fully equalize between times t_A and t_B. Further, the temperature at the bottom of the crucible is highest and there is a temperature gradient from the bottom to the top of the crucible where the temperature is lowest.
5. At B, pockets of Alloy 1 and Alloy 2 exist.

Stage II ◊ Segment BCD

1. Equalization of all temperature fields in the melt, especially the gradient from the bottom to the top of the mixture as described in event 4 of Stage I. The temperature equalization takes place in the segment BC and denoted as Stage IIa in Figure 1.
2. Simultaneous heat loss from the mixture due to ambient conditions.
3. Growth of the nuclei of primary Al phase from Alloy 1 formed in Stage 1. Figure 2(a) shows the compositional and temperature fields ahead of a growing nucleus at the end of stage I at point B. Figure 2(b) is the compositional and temperature fields observed ahead of a growing nucleus in conventional casting process.

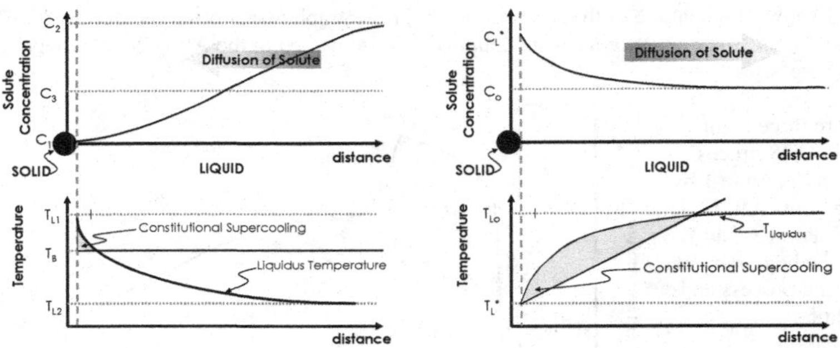

(a) (b)

Figure 2. Solidification conditions in the liquid ahead of the growing primary Al phase in CDS and conventional casting process at the beginning of solidification. The solute concentration and temperature profiles are presented for (a) CDS and (b) conventional casting. The constitutional supercooling for CDS is negligible compared to conventional casting process. The critical difference lies in the direction of solute diffusion at the beginning of solidification.

The main difference between the growth in stage II of the CDS process and conventional casting process is that the solute diffusion occurs towards the growing solid phase in the former and away from the growing solid phase in the latter. This leads to a solute concentration profile is stage II of CDS opposite in nature to that observed in a conventional casting process. The trace of the liquidus and temperature gradient of the melt in CDS and conventional casting is shown in Figures 2(a) and 2(b), respectively. It can be observed that the constitutional supercooling is barely available in the liquid ahead of the solidifying phase in CDS whereas in conventional casting the constitutional super cooling is much larger. Hence, at point B in the CDS process, the solid phase will begin to grow with a planar front in stage II of CDS and dendritic front in conventional casting process.

Figure 3 shows the concentration and temperature profiles in the liquid ahead of the solidifying phase in Stage II of CDS process in the line segment BCD. Between point B and D the solute atoms will diffuse towards the Solid/Liquid (S/L) interface until the solute atoms equalize sometime at or before time t_D. During this time the temperature gradient ahead of the S/L interface is shown by the lines G in the temperature profile of Figure 3. Until point C the temperature gradient ahead of S/L interface is zero and between C and D a low but positive temperature gradient develops. The constitutional supercooling in the liquid ahead of the S/L interface between B and D is defined by the rate at which the relationship between the temperature gradient, G and the liquidus temperature of the liquid at the S/L interface. Equations (1a), (1b) and (1c) show the three possible relationships between G and the liquidus temperature.

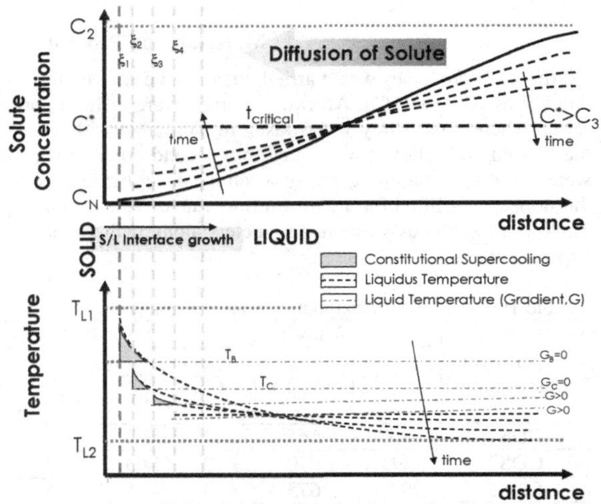

Figure 3: Schematic of the solute composition and temperature fields in the liquid ahead of the growing solid phase in Stage II of CDS.

T_L in Equation 1 is the liquidus temperature. The left hand terms in Equation 1 is effected by the diffusion rate of the solute atoms in the liquid (solute equalization rate) and the right hand term is effected by the solidification rate of the casting process (heat extraction). Equation 1(a) can be observed in a slow solidification rate casting processes (sand cast or an alloy wherein the diffusion coefficient of the solute in the liquid is small. Equation 1(c) can be observed in a casting process with high solidification rate (die casting) or an alloy with high diffusion coefficient of the solute atoms in the liquid.

$$\frac{d(T_L)_\xi}{dt} \leq \frac{d(T)_\xi}{dt} \quad \Diamond \quad \text{Globular Morphology of Primary Phase} \qquad (1a)$$

$$\frac{d(T_L)_\xi}{dt} \overset{(\text{Marginally})}{>} \frac{d(T)_\xi}{dt} \quad \Diamond \quad \text{Rosette Morphology of Primary Phase} \qquad (1b)$$

$$\frac{d(T_L)_\xi}{dt} >> \frac{d(T)_\xi}{dt} \quad \Diamond \quad \text{Dendrite Morphology of Primary Phase} \qquad (1c)$$

It is preferred that the solute equalization shown by C^* in Figure 3 occurs at or close to the point D. Subsequent to solute equalization, the profile of the solute and the temperature will begin to look similar to that of the conventional casting process shown in Figure 2(b) and the main difference being that the amount of nucleation in CDS is far greater than that in the conventional casting process.

Stage III – Segment DE

At point D, there are innumerable nuclei and non-dendritic solid phase distributed evenly in the liquid (if Equations (1a) and (1b) are in effect), a nearly constant solute field and a low to negligible temperature gradient in the liquid ahead of the S/L interface. This scenario is similar to that found in conventional rheocasting process and will yield a non-dendritic primary phase morphology in the cast microstructure [9-10, 14].

Materials and Procedure

All the CDS processes were carried out with the Al-Cu alloy system with Alloy 1 as pure Al and Alloy 2 as Al-33wt%Cu. All raw materials were of high purity. Figure 4 shows the experiment set-up wherein the Alloy 1 was taken in a crucible with a hole (9mm diameter) and a stopper at the bottom and Alloy 2 was taken in a second crucible with two thermocouples. Temperatures were monitored continuously and when the desirable temperatures T_1 and T_L (=T_2) were reached, the stopper was lifted for a uniform flow rate of Alloy 1 mix with Alloy 2 in the bottom crucible. Table 1 shows the experiment parameters along with the designations used to refer to the various experiments.

Table 1: Experiment Designations and Parameters

Experiment Designation	Alloy 1			Alloy 2			m_r
	Mass (g)	T_1 (°C)	T_{L1} (°C)	Mass (g)	T_2 (°C)	T_{L2} (°C)	
CDS1	290.4	665	649	47.1	555.4	547	6
CDS2	293	670	649	47.3	554	547	6
CDS3	292.1	675	649	47.5	555	547	6
CDS4	291.1	683	649	47.3	553	547	6
CC663	Conventional Solidification at T_{pour} = 663 °C						

In CC663 (conventional solidification), the resultant alloy obtained from each of the experiments CDS1 to CDS6 were re-melted and poured (at 663 °C) into a empty crucible at 555 °C and with two thermocouples as shown by the setup in Figure 4.

Evidences Substantiating CDS Mechanism

In this section, the evidences from experiment results will be presented to substantiate the hypothesis for the mechanism of CDS process presented earlier.

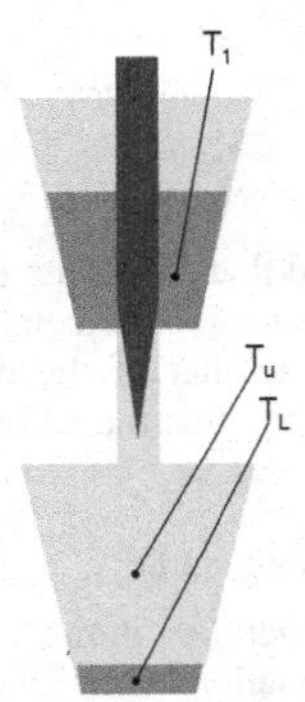

Figure 4. Experiment Set-Up

Figure 5(a) shows the typical thermal analysis data during solidification for experiments CDS2 and CC663 given in Table 1. The various state points of the three stages shown in Figure 2 can be observed in Figure 5. Point A shows the start of mixing Alloy 1 in Alloy 2, point B is the maximum temperature reached at the end of mixing, point C is the end of thermal equilibration in Alloy 3. It can be observed that the thermal data for CC663 reaches a higher temperature than CDS2 showing that the Alloy 2 plays a critical role in the mixing stage. Further it can be observed in the inset of Figure 5(a) that the nucleation of primary phase in CC663 occurs at T_{L3} and the final nucleation of the CDS2 occurs at T_{C*} (point D) showing evidence of Al nucleation in CDS2 during mixing stage (before point D) and thus increasing the average composition of the inter-dendritic liquid at point D (see Figure 3). Figure 5(b) shows the typical thermal data during solidification for T_L and T_U where in the thermal equilibration can be observed at point C. In Figure 5, the time interval for segment AB is about 2 s, BC is about 0.6 s and CD is about 0.6 s as well. Figure 6 shows thermal data superimposed on the rate of change

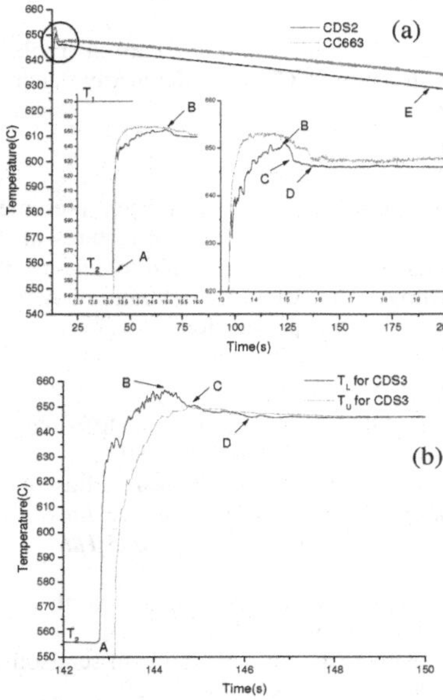

Figure 5. Typical thermal analysis data during solidification. (a) CDS2 and CC663 showing the details of the states in the three stages as shown in Figure 1 and (b) typical profiles of T_L and T_U in CDS3 showing the mixing stage and the point C where the thermal fields in the crucible equilibrate.

of mass ratio, Cu concentration and instantaneous liquidus temperature during the mixing process given by line segment AB in Figure 1 and 5. We observe that t_{mix} = 2 s and the instantaneous liquidus temperature between 0 and 2 s is always less than the alloy mixture temperature. However, T_{L1} is always greater than the instantaneous temperature; thus, facilitating nucleation of primary Al phase from Alloy1.

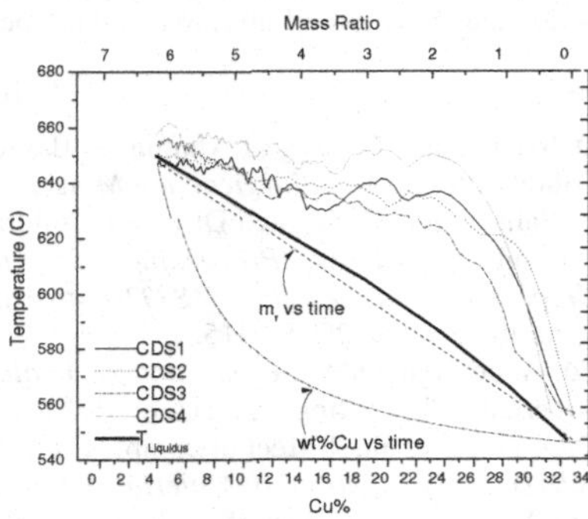

Figure6. The thermal data of all the experiments in Table 1 superimposed on the rate of change of mass ratio, solute concentration and the instantaneous liquidus temperature of the mixture.

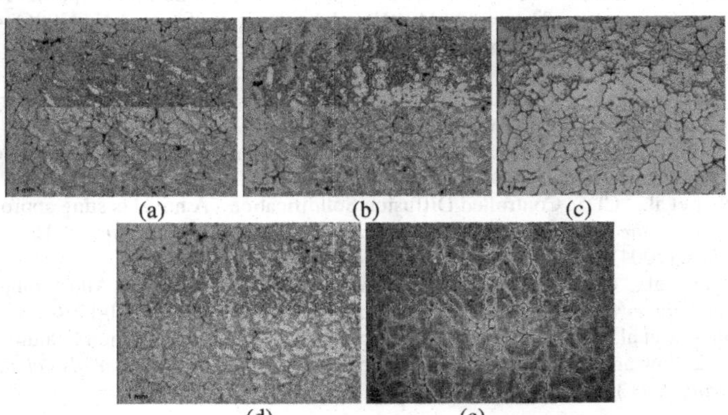

Figure 7. Typical microstructures. (a) CDS1, (b) CDS2, (c) CDS3, (d)CDS4 and (e)CC663.

Figure 7 shows typical microstructures (optical) obtained from the experiments in Table 1. Figure 7 shows the gradual change from the globular structure in CDS1 to the nearly dendritic structure in CDS4 and the dendritic structure in CC663.

Summary

A viable hypothesis to describe the mechanism of the CDS process has been presented. Evidences from experiments have been presented to substantiate certain critical features of the hypothesis. Owing to space constraints, further experimental evidences gathered to quantify the effect of critical parameters such as temperatures of Alloy 1 and Alloy 2, mass ratio of Alloy 1 and Alloy 2 and the rate of mixing could not been presented in this publication.

References

1. D. Saha et al., "Casting of Aluminum-Based Wrought Alloys Using Controlled Diffusion Solidification," *Metallurgical and Materials Transaction A*, 35 (2004),2147-180.
2. D. Saha et al., "Controlled Diffusion Solidification – Maufacturing Net Shaped Al Based Wrought Alloy Parts," *Proceedings of the John Campbell Honorary Symposium Edited by P. Crepeau and Tiryakioklu, TMS (The Minerals, Metals & Materials Society), 2005 TMS Annual Meeting*, (2005), 415.
3. Ed. A. de Figueredo , *Science and Technology of Semi-Solid Metal Processing*, (North American Die Casting Association (NADCA), 2001), Rosemont, IL, USA.
4. Q.Q. Zhang et al., "Effect of compression ratio on the microstructure evolution of semisolid AZ91D alloy," *Journal of Materials processing Technology*, 184 (2007), 195-200.
5. J.G. Wang et al., "Effect of predeformation on the semisolid microstructure of Mg–9Al–0.6Zn alloy," *Materials Letters*, 58 (2004),3852-3856.
6. Y. Birol, "A357 thixoforming feedstock produced by cooling slope casting," Journal of Materials processing Technology,May,186, n1-3 (2007), 94-101.
7. Q.D. Qin et al., "Semisolid microstructure of Mg2Si/Al composite by cooling slope cast and its evolution during partial remelting process," *Materials Science and Engineering*, A444(2007), 99-103.
8. P.K. Seo, D.U. Kim, and C.G. Kang, "Effects of die shape and injection conditions proposed with numerical integration design on liquid segregation and mechanical properties in semi-solid die casting process ,"*Journal of Materials Processing Technology*,176(2006), 45-54.
9. H.V. Atkinson, "Modelling the semisolid processing of metallic alloys ,"*Progress in Materials Science*, 50 (2005), 341-412.
10. D. Apelian, Personal Communications, Metal Processing Institute (MPI), Worcester Polytechnic Institute (WPI), Worcester, MA, USA, (2007).
11. D.Saha et al., "Controlled diffusion solidification - Manufacturing quality net shaped Al based wrought alloy parts," *JOM*, 56(2004), 302.
12. S. Shankar et al., "CDS: Controlled Diffusion Solidification - A novel casting approach," *Proceedings of the 8th International Conference on Semi-Solid Processing of Alloys and Composites*, (2004), 629-636.
13. D. Apelian et al., "CDS Method for Casting Aluminum-Based Wrought Alloy compositions: *theoretical framework," Materials Science Forum*, 519-521(2006), 1771-1776.
14. B.K.Dhindaw et al., "Microstructure development and solute redistribution in aluminium alloys under low and moderate shear rates during rheo processing," *Materials Science and Engineering* A 413-414(2005),156-164.

Shape Casting: The 3rd International Symposium
Edited by: John Campbell, Paul N. Crepeau, and Murat Tiryakioğlu
TMS (The Minerals, Metals & Materials Society), 2009

Favorable Alloy Compositions and Melt Temperatures to Cast 2XXX and 7XXX Al alloys by Controlled Diffusion Solidification (CDS).

Peyman Ashtari, Gabriel Birsan and Sumanth Shankar

Light Metal Casting Research Centre (LMCRC),
Department of Mechanical Engineering,
McMaster University, Hamilton, ON, Canada

Keywords: CDS, Controlled Diffusion Solidification, 2024, 7005, 7075, Rheocasting

Abstract

Controlled diffusion solidification (CDS) is an innovative rheocasting (Semi-solid) processing route to obtain a cast part with a non-dendritic morphology of the primary Al phase. The process involves mixing two alloy melts with specific individual compositions and temperatures to produce the desired final alloy by mixing and immediately casting in a mold. The process enables the shape casting of Al based wrought alloy along with their superior cast properties and performance. The present work defines process conditions to enable shaped casting of the 2XXX and 7XXX series of Al based wrought alloys, specifically, 2024, 7005 and 7075 alloys.

Nomenclature

Alloy 1	Pre-cursor alloy with higher thermal mass (higher temperature and higher mass).
Alloy 2	Pre-cursor alloy with lower thermal mass.
Alloy 3	Resultant mixed alloy.
T_{L1} T_{L2} and T_{L3}	Liquidus temperature of Alloy 1, Alloy 2 and Alloy 3, respectively.
T_1, T_2 and T_3	Melt Temperature of Alloy 1, Alloy 2, and Alloy 3, respectively.
m_1 and m_2	Mass of Alloy 1 and Alloy 2 respectively.
m_r	Mass ratio of Alloy 1 and Alloy 2 (Alloy 1: Alloy2).

Introduction

In our previous paper published in this proceeding [1], we have elaborately discussed the mechanisms involved during the mixing and solidification stages of the CDS process. CDS is a novel rheocasting route to enable high integrity shaped casting of Al cast and wrought alloys alike. CDS is a cost effecting route wherein conventional casting equipment can be used in both the high pressure die casting and permanent mould casting processes with the installation of an additional melting/holding furnace. The unique feature of the CDS process is that it enables the shaped casting of Al based wrought alloys along with their superior properties and performances [2]. The key feature of CDS process that makes it unique amongst the conventional rheocasting routes is that it is easier to obtain the resultant alloy at the required temperature in the semi-solid state and there is no intermediate holding time to equalize the temperature and the microstructure in the semi-solid region.

In the CDS process, two alloys of specifically chosen solute compositions and melt superheat temperatures are mixed such that the alloy with the higher thermal mass is poured into the alloy with the lower thermal mass to obtain a resultant mixture in which the temperature and the solute composition fields equalize in a controlled manner; and the resultant mixture is poured into a

mold to yield a shaped casting with a non-dendritic morphology of the primary Al phase in the solidified microstructure [1]. The critical process parameters in the CDS process are the compositions of the two precursor alloys, melt superheat temperatures of two precursor alloys and rate of mixing of the two alloys. The compositions and melt superheat temperatures of the two precursor alloys define the mass ratio and temperature difference (difference in thermal mass) between the two alloys. There are many possible combinations of the two precursor alloys with defined composition and melt temperatures to yield one specific resultant alloy.
In this study we wish to elaborate on the process of choosing favorable precursor alloy compositions and melt temperatures to yield a non-dendritic primary Al morphology in the solidified structure of the CDS process. The mechanism presented in our previous paper will be used to define the alloy specifications [1]. Three Aluminum wrought alloy compositions were used in this study: *2024, 7005 and 7075.*

The 2024 wrought alloy is also termed 206 in the cast alloy family and this can be used for structural automotive and aerospace applications. Shape casting of this alloy is tedious because of the extensive hot tearing issues. The 7XXX series of alloys are a part of Al-Zn family of alloys. These alloys have shown the best mechanical properties and performances among all Al alloys. Specifically 7075 is a popular 7XXX alloy.* The 7005 was chosen to cast a 7XXX series without Cu content and the 7075 has up to 2 wt% Cu. The 7005 alloy is weldable and exhibits high corrosion resistance; and it is used in the ground transportation industry, and the 7075 alloy with its high strength to weight ratio is used for aircraft structures.

Materials and Procedure

All the alloys were prepared using commercial purity raw materials (>99 % pure). The nominal compositions of 2024, 7005 and 7075 alloys are presented in Table 1. All compositions presented in this study are given in weight % of the respective solute element and all temperatures are in °C.

Table 1. Nominal composition (weight %) of 2004, 7005 and 7075†.

Alloy	Cu	Zn	Mg	Mn	Cr	Al
7005	-	4-5	1-1.8	0.2-0.7	0.06-0.2	Bal.
7075	1.2-2	5-6	2.1-2.9	-	0.2-0.28	Bal
2024	3.8-4.9	-	1.2-1.8	0.3-0.9	-	Bal

Table 2 presents the experiment parameters for the CDS casting of the alloys shown in Table 1. The alloy compositions presented in Table 2 were evaluated by Glow Discharge Optical Emission Spectrometry (GDOES). The Sample Identification (ID) in this study is also presented in Table 2. The first four numerals in Sample ID represent the final alloy required, the alphabet 'R' stands for Recipe and the last alphabet distinguishes the recipes from each other. The recipes for CDS presented in Table 2 have been obtained after an extensive review of various isopleths in the respective multi-component phase diagrams of the alloys. The Al-Cu-Mg phase diagram was used for the 2024 alloy and Al-Zn-Mg-Cu phase diagram was used for the 7XXX alloys. Figure 1 shows typical isopleths in respective alloy phase diagrams showing the experiment conditions in Table 2.

* http://www.Britanica.com/EBchecked/topic/18071/aluminum-processing/81517/Wought-alloys
† http://www.matweb.com

Table 3 shows the experiment conditions for the CDS recipes shown in Table 2. In Table 3, the temperatures T_B and T_D represent the peak temperature reached upon mixing and the final nucleation temperature of Alloy 3 as described in our previous manuscript [1]. Between 7005Ra and 7005Rb, the only difference in the recipes is the higher melt superheat T_1 in 7005Ra. 7005Rc has different compositions for Alloy 1 and Alloy 2 compared to 7005Ra and 7005Rb. The difference between the two recipes of 7075, 7075Ra and 7075Rb, are the compositions of Alloy 1 and Alloy 2. In Tables 2 and 3, all the alloy recipes were cast using the experiment setup presented in our previous manuscript [1] except the 2024Rb which was cast in a tilt-pour permanent mold process with a total pouring weight of 2.3 kg. All the alloys in Table 2 were re-melted after the CDS process and cast into an empty crucible to obtain the microstructure of the conventionally solidified alloy. The method for conventional casting is presented in our previous manuscript [1]. Figure 2 shows the two halves of the mild steel mold used in the tilt-pour casting process and the cast component with 2024Rb. In the tilt-pour casting of 2024Rb, the Alloy 1 and Alloy 2 shown in Table 2 were melted in two separate electric resistance furnaces maintained at 20 °C melt superheat temperatures, respectively. The pouring cup was pre-heated to about 500 °C and the casting mold was preheated to about 400 °C. About 330 g of Alloy 2 was poured by a ladle into the pouring cup of the tilt machine followed by 1.97 kg of Alloy 1 onto the Alloy 2 in the pouring cup. The tilting process was commenced immediately after Alloy 1 was poured in the pouring cup. The tilting process lasted about 15 s. The cast part was ejected after about a 3 min waiting period.

Table 2. Recipes for alloy design for CDS casting. Temperatures are in °C.

Sample ID	Alloy 1	Alloy 2	T_{L1}	T_{L2}	ΔT_L	M_r	Alloy 3	T_{L3}
2024Ra	Al-1.5Mg	Al-33Cu-1.5Mg	651	543	114	6	Al-6Cu-2Mg	640
2024Rb	Al-1.5Mg	Al-33Cu-1.5Mg	651	543	114	6	Al-4.5Cu-2Mg	640
7005Ra	Al	Al-25Zn-5.6Mg	660	573	87	4	Al-5Zn-1.2Mg	643
7005Rb	Al	Al-25Zn-5.6Mg	660	573	87	4	Al-5Zn-1.2Mg	643
7005Rc	Al-1.5Mg	Al-25Zn-1.5Mg	651	608	43	4	Al-5Zn-1.4Mg	642
7075Ra	Al-3Mg	Al-16Cu-36Zn	644	530	114	6	Al-2.5Cu-5.6Zn-2.7Mg	634
7075Rb	Al-6.5Zn-3Mg	Al-36Cu	644	530	114	6	Al-2.5Cu-5.6Zn-2.7Mg	634

Table 3. Experiment parameters for CDS casting of the alloys in Table 2.

Sample ID	T_1 (°C)	T_2 (°C)	T_3 (°C)	m_1 (g)	m_2 (g)	Casting Process
2024Ra	660	548	645	300	50	Crucible
2024Rb	656	554	638	1970	330	Tilt-Pour Permanent Mold
7005Ra	674	575	652	200	50	Crucible
7005Rb	665	582	643	200	50	Crucible
7005Rc	664	618	647	200	50	Crucible
7075Ra	648	544	655	300	50	Crucible

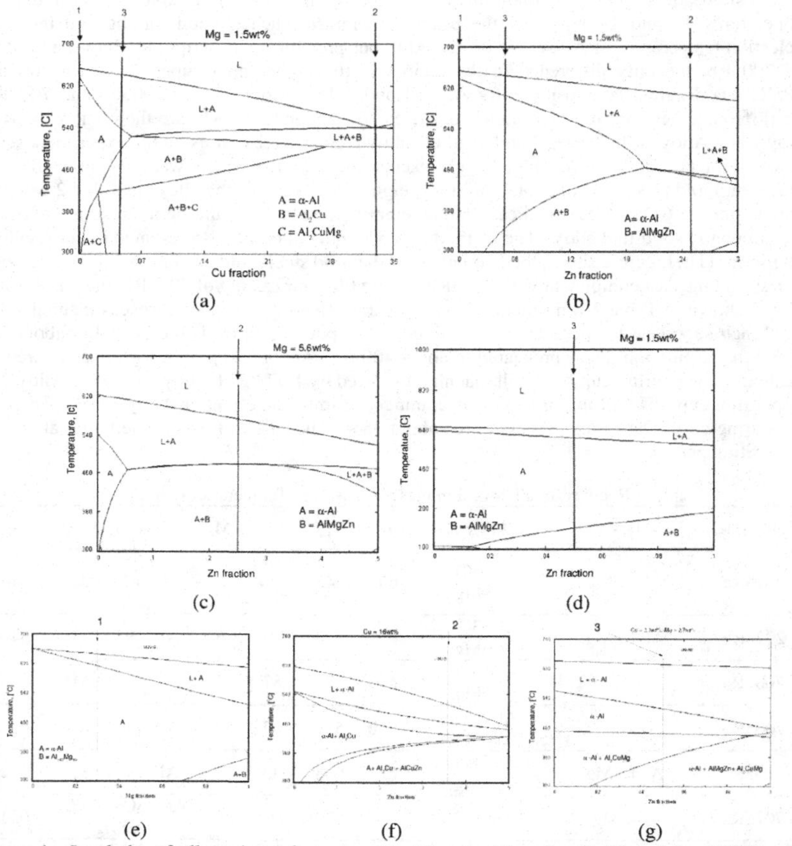

Figure 1. Isopleths of alloy phase diagrams. (a) Alloy 1, Alloy 2 and Alloy 3 for 2024Ra and 2024Rb, (b) Alloy 1, Alloy 2 and Alloy 3 for 7005Rc, (c) Alloy 2 for 7005Ra and 7005Rb, (d) Alloy 3 for 7005Ra and 7005Rb, and (e) through (g) show Alloy 1, Alloy 2 and Alloy 3 for 7075Ra.

Results and discussion

Figures 3(a), 3(b) and 3(c) show typical microstructures for 2024Ra cast with CDS, 2024Rb cast with CDS and 2024Ra alloy cast by conventional solidification. It can be observed that the 2024Ra and 2024Rb are both non-dendritic microstructures whereas the conventionally solidified microstructure is dendritic. The casting obtained from 2024Rb with the tilt-pour permanent mold process was sound and is shown in Figure 2 (c). Figure 2(d) shows the thermal data recorded during CDS and conventional solidification of 2024Ra alloy.

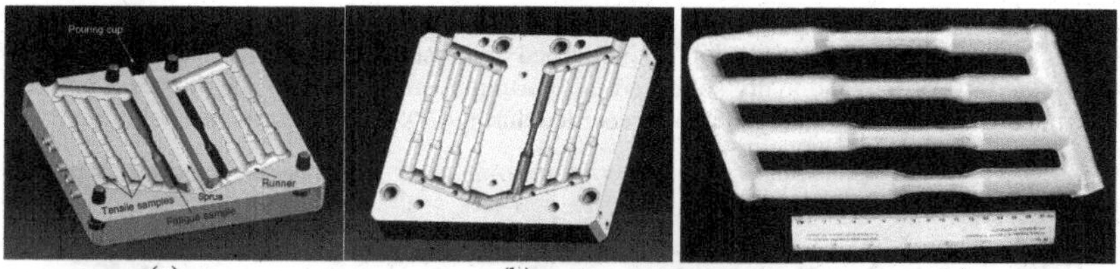

(a) (b) (c)

Figure 2. Mild steel mold used in tilt-pour casting process for CDS and cast part. (a) and (b) Two halves mold , and (c) the cast component

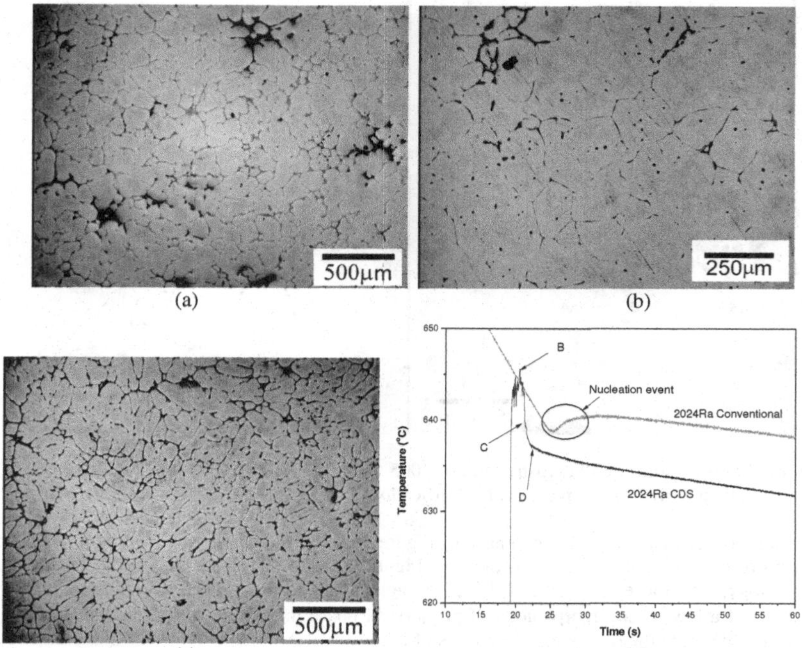

(a) (b)

(c) (d)

Figure 3. Typical solidified microstructures of 2024 alloy. (a) 2024Ra CDS, (b) 2024Rb CDS (c) 2024 conventional solidification. (d) Thermal analysis data collected during CDS in which the points B, C and D are explained in Khalaf et al [1].

In Figure 3(d), the point B is when the peak temperature is reached during mixing, point C is when the temperature equalization is completed and point D is when the solute equalization is completed and results in the final nucleation of the inter-dendritic liquid [1]. A favorably non-dendritic microstructure was obtained during solidification in the crucible as well as in the tilt-pour casting process for the 2024 alloy.

Figures 4(a) through 4(d) show typical solidified microstructures for the 7005Ra, 7005Rb, 7005Rc and 7005 conventionally solidified alloy samples, respectively. Figures 5(a) and 5(b) show the thermal data for the 7005 alloys recorded during CDS and conventional solidification.

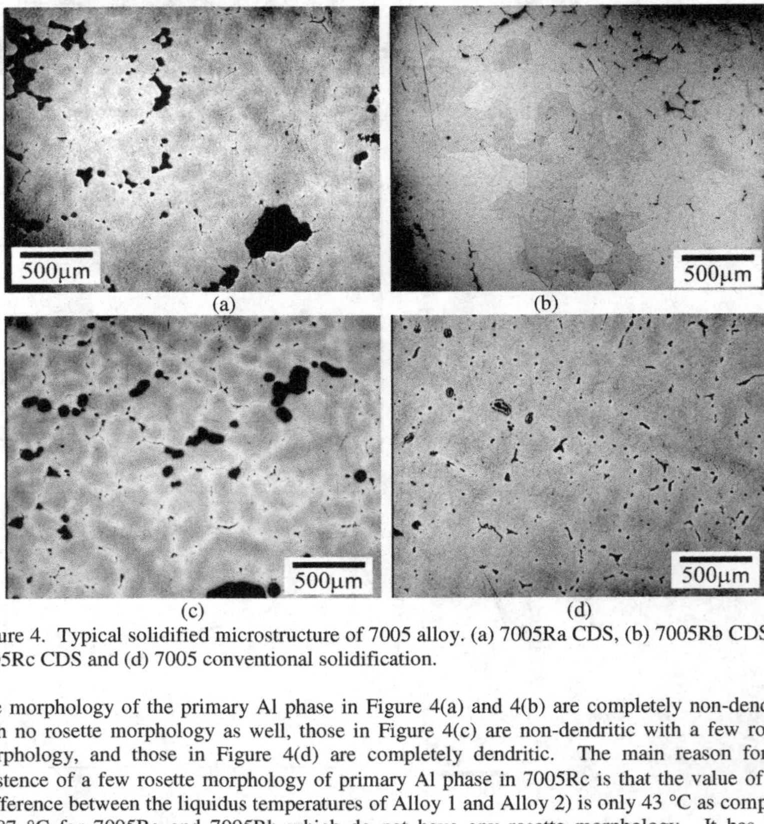

Figure 4. Typical solidified microstructure of 7005 alloy. (a) 7005Ra CDS, (b) 7005Rb CDS, (c) 7005Rc CDS and (d) 7005 conventional solidification.

The morphology of the primary Al phase in Figure 4(a) and 4(b) are completely non-dendritic with no rosette morphology as well, those in Figure 4(c) are non-dendritic with a few rosette morphology, and those in Figure 4(d) are completely dendritic. The main reason for the existence of a few rosette morphology of primary Al phase in 7005Rc is that the value of ΔT_L (difference between the liquidus temperatures of Alloy 1 and Alloy 2) is only 43 °C as compared to 87 °C for 7005Ra and 7005Rb which do not have any rosette morphology. It has been observed that the greater the value of ΔT_L, the lesser the existence of rosette morphology of the primary Al phase in the CDS solidified structure and a greater value of ΔT_L will present a greater freedom to choose higher melt superheats for the two alloys. Hence, it is suggested that 7005Ra recipe is the best among the three recipes investigated to cast 7005 alloy via CDS process.

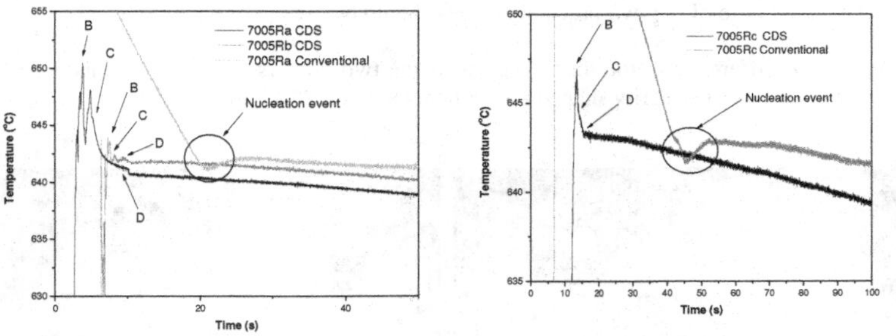

(a) (b)

Figure 5. Thermal data during solidification (a) 7005Ra and 7005Rb, and (b) 7005Rc.

Figures 6(a) and 6(b) show typical solidified microstructures for the 7075Ra alloy in CDS and conventional solidification, respectively. The morphology of the primary Al phase of the 7075Ra alloy in CDS microstructure is non-dendritic without any rosette morphology and that in the conventionally solidified microstructure is dendritic. Figure 6(c) shows the thermal data during solidification of 7075Ra using CDS and conventional solidification. Although the maximum temperature (state point B) is below T_{L3} the composition and microstructure of Alloy 3 were uniform. In Figure 6(c), the low temperature of state point B seems to be favorable from the microstructure view point; however, it would be beneficial to have state point B higher than T_{L3} to enable better castability of the alloy. In the casting of 7075Ra with CDS, the temperature of Alloy 1 (T_1) can be as high as 20 °C melt superheat. Further work is being carried out to establish an optimum temperature of T_1 for this alloy recipe to enable sound shaped castings. Alloy 7075Rb was not cast with CDS because the ΔT_L for this alloy was 76 °C compared with 114 °C in the 7075Ra alloy recipe. It was observed in 7005 alloys (Figure 4) that the smaller ΔT_L is not favorable for CDS. It is suggested that 7075Ra recipe is superior to 7075Rb recipe to cast the 7075 alloy with CDS and the temperature T_1 (Table 3) can be even 20 °C higher than that used in this study.

The favorable recipes in Table 2: 2024Ra, 7005Ra and 7075Ra are currently being cast in the tilt-pour permanent mold process to evaluate the mechanical properties of the cast components.

Summary

CDS is a viable and low cost process to obtain shaped castings of Al alloys, especially the Al wrought alloys. Each Al alloy can have many recipes to cast with CDS. However, only a few of those recipes are favorable with respect to both energy utilization and casting quality. In this study, favorable recipes have been presented to cast shaped components using the 2024, 7005 and 7075 alloys via CDS process. The following guidelines can be adopted to obtain favorable casting recipes for CDS:

- The difference of more than 80 °C is preferred between the liquidus temperatures of the two precursor alloys.
- The peak temperature upon mixing the two precursor alloy should be higher than the liquidus temperature of the resultant alloy to facilitate the casting process.

- The mass ratio of the two precursor alloys should be greater or equal to 3.

In sum, a large difference in the thermal mass of the two precursor alloys presents a favorable condition to cast a high integrity shaped component using the CDS process.

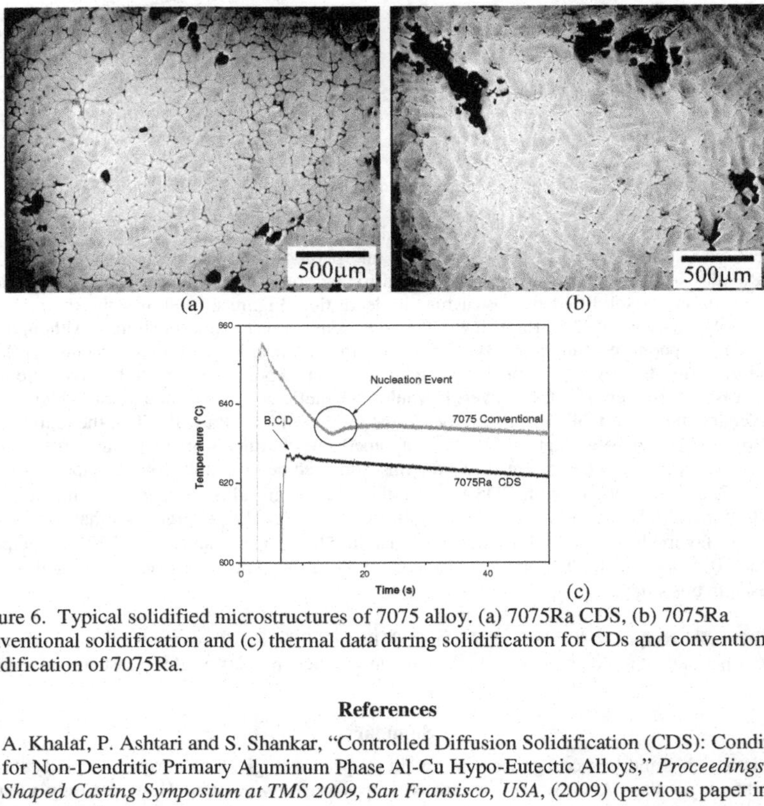

Figure 6. Typical solidified microstructures of 7075 alloy. (a) 7075Ra CDS, (b) 7075Ra conventional solidification and (c) thermal data during solidification for CDs and conventional solidification of 7075Ra.

References

1. A. Khalaf, P. Ashtari and S. Shankar, "Controlled Diffusion Solidification (CDS): Conditions for Non-Dendritic Primary Aluminum Phase Al-Cu Hypo-Eutectic Alloys," *Proceedings of Shaped Casting Symposium at TMS 2009, San Fransisco, USA*, (2009) (previous paper in this proceeding).
2. S. Shankar, D. Saha, M. M. Makhlouf and D. Apelian, "Casting of wrought aluminum-based alloys," *Die Casting Engineer*, 48 (2) (2004), 52-54+56-57.

Shape Casting: The 3rd International Symposium
Edited by: John Campbell, Paul N. Crepeau, and Murat Tiryakioğlu
TMS (The Minerals, Metals & Materials Society), 2009

THE APPLICATION OF POSITRON EMISSION PARTICLE TRACKING (PEPT) TO STUDY THE MOVEMENT OF INCLUSIONS IN SHAPE CASTINGS

W. D. Griffiths[1], Y. Beshay[1,2], D. J. Parker[3], X. Fan[3] and M. Hausard[4]

[1]Department of Metallurgy and Materials Science, University of Birmingham, Edgbaston, Birmingham, United Kingdom. B15 2TT.
[2]Beshay Steel, Heliopolis, Cairo, Egypt.
[3]School of Physics and Astronomy,
University of Birmingham,
Edgbaston, Birmingham, United Kingdom. B15 2TT.
[4]Formerly at the School of Physics and Astronomy, University of Birmingham, at the time the work was carried out, but now at the Centre de Calcul de l'Institut National de Physique Nucléaire et de Physique des Particules, Domaine scientifique de la Doua, 27 bd du 11 Novembre 1918, 69622 Villeurbanne Cedex, France.

Keywords: Inclusions, shape casting, Positron Emission Particle Tracking, PEPT, radioactivity.

Abstract

Positron Emission Particle Tracking, (PEPT), was used to track radioactive particles entrained into liquid metal during casting. The purpose of these experiments was to test the technique for its application to the study of inclusion movement in castings, and so provide a method for validation of computer simulations of inclusion behaviour. Two types of experiments were carried out. In the first, Al alloy plate castings were made in resin-bonded sand moulds, into which were entrained radioactive alumina particles of size 355 to 710 μm. In the second type of experiment smaller alumina and resin particles, around 60 to 100 μm in size, were entrained into a low melting point In-based alloy, (Field's Metal), cast at 80°C into an acrylic die. In each experiment the particle locations were recorded in real time, using a positron detection camera. The particle paths were obtained for each casting and the reproducibility of the technique examined.

Introduction

Inclusions in castings act as sources of fatigue cracks and reduce the life of the cast product. This is true for both shape casting of metals, DC casting of light alloys, and continuous casting of iron and steel. These inclusions are often exogenous, having their source outside of the metal. Examples of sources for these include refractories undergoing wear, tool coatings, particles from the mould-making process trapped inside the mould, or particles released from the mould surface during filling, (such as might occur in the case of sand casting). Indigenous inclusions refer to those inclusions that occur inside the liquid metal, perhaps due to *in situ* chemical reactions, such as alumina in Al-killed steel or MnS [1], or due to physical effects such as agglomerated TiB_2 particles in grain refined Al alloys.

It is necessary to know more about the behaviour of such inclusions in the liquid metal as the mould is being filled, in order to improve cast products. In shape castings a range of inclusion-removal strategies are often employed. These include running system designs intended to prevent liquid metal containing inclusions from entering the mould cavity, and ceramic foam filters intended to trap inclusions within their pores. Ceramic foam filters are also used in the DC casting of Al, being placed in the launder before the mould. The study of the behaviour of inclusions in castings has been facilitated by models of particle movement accompanying the computer simulation software now used widely in the industry [2]. However, there is currently no known method available to validate these models.

The results in this paper relate to the application of Positron Emission Particle Tracking (PEPT) to the study of the movement of inclusions in shape castings. In this technique a radioactive particle was introduced during the casting process, and its movement followed by detection of its decay products. PEPT has previously been used to study flows in processing operations that involve mixing and movement of pastes and granular solids [3,4], and the technique should therefore be as useful in the study of inclusion movement in cast liquid metals. In previous work radioactive alumina particles were added to a low-C steel cast in investment moulds [5]. However the risk of damage from the liquid steel to the detectors used to locate the radioactive inclusion meant that the casting operation took place separately, and only final inclusion locations were obtained. Also, in these experiments comparatively large particles of alumina were introduced, of size 355 to 710 μm. From the initial and final locations of the radioactive inclusion some information about its behaviour in between could be obtained, and a comparison with models of inclusion movement attempted. These initial results showed that, while the Positron Emission Particle Tracking (PEPT) technique could be used in shape casting, the particle location results were only occasionally reproducible, probably because of unreproducibility in the casting technique itself.

This paper describes the latest experiments using PEPT with shape castings, with two types of experiments having been carried out. In one, a portable detector system was used in conjunction with the sand casting of an Al alloy, again using relatively large particles. In the second experiment smaller entrained radioactive particles were used, more representative of the size of inclusions to be found in castings, and the experiments were carried out with a low melting point alloy cast into an acrylic die. In both experiments the particle track was now determined continuously, rather than simply identifying the initial and end locations as in previous work [5].

Experimental Procedure

Aluminium Sand Casting

The PEPT technique used in this case employed positron-emitting radioactive particles prepared in the University of Birmingham 35MeV cyclotron by direct bombardment of alumina particles with ^{3}He nuclei. This resulted in the conversion of some of the oxygen to radioactive ^{18}F, which has a half-life of 110 minutes and decays by positron emission. The emitted positron travels within the local material until it meets an electron, resulting in mutual annihilation. This annihilation produces two collinear γ-rays which travel at almost exactly 180° to each other, which are then detected using bismuth germanate-based scintillators placed around the experiment location [6]. The γ-ray signals from several such annihilation events are determined and passed through an algorithm to eliminate signals that are likely to be spurious or otherwise unuseable. The source of the γ-ray emissions, and hence the location of the radioactive particle, can be determined within an error of a few mm by triangulation of the γ-ray pairs.

To use this approach in casting, simple plates were made in an Al-7Si-0.3Mg alloy, cast in resin-bonded sand moulds, with a pouring temperature of about 730°C. The design of the plate and its running system has been shown in Figure 1(a), with the plate having dimensions of 200 mm in length, 100 mm in height, and 15 mm in thickness. The moulds were cast using a graphite stopper initially placed at the top of the downsprue in the pouring basin, so as to make the casting process as reproducible as possible. Casting took place within a modular arrangement of four blocks of the γ-ray detectors arranged in a square, as shown in Figure

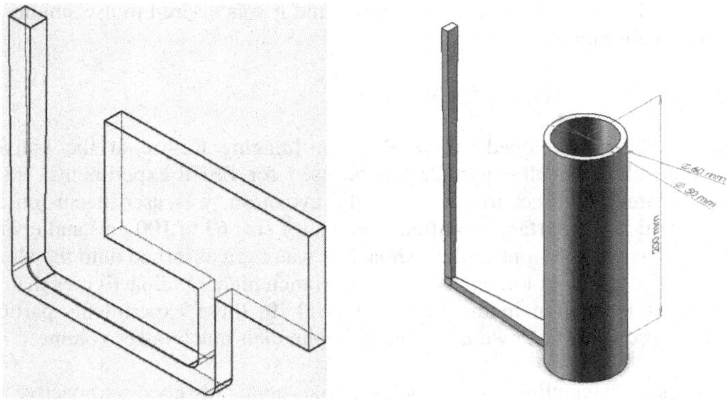

(a) (b)

Figure 1. Sketches of the castings used in the experiments. (a). The plate casting made in a resin-bonded sand mould with an Al alloy. (b). The cylindrical tube casting made in an acrylic die, cast with a low melting point alloy.

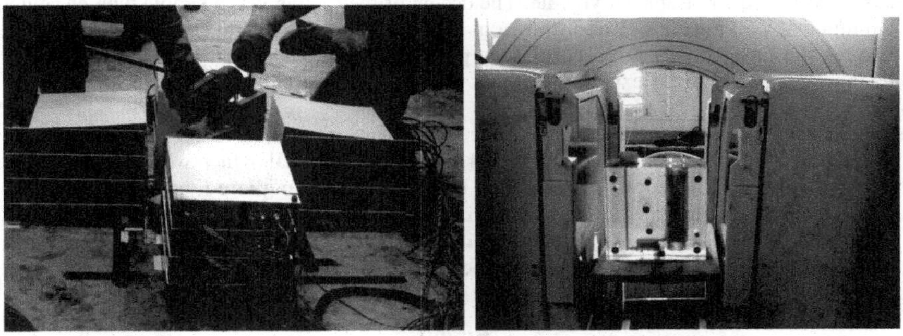

(a) (b)

Figure 2. (a). The arrangement of the γ-ray detectors around the sand mould, as it was being cast. (b). The acrylic die between the faces of the ADAC Forte γ-ray positron camera, ready to be cast.

2(a). Five castings were made in this way, and the particle tracks expressed in x, y and z coordinates, determined every few milliseconds.

Radioactive alumina particles of size 355 to 710 μm were prepared using the procedure of direct bombardment in the cyclotron described earlier. This resulted in particles having an initial radioactivity, measured at the time of their selection, of between 90 and 200 μCi. The minimum radioactivity a particle should have depended on many factors that vary from experiment to experiment, such as the thickness and density of the material through which the γ-rays have to pass before detection. Consideration of the half-life of ^{18}F indicated that the particles used here should have been detectable for around 3 hours. To introduce the radioactive particles into the liquid metal they were lightly glued to the head of a pin which was then inserted through the mould wall into the central part of the downsprue, 10 mm below the upper face of the mould. Once the liquid metal entered the downsprue the glue should have been vaporised, releasing the radioactive particle into the liquid metal.

This experiment used relatively large particles, in order to obtain a level of radioactivity that was sufficient to allow detection. However, the size of the particles would have represented only the largest inclusions to be found in castings, and it was desired to use smaller particles to represent the smaller inclusions also.

Casting Fields Metal in an Acrylic Mould

A technique has been developed at the Positron Imaging Centre at the University of Birmingham that allows smaller particles to be used for PEPT experiments. Radioactive water, again prepared by direct irradiation in the cyclotron, was used in an ion-exchange technique with particles of surface-modified alumina (of size 63 to 100 μm) and resin (of size 63 to 73 μm). In this case ^{18}F ions in the radioactive water are adsorbed onto the alumina and the resin particles, resulting in ion-exchange, giving much higher radioactivities than could be obtained by direct irradiation in the cyclotron [7,8]. In these experiments particles with radioactivities of 100 to 600 μCi were obtained, despite their much lower volume.

Since the process of labelling the particles relied upon adsorbed radioactive ions the experiments must take place at lower temperatures if the radioactive labelling of the particles was to be preserved. This precluded the use of any of the normal cast metals, so in these experiments Field's Metal was used, (51wt.%In, 32.5wt.%Bi and 16.5wt.%Sn, with a melting point of 62°C), cast at about 85°C. Advantage was taken of the low melting point involved to make castings in a reusable acrylic die. The design of the casting was a simple tube of height 200 mm and wall thickness 5 mm, shown in Figure 1(b), and the casting operation was done between the faces of an ADAC Forte γ-ray positron camera, as shown in Figure 2(b), since the low temperature of the experiment meant minimal risk of damage to the detector.

The radioactive particles, either alumina or resin, were introduced to the metal stream by placing them in a depression formed in a thin strip of Fields Metal, which was placed over the central part of the downsprue opening, with the particles kept in place by a light film of oil. Upon casting the liquid metal into the pouring basin the strip melted, releasing the particle into the liquid metal with which it travelled into the running system and the mould cavity. In this case the average error in the particle location should have been less than 2 mm, which was a great improvement in the error of location in previous work with steel castings [5].

Results

Aluminium Sand Castings

Examples of two particle tracks from the sand moulds cast with the Al-Si-Mg alloy have been shown in Figure 3. Although data on the particle position was collected in the x, y and z-directions, only the x and y coordinate data possessed sufficient resolution for display so the particle tracks in the plate castings have been shown in two-dimensions only. In both cases the particles were swept down the downsprue as the casting was being filled, passed along the horizontal runner bar, and were carried through the vertical ingate into the plate casting. In one case, (Figure 3(a)), the particle was swept to the right, but eventually followed an oscillating path upwards to come to rest in the upper right-hand corner of the plate. In Figure 3(b), the particle was swept to the left, and came to rest in the left-hand side of the plate, just below the mid-point.

Although apparently different, the particle tracks in Figure 3 do possess some similarities. The plate is symmetrical about a vertical line passing through the centre of the ingate, so whether the particle passes to the right or left is not significant. After emerging from the ingate the particle was carried upwards, and then downwards and away from the ingate close to the bottom of the plate, and then upwards again. Its final location differed because of the difference in the subsequent movement, and this may have been due to the differing degrees of solidification that had occurred in the vicinity of the particle location.

However, not all of the particles behaved in this way. In the example shown in Figure 4(a) the particle has remained in the horizontal runner bar, possibly because it became attached to the mould wall. In Figure 4(b) the particle remained just below the downsprue opening, close to its point of introduction, for reasons unknown.

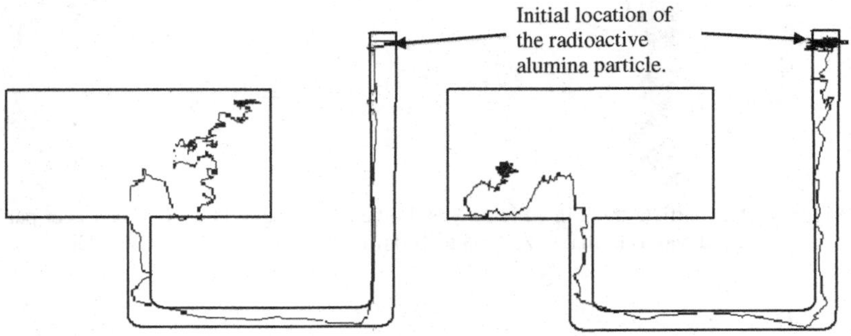

Figure 3. Two particle tracks for alumina particles entrained in the Al alloy plate casting. (a) size = 355 to 425 μm. (b). size = 425 to 710 μm.

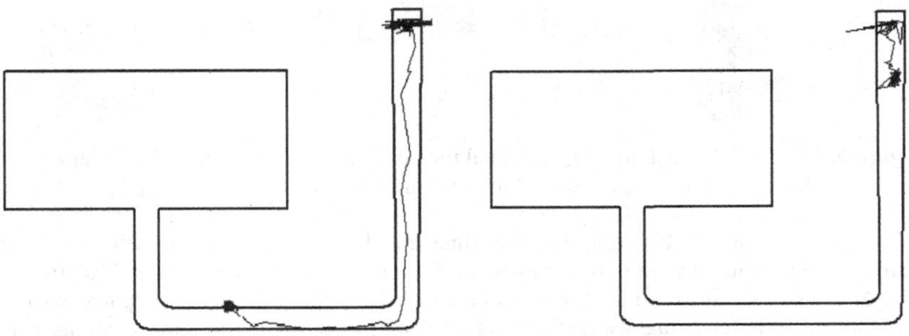

Figure 4. Two particle tracks for alumina particles entrained in the Al alloy plate casting. Both entrained particles were of size 355 to 425 μm.

Casting Fields Metal in an Acrylic Mould

Figures 5 to 8 show the tracks of entrained radioactive particles of alumina and resin in the low melting point alloy, Fields Metal, cast in the acrylic die. Both isometric and plan views have been shown. In this case the resolution was sufficient to see the particle movement in three dimensions. In Figure 5, the particle track is shown descending the downsprue, making its way along the runner bar, entering the tubular casting and becoming trapped at the base of the tube about 140° from the point of entry. The zig-zag path shown in the plan view in the Figure is due to the error of several mm with which the particle location has been determined; the actual particle path was probably smoother. In Figure 6, the behaviour of the particle track was similar, despite being cast at a higher temperature of 110°C, except this time the particle travelled around the opposite side of the tube. The small loop at the top of the downsprue indicated that the radioactive particle was briefly washed around the pouring basin before entering the downsprue.

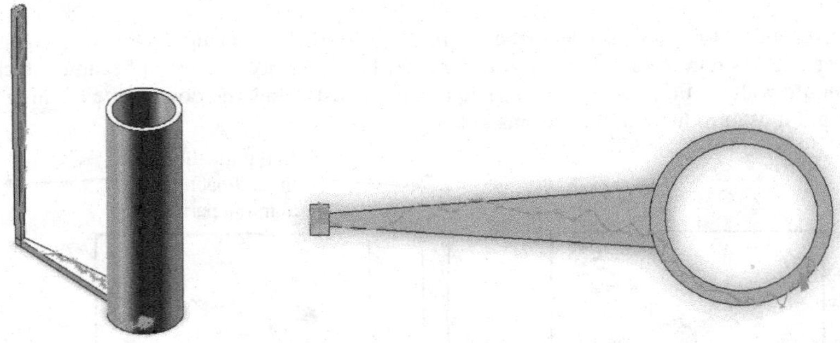

Figure 5. The particle track in a Fields Metal tube casting, cast at 85°C. The tracer particle used was alumina, (size = 63 to 100 μm), with an activity of 274 μCi.

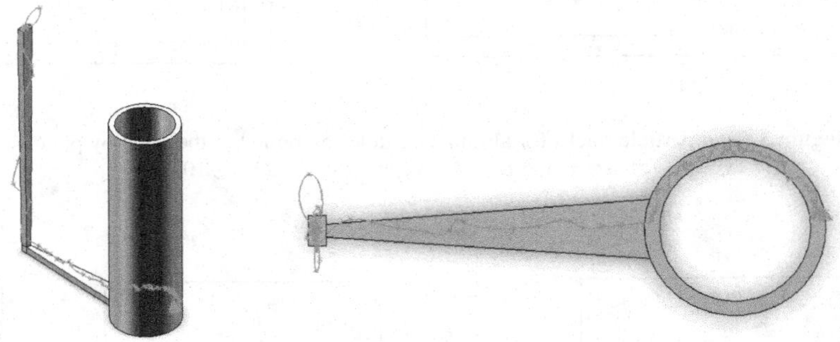

Figure 6. The particle track in a Fields Metal tube casting, cast at 110°C. The tracer particle used was alumina, (size = 63 to 100 μm), with an activity of 168 μCi.

In Figure 7 the entrained particle has this time travelled upwards, at the rear of the tube, relative to the ingate, and has come to rest in the top part of the tube, about 270° from the point of entry in the ingate. Figure 8 shows a good example of the entrained particle (in this case a resin particle) being washed around the pouring basin for several rotations before entering the downsprue.

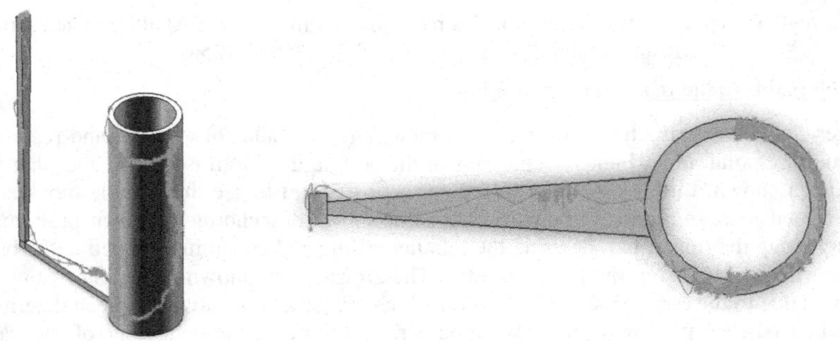

Figure 7. The particle track in a Fields Metal tube casting, cast at 87°C. The tracer particle used was alumina, (size = 63 to 100 μm), with an activity of 99 μCi.

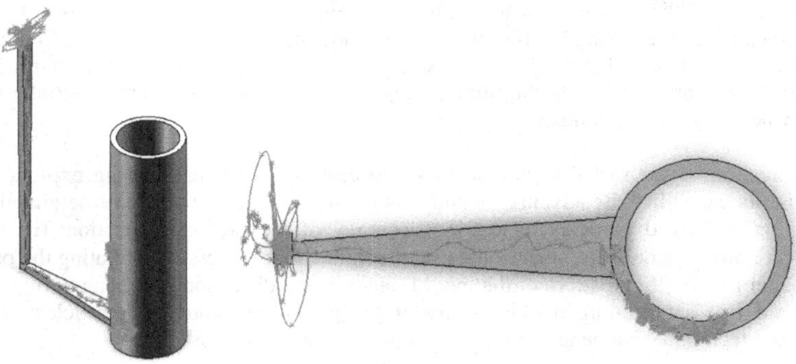

Figure 8. The particle track in a Fields Metal tube casting, cast at 87°C. The tracer particle used was resin, (size = 63 to 73 μm), with an activity of 418 μCi.

Discussion

Positron Emission Particle Tracking (PEPT) has been shown to be a useable technique for the study of the movement of inclusions in shape casting, demonstrated by two practical examples, the casting of Al in a sand mould, and the casting of a low melting point alloy in an acrylic mould. With the Al castings the radioactive particles were made so by direct bombardment in a cyclotron, giving activities of less than 200 μCi with particle sizes of around 0.5 mm. To reduce the particle size to 100 μm or less (approaching the typical size of inclusions in cast metals), would reduce the particle volume by more than two orders of magnitude, with an associated reduction in radioactivity for particles prepared by bombardment. The ion-exchange mechanism used for the particles used in the Field's Metal experiments allowed the use of radioactive particles that were less than 100 μm in size, but possessed radioactivities of up to three times greater, (600 μCi). The PEPT technique could therefore, in principle, be used to test and validate inclusion movement models in commercial modelling packages of the casting process. The technique could also be used to test inclusion removal strategies in cast metals.

However one obstacle to the application of PEPT in this way is the lack of reproducibility observed in the particle tracks and their final location, in both types of experiments. Both experiments were carried out in such a way as to minimise the variability in the casting procedure. Some reproducibility of results, especially if the symmetry of the casting geometries is considered, can be seen, (compare Figures 3(a) and 3(b), and Figures 5 and 6). Nonetheless, as shown in Figures 3 and 4, and, say, 6 and 7), large differences in the final particle locations were also seen.

This variability may be accounted for in two ways. Solidification would obviously begin at the cold mould surfaces. If the particle track takes it close to the mould wall, it could become entangled in the growing solidification structure and its movement would be hampered. In the thin-walled plates and tubes used here, if the radioactive particle followed a track that took it into the central regions of the castings, away from the solidification occurring at the mould walls, it would continue to move until this region solidified. In Figure 7, for example, the track of the particle suggests that it travelled up the liquid central part of the tube in order to reach the upper part of the casting, in contrast to the other experiments.

A second factor in interpreting the results is that it is not clear exactly when the entrained particle actually passes through the liquid metal surface and enters the bulk liquid metal. It was presumed that the method of insertion means that the particle was entrained almost immediately, and the results shown in Figures 6 and 8 the particle travelled around the pouring basin before entering the running system, showing that it was entrained. Conversely, such results as shown in Figure 4(a) suggest that the entrained particle may perhaps, in this case, have been carried along at the liquid metal surface to become attached onto the mould surface when it came into contact.

The application of the PEPT technique to shape casting is still undergoing exploration to determine its capability, its advantages and its limitations. Future work will be aimed at (i) locating the entrained particles to test the accuracy of the predicted location, (ii) testing inclusion removal techniques, such as the use of ceramic foam filters, (iii) relating the particle tracks better to the filling pattern of the moulds, and the solidification occurring during mould filling, and (iv) determining the feasibility of using multiple entrained particles to show whether the technique can be used to detect inclusion agglomeration.

Summary

Positron Emission Particle Tracking in metal casting is a technique with the potential to enhance understanding of inclusion behaviour in metals, and can help to develop commercial modelling packages. In these experiments it has been used to reveal the movement of particles entrained in cast Al alloy plates with the use of relatively large inclusions, of size up to 700 μm. However, by using a low melting point alloy and ion exchange techniques to label particles, the tracks of inclusions an order of magnitude smaller (60 to 100 μm) can be determined.

Acknowledgements

The authors would like to gratefully acknowledge the technical assistance of Mr. A. Caden of the School of Metallurgy and Materials Science of the University of Birmingham. YB would like to gratefully acknowledge the provision of funds from Beshay Steel, Heliopolis, Egypt.

References

1. R. Kiessling: "Non-Metallic Inclusions in Steel, Part III", 1989, London, The Institute of Materials.
2. MAGMA Gieβereitechnologie GmbH, Kackertstrasse 11, 52072 Aachen, Germany. http://www.magmasoft.de
3. D. J. Parker, C. J. Broadbent, P. Fowles, M. R. Hawkesworth and P. A. McNeil: *Nucl. Instrum. and Meth.*, 1993, A326, 592-607.
4. D. J. Parker, T. W. Leadbeater, X. Fan, M. N. Hausard, A. Ingram and Z. Yang: *Meas. Sci. Technology*, 2008, **19**, article number 094004.
5. W. D. Griffiths, Y. Beshay, D. J. Parker and X. Fan: *J. Mat. Sci.,* in press.
6. A. Sadrmomtaz, D. J. Parker and L. G. Byars: *Nucl. Instr. and Meth. A*, 2007, 573, 91-94.
7. X. Fan, D. J. Parker, M. D. Smith: *Nucl. Instr. and Meth. A*, 2006, 558, 542-546.
8. X. Fan, D. J. Parker, M. D. Smith: *Nucl. Instr. and Meth. A*, 2006, 562, 345-350.

Shape Casting: The 3rd International Symposium
Edited by: John Campbell, Paul N. Crepeau, and Murat Tiryakioğlu
TMS (The Minerals, Metals & Materials Society), 2009

MICROSTRUCTURAL AND SURFACE CHARACTERISTICS OF LEAD FREE BISMUTH BRONZE PRODUCED THROUGH THE FROZEN MOLD CASTING PROCESS

Shuji Tada, Hiroyuki Nakayama, Toshiyuki Nishio, Keizo Kobayashi

National Institute of Advanced Industrial Science and Technology
2266-98 Anagahora Shimoshidami Moriyama-ku Nagoya, Aichi 463-8560, JAPAN

Keywords: Bronze casting, Lead free, Bismuth, Casting, Frozen mold, Colloidal silica, Microstructure, Cast surface

Abstract

The frozen mold is a kind of sand mold which is produced by freezing the mixture of sand and water. The frozen mold casting process has the possibility to reduce the environmental load and the rapid cooling effect on cast products is expected. The effect of cooling rate on the microstructure of bronze castings was investigated. The frozen mold indicated better cooling property compared with conventional green sand mold. The microstructure of bronze castings produced through the frozen mold casting process was refined in the thinner sample but the quenching effect did not work well in the thicker sample. The surface condition of bronze cast was also examined. The surface of bronze castings produced using frozen mold consisting of only sand and water was rather rough. The surface roughness, however, was improved by adding colloidal silica solution into the sand mixture.

Introduction

The frozen mold casting process, sometimes called the Effset process, was invented in United Kingdom in the early 1970's [1]. The frozen mould is a type of sand mold basically consisting of only sand and water and is prepared by freezing this mixture. This new casting process promises industrial advantages such as the reduction of the environmental load and the improvement of working conditions. Fig. 1 indicates the schematic illustrations of the conventional casting process utilizing a traditional sand mold with organic binders and the frozen mold casting process. In the conventional method, shake-out and reclamation processes are mandatory required to complete the casting work. These processes often cause poor working conditions such as vibration, noise and dust. In contrast, the frozen mold casting process enables to improve the working conditions, since the above processes are basically not necessary in this process. The mold usually decays spontaneously with thawing. The reclamation process is basically not required because the frozen mold generally contains no other materials than sand and water.

Several studies of this technique have been reported as to freezing method [2], mechanical properties of frozen mold [3-5], and its deformation during freezing [6] in Japan. This new technique, however, was not put into practical use due to the high cost. It was necessary to use liquid nitrogen to produce a frozen sand mould. Recently, the advanced freezing method in which the cold air was ventilated through the sand mold was developed [7-8]. This method enabled to provide frozen mold in short time and at low cost, so the industrial try of frozen mold casting process has been actually started.

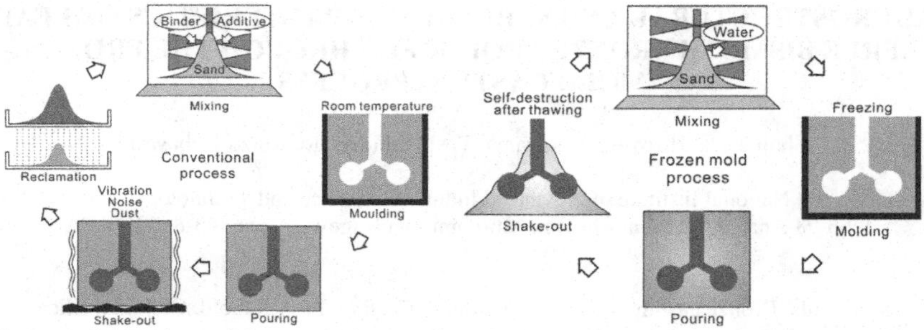

Figure 1. Comparison of the conventional casting process and the frozen mold casting process.

In recent years, the desire for practical application of lead-free bronze castings has been increasing rapidly as people become more health conscious. Bismuth is one candidate for replacing lead. But it is very important to reduce bismuth amount used to accelerate substitution, since it is an expensive material. As previously described, the frozen mold casting process has the big industrial advantages. Also the improvement of mechanical properties utilizing the quenching effect due to a cryostatic mold can be expected. However, the effect of frozen mold on the microstructure of lead free bismuth bronze has not yet been investigated. At the same time, the surface quality of bronze cast produced by a frozen mold tends to drop compared to the conventional products. In present study, in order to promote lead-free in bronze castings, the microstructures of bismuth bronze produced through the frozen mold casting process were examined. The additional technique in the frozen mold casting process to improve the surface quality of bronze cast was also investigated.

Experimental Procedure

Preparation of Frozen Mold

The artificial silica sand, R6, was used to provide sand molds in this study. The grain size distribution and the microscopic profile of this sand are shown in Table 1 and Fig. 2, respectively. As shown in Fig. 3, two types of sand molds were prepared. One was a stepped mold (S series) applied for the microscopic examination of produced bronze castings and the other (D series), which was corresponding to a disk form, was used for the evaluation of cast surface quality. Several kinds of sand mixtures with different compositions were prepared to evaluate the effect of frozen mold on the microstructure and surface characteristics of bronze cast. The colloidal silica solution which contained 30 or 50 mass% fine silica particles with diameters between 10 and 30 nm was added for the purpose of improving mold strength. In some

Table 1. Grain size distribution (mass%) and grain fineness number of silica sand.

Micron	425	300	212	150	106	75	AFS.FN
R6	0.2	5.6	54.8	34.8	4.4	0.2	58.7

Figure 2. Microscopic profile of silica sand.

molds, coating treatments for the mold surface were conducted. Fine silica and alumina particles were pre-coated to the mold surface prior to freezing and fine boron nitride was sprayed on the surface after freezing. Fabricating conditions of sand molds are listed in Table 2. All molds were frozen by keeping in a freezer held below 250 K for more than 50 ks.

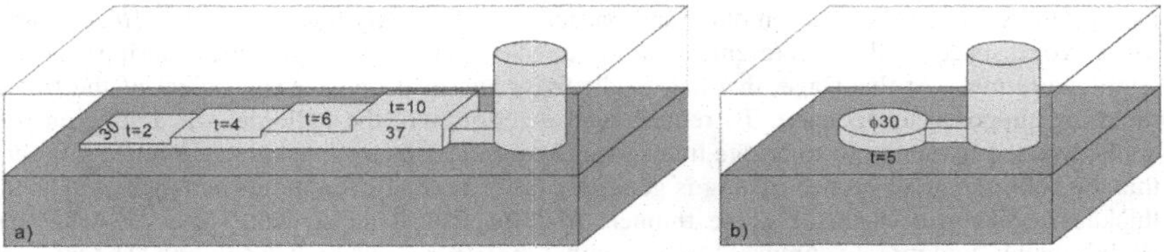

Figure 3. Dimensions of sand molds, a) stepped mold, b) disk mold.

Casting

Lead-free bismuth bronze ingot was used for the casting work in this study. The bismuth content of this ingot was 2.25 mass%. The chemical compositions of this material are shown in Table 3. Melting was conducted in a high frequency electric induction furnace. Fig. 4 indicates the heating diagram. Ingots were heated to 1200 °C and the heating was once stopped at this temperature. When the melt temperature dropped at 1160 °C, 0.02 mass% phosphor copper was added and stirred for deoxidization. After holding the molten bronze at 1160 °C for 240 s, it was poured into a mold at the desired temperature.

Table 2. Fabricating conditions of sand molds.

Mold	Sand	Water	Silica	Remarks
F01	95.00	5.00		
F02	93.00	4.90	2.10	
D01	95.00	5.00		
D02	92.86	5.00	2.14	
D05	92.86	5.00	2.14	Silica coat
D06	90.00	5.00	5.00	Alumina coat
D07	90.00	7.00	3.00	
D08	86.00	7.00	7.00	
D09	90.00	7.00	3.00	BN spray

Cu	Bi	Sn	Zn	Pb	Fe	Ni	Sb	P
85.70	2.25	4.38	7.40	0.02	0.02	<0.005	0.20	0.021

Evaluation of Frozen Mold Casting

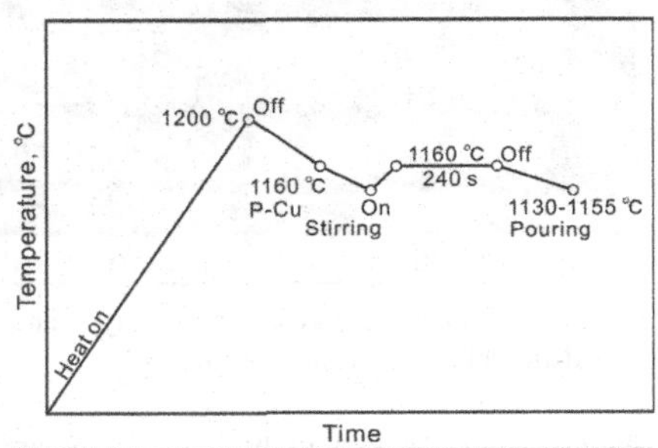

The cooling behavior of a bronze casting was monitored in the stepped sample. Chromel-alumel thermocouples were arranged at four positions, corresponding to the center of each thickness. Microstructures at each thickness were observed after the etching treatment. The relationship between the cooling rate and microstructure was examined in order to evaluate the advantage of the frozen mold casting process. After the shot-peening treatment, bronze disks were cut into two pieces along the radial direction. The profile of its cross section was observed in an optical microscope to evaluate surface quality.

Results and Discussion

Microstructure of Bronze Castings

Fig. 5 shows the cooling curves of stepped lead-free bismuth bronze obtained from the casting by frozen molds and a green sand mold. Sand particles are generally bound by ice in frozen mold. Many void spaces will be present in a mold when ice thaws following pouring of high temperature melt. At this stage, thermal conductivity and heat transfer coefficient of the frozen mold are supposed to decrease. To reduce such deterioration, the colloidal silica solution was applied to one frozen mold to bridge interspaces between sands by fine silica particles. It is seen that the cooling rate of bronze casting is generally faster in the frozen molds and increased as the thickness of bronze casting became thinner. And the frozen mold with 5 mass% water has slightly better cooling rate compared to that with 7 mass% colloidal silica solution. The addition of colloidal silica, unfortunately, did not accelerate the cooling rate. This seems to be caused by the fine silica particles which inhibited the ventilation property of the mold. The heat transfer coefficient might be degraded by vaporized water stuck between the mold and the cast material.

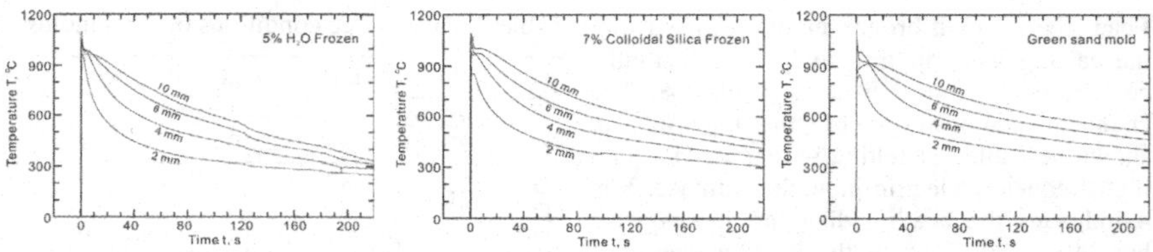

a) 5 mass% water frozen b) 7 mass% colloidal silica frozen c) Green sand mold

Figure 5. Cooling curves of stepped bronze at each thickness for a) 5 mass% water frozen mold, b) 7 mass% colloidal silica frozen mold and c) green sand mold.

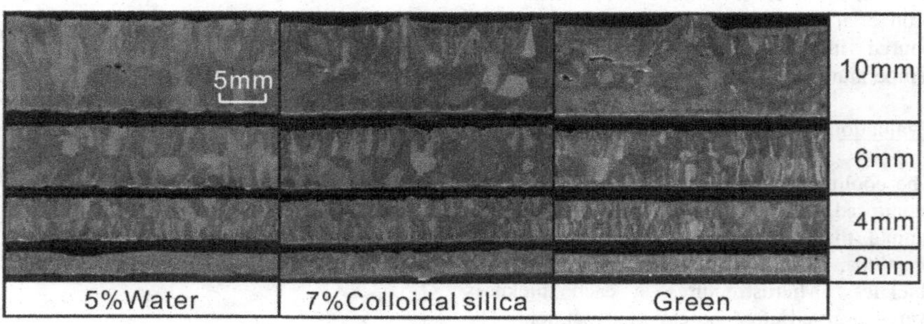

Figure 6. The effect of casting mold type and sample thickness on the microstructure of lead-free bismuth bronze castings.

Microstructures of various bronze castings are shown in Fig. 6. Almost all samples represent columnar structure but the 2 mm thick section obtained by the frozen mold with 5 mass% water, which had the fastest cooling rate, shows a fine equiaxed structure. This result suggests that the frozen mold has the potential to refine the structure of bismuth bronze casting by rapid cooling. This advantage is now limited to thin products, but the frozen mold casting process has the potential for structural refinement of thick sections provided the thermal conductivity of a frozen mold is improved.

Surface Quality of Bronze Castings

Fig. 7 indicates the surface characteristics and section profiles of lead-free bismuth bronze disks produced by the frozen mold casting process and the conventional casting with green sand mold. It is obvious that the surface quality of bronze disk produced by a frozen mold becomes worse compared with that obtained by a green sand mold. The strength of a frozen mold drops remarkably when it thaws, so there is a possibility that the condition of mold surface gets rough when the melt is poured at high temperatures. It has been reported that the strength of a frozen mold after thawing could be improved by the addition of colloidal silica solution to sand mixture [9]. The effect of the application of colloidal silica solution to sand mixture for frozen mold on the surface quality of cast products was investigated.

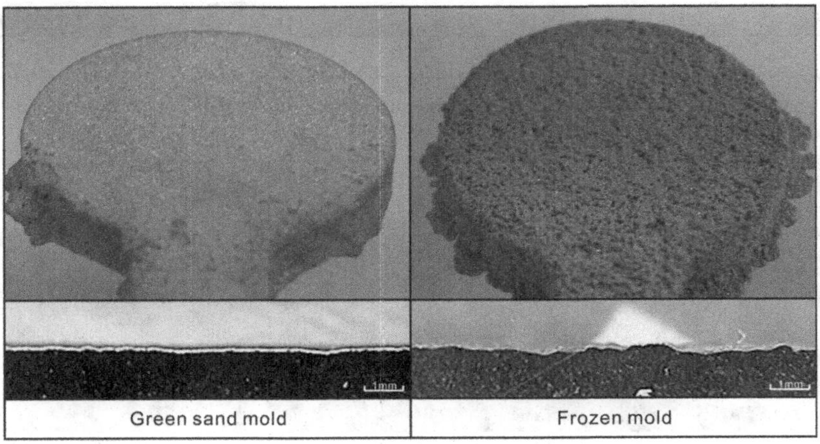

Figure 7. Surface aspects and section profiles of bronze disks produced by green sand mold and frozen mold.

Examples of surface quality and section profile of bronze disk produced by the frozen molds with various amounts of colloidal silica additions are shown in Fig. 8. The surface condition was improved by adding colloidal silica solution. This seems to be caused by the maintenance of the mold surface even after thawing by the addition of fine silica particles. The surface quality became better when the additive amount of fine silica particles increased from 2.14% to 3.0% but no obvious improvement was observed with further additions. As previously described, the addition of fine silica particles to the sand mixture sometimes inhibits the ventilation property of the mold. It is important to find the optimum amount taking into consideration a balance between the strength and air permeability.

Fig. 9 shows the results of coating treatment for frozen mold. In the case of pre-coating, both silica (D05) and alumina (D06) coatings caused the difficulty in mold release due to sticking of the additive agents to the molding flask, resulting in the desolation of mold surface. The application of hydrophobic material to the flask has the possibility of removing this hurdle. The BN coating to the frozen mold as a mold wash also roughened mold surface since the surface thawed by spraying. It seems to be desirable to use solid state material for mold wash to prevent mold surface from thawing.

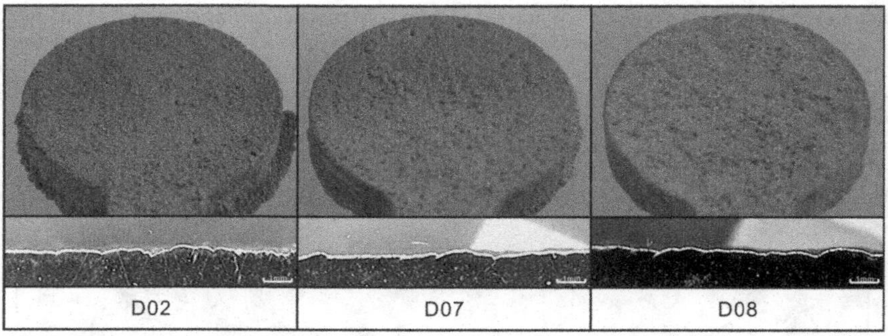

Figure 8. The effect of adding amount of colloidal silica to frozen mold on the cast surface of bronze disk.

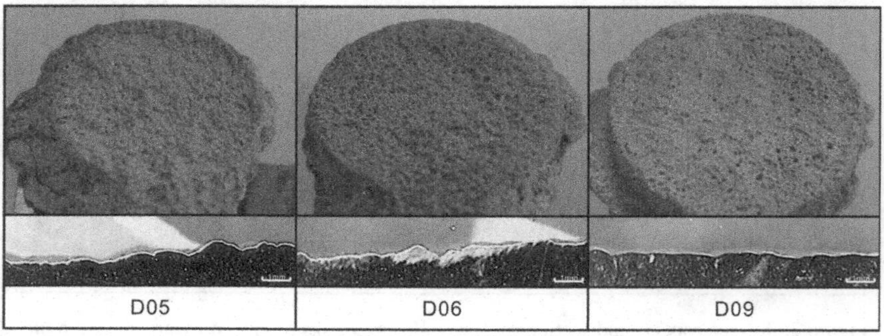

Figure 9. The effect of coating treatments for frozen mold on the cast surface of bronze disk.

Concluding Remarks

In this study, the effect of the frozen mold casting process on the microstructure and surface quality of lead-free bronze castings was investigated. The results are summarized as follows.

1) The frozen mold enables to increase the cooling rate of bronze castings and has the potential to refine the structure. But the addition of colloidal silica to the frozen mold does not improve the cooling property. It is important to find out the heat transfer behavior between the frozen mold and the cast product in order to use the quenching effect of frozen mold casting process at its full value.

2) The surface quality of bronze castings produced by the frozen mold is usually not equal to that by the general green sand mold. The quality, however, can be improved by adding colloidal silica solution into sand mixture. It is important to find the optimum amount while considering the balance between the strength and air permeability of the frozen mold.

3) The coating treatments for the frozen mold did not work well to improve surface quality of bronze castings. In the pre-coating before freezing, it was rather difficult to release sand mold from the mold flask. The post-coating caused the thawing of mold surface. The application of hydrophobic material to the flask and solid type mold wash may be effective in further development of the frozen mold casting process.

REFERENCES

1. C. Moore and D. Beat, "Effset-metallurgy, sand technology and economics", Foundry Trade J. May 1979, 1049-1063.

2. "Imono handbook", 4th revised edn, ed. Japan Foundrymen's Society, 205, 1986, Tokyo, Maruzen.

3. Kiyoshi Kita, Hahuki Nino and Masatake Tominaga, "Characteristics of frozen mold", IMONO, 1980, 52, 28-33.

4. Susumu Minowa, Hideaki Ohta and Mitsuo Ninomiya, "Behavior of water in freezing molds", IMONO, 1980, 52, 530-535.

5. Susumu Minowa, Mitsuo Ninomiya, Hideaki Ohta and Takeshi Takayanagi, "Transverse strength of frozen mold sand", IMONO 1982, 54, 309-313.

6. Susumu Minowa, Mitsuo Ninomiya and Hideaki Ohta, "Deformation and pattern drawing force of sand mold at freeze", IMONO, 1981, 53, 15-19.

7. Yoshinobu Fukuda, "Development of the casting system using a rapid frozen molding process", J. Japan Foundry Society Inc, 2006, 2, 919-925.

8. Hideto Matsumoto, Yoshiyuki Maeda and Yoshinobu Fukuda, "Introduce casting plant for frozen molding process", J. JFS, 2008, 80, 370-374.

9. Shuji Tada, Yuya Makino, Toshiyuki Nishio, Keizo Kobayashi and Ken Aoyama, "The compressive strength of frozen mold with colloidal silica addition", J. JFS, 2008, 80, 531-535.

Shape Casting: The 3rd International Symposium
Edited by: John Campbell, Paul N. Crepeau, and Murat Tiryakioğlu
TMS (The Minerals, Metals & Materials Society), 2009

PRESSURE MOLD FILLING OF SEMI-SOLID DUCTILE CAST IRON

B. Heidarian[1], M. Nili-Ahmadabadi[1, 2], M. Moradi [1], J. Rasizadehghani[1]

[1]School of Metallurgy and Materials Engineering, University of Tehran;
North Karegar St., Tehran, Iran
[2]Center of Excellence for high performance material, University of Tehran;
North Karegar St., Tehran, Iran

Keywords: Semi- solid, ductile iron, high pressure mold filling, Graphite deformation

Abstract

The processing of metals in the semi-solid state is becoming an innovative technology for the production of globular structure and high quality cast parts. Ductile irons because of spherical graphite have specific properties such as good mechanical properties, strength and toughness together and suitable castability. This engineering alloy along with growing application has several shortcoming which have limited its applications such as non-formability, dendritic structure and alloying element segregation, micro-porosity resulted from solidification mode and fabrication of thin section parts. It seems that replacing dendritic structure with globular structure and thixoforming results in improving of mechanical properties, controlling of alloying element segregation, decreasing of micro-porosity and increasing of ability to thin section filling. This paper reports successful high pressure, semi-solid mold filling of ductile iron containing Mn and Mo. Filling properties, fluidity, liquid segregation and defects like shrinkage and cracks are characterized.

Introduction

Ductile irons have unique properties to the other types of cast irons, such as good mechanical properties, strength and toughness with suitable castability. However, the use of ductile iron for light weight automotive components has been limited in the past by the problems of the foundries to produce thin wall (3 mm and less) parts. Also, this engineering material cannot be hot worked—like steel—to produce thin walled parts because of deformation of the spheroidal graphite and reduction of mechanical properties [1].

The need for fuel economy drives weight reduction, for instance, by converting from ductile iron to aluminum and powder metal. However, if the yield stress/cost ratio of the various materials is considered ductile iron is the winner in the most of the time [2].

Dendritic structure and the microporosity are intrinsic to solidification of ductile irons. Therefore, semi-solid processing seems to be the proper solution to improve the ductile iron structure and the mechanical properties without deformation in graphite spherioids. It is worthwhile to note that graphite deformation more than 15% decreases mechanical properties dramatically; 40% and more graphite deformation decreases mechanical properties more than 50% due to stress concentration at sharp edges of deformed graphite [3,4]. In addition, semi-solid processing of ductile iron offers the capability to produce near net shape and thin section parts.

Prior research has reported semi-solid forming of gray cast iron[5,6]. In the case of ductile iron the process is more complicated because of degeneration of graphite during extended conventional semi-solid processing (such as mechanical stirring) and deformation of graphite during thixo-forming. With the development of the inclined cooling plate technique It is possible to prevent Mg fading in ductile iron, which may happen in conventional semi-solid processing of ductile iron [7,8]. Thus this process offers better structural control and reduced complexity.

Having an alloy with suitable freezing range and low liquid fraction sensitivity* is the most important parameter to semi-solid forming of the ductile irons, According to previous works although hypoeutectic ductile iron has a long freezing range, it is not suitable for semi-solid forming due to solution of graphite during re-heating[9]. This phenomenon was also observed in hypoeutectic gray iron [5,6]; but in hypereutectic alloy, graphite spheroids remain in the microstructure during reheating and after semi-solid forming. This present work is an attempt to study thixo-forming of hypereutectic ductile iron alloyed with Mn and Mo. Special attention was paid to control microstructure and mechanical properties of products.

Experimental procedure

Semi solid Casting

Ductile cast iron with the chemical composition shown in Table 1 was prepared in a medium frequency induction furnace. Magnesium was added by the sandwich method with 2.5% Fe-Si-Mg as inoculant and 0.4% Fe-Si as post inoculant.

Table 1 Feedstock chemical composition (%)

Element	C	Si	Mn	Mo
Content	3.6	2.8	1	0.5

Hyper-eutectic ductile iron containing Mn and Mo produced by inclined cooling plate has been used for thixo-forming. Processing of semi-solid feedstock of ductile iron was carried out with inclined cooling plate method.

A 1000×50×20 mm water cooled copper plate served as an inclined cooling plate (Fig. 1) coated with boron nitride. Important variables such as superheat of the melt, melt flow rate, and temperature, length and angle of the inclined cooling plate were selected according to the previous studies [8, 10]. The slurry after flows down an inclined cooling plate collected in a sand mold.

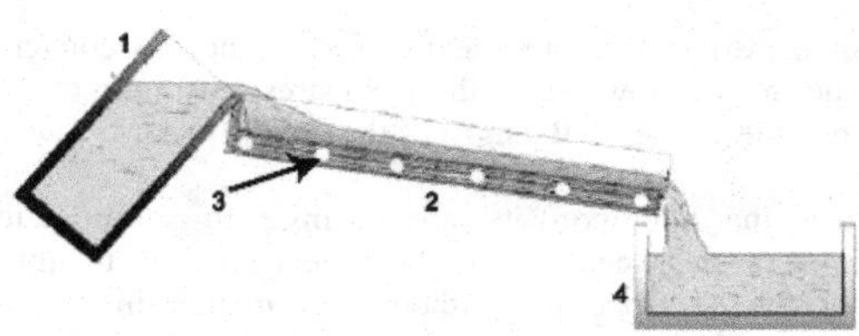

Fig.1. Schematic of inclined cooling plate process

* Liquid fraction sensitivity is the curve slope of liquid fraction versus temperature which shows liquid fraction variation as a function of temperature.

(1-pouring system; 2-inclined cooling plate; 3-water cooling system; 4-mold)

Reheating process

Feed stock samples were reheated in graphite tube in a resistance furnace under an Ar atmosphere to 1150 °C for 40-45 min to achieve about 50% liquid fraction.

Semi-solid forming

Semi-solid forming was carried out at 3 temperatures (450, 500 and 600°C) in a step die having interchangeable gates with different V/A moduli. The die was quenched and tempered H13 tool steel coated with TiO_2, shown in Figs. 2 and 3. This die was designed with stepped sections for fluidity testing of ductile irons, influence of the wall thickness and the metal velocity on the filling behavior, phase segregation and microstructure in the semi-solid state. Metal after pass through the gate entered the 30 mm step and was pushed to the thinner steps in the both side of gate. A wedge part was included for studying flow behavior in longer length.

Metal was introduced into the die with a hydraulic press with maximum ram speed of 100 mm/sec and 12 ton force. At the final step of forming process, the same as high pressure die casting techniques, pressure was maintained for 5 s as solidification completed.

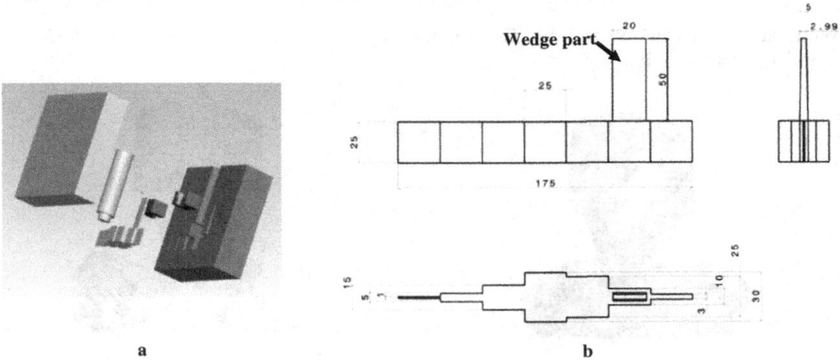

Fig. 2. Schematic of step die showing stepped sections 1, 5, 10, 15, 25, and 30 mm thick used to assess fluidity.

Optical microscopy (ZEISS model) and image analyzer software (UTHSCSA image tool Ver. 1.27) used for micro-structural study (calculation of graphite's shape factor and liquid fraction). Shape factor of graphite particles were calculated according to Eq. 1 [11] and using images taken at 100X magnification. Images were taken from the top, middle, bottom, left, center and right of each step after polishing.

$$SF = \frac{N}{\sum_{1}^{N} \frac{P^2}{4\pi A}} \quad (1)$$

Polished samples were color etched with a solution of sodium hydroxide (NaOH), potassium hydroxide (KOH) and picric acid for liquid segregation study. Liquid fraction was also calculated according to the images taken from the top, middle, bottom, left, center and right of each step.

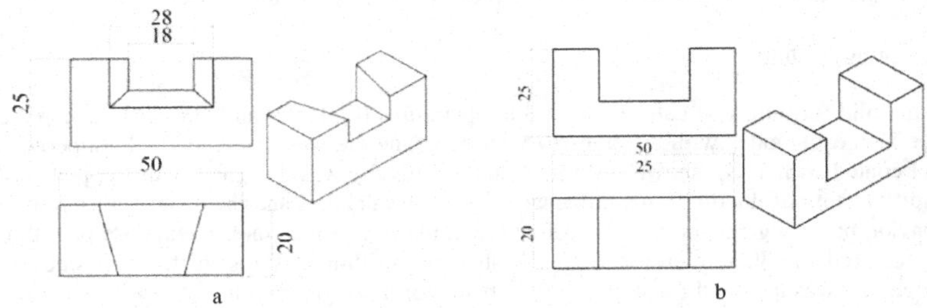

Fig. 3. Gate dimensions a) gate No.1, b) gate No.2

Result and discussion

With gate No. 1 and 450 and 500°C die temperature 1-10 mm steps is not fill and 15-30 mm steps filled partially, Fig. 4.

a b

Fig.4. Thixo-formed samples with gate No.1 at die temperature a) 450°C, b) 500°C

It is expected that solidification occurred in the gate and prevented complete filling. Low ram speed and consequent long fill time led to premature solidification. Filling time must obviously be less than solidification time. To this end ram speed and die temperature should be increased. Inasmuch as incomplete filling is due to short solidification time, the die temperature and gate modulus were increased. Fig. 5 shows thixo-formed specimens at 600°C die temperature and gate No.2. The sample filled completely and all of the steps (30 to 1mm) were filled without missrun or visible defect. So, with increasing die temperature, solidification time increases and leads to prevent of early solidification and complete filling of the cavity.

a b

Fig.5. Thixo-formed samples with gate No.2 and die temperature 600°C.
a) Before and b) after trimming.

Fig. 6 shows the microstructure of thixo-formed sample shown in Fig.6. Except for the 30 mm step, all of the other steps are defect free and graphite remains spheroidal.

Fig. 7 shows un-etched structure and graphite shape of thixo-formed samples. Except for the 30 mm step, all of the other steps are defect free and graphite remains spheroidal.

Defects were seen in front of the gate at the bottom of the 30 mm step (fig. 8). It is expected that non-globular particles and segregation of solid-liquid during reheating process are responsible for defect formation in this step. As a consequence of this inhomogeneity, segregated liquid solidified in the bottom of the part with dendritic structure. The interface of solidified liquid and semi-solid encourages crack and porosity formation.

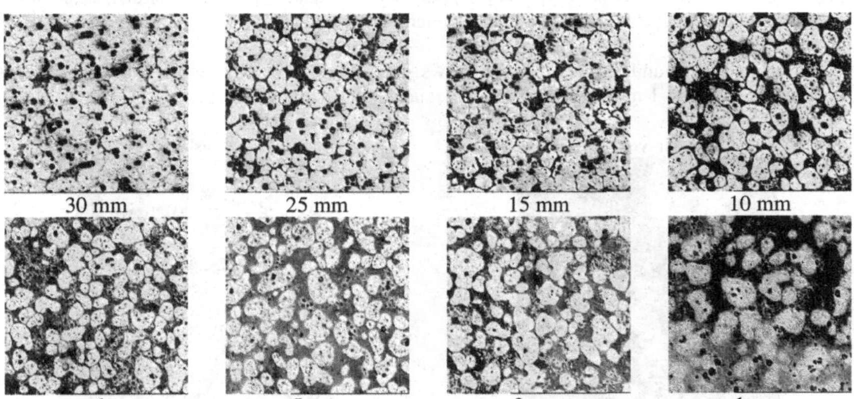

Fig.6. Microstructure of thixo-formed samples with gate No.2 and die temperature 600°C.

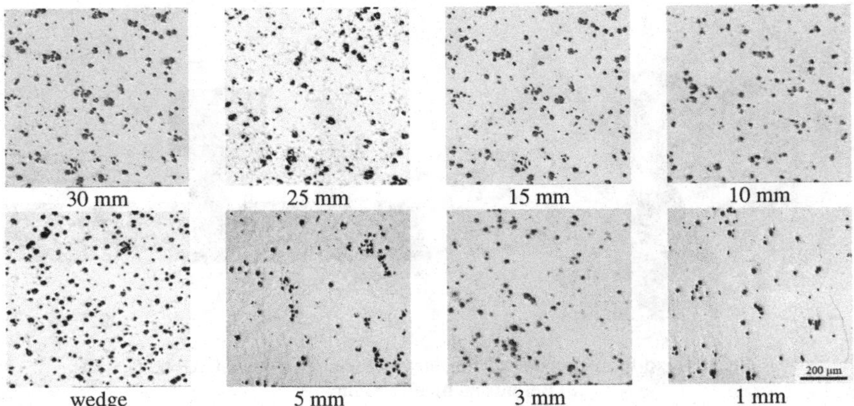

Fig.7. Microstructure of thixo-formed samples with gate No.2 and die temperature 600°C.

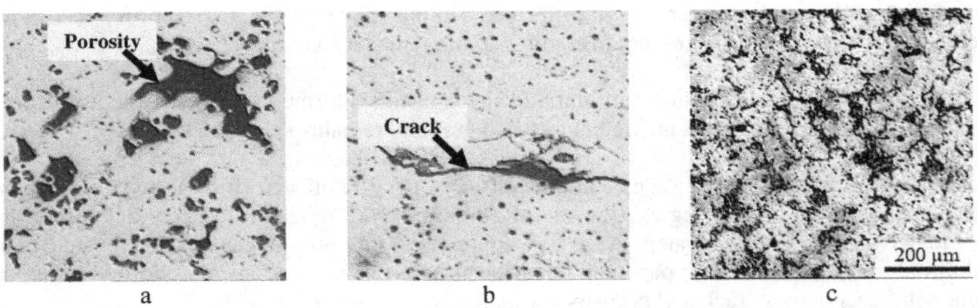

Fig.8. Formation of defect in 30 mm step, a) porosity, b) crack, c) dendritic solidification. with gate No.2 and die temperature 600°C.

According to Fig.5a and 9 ductile irons shows good fluidity in the semi-solid state and has capability of filling to 1 mm thickness and less, such as flash and vents (Fig.9).

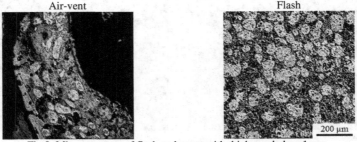

Fig.9. Microstructure of flash and vents with thickness below 1 mm.

Table 2 shows that graphite shape factor in the different steps changes little, varying less than 15%. This matter clearly shows the advantage of thixo-forming in comparison to the other deformation process of ductile iron. It must be considered that in 30, 25 and 15 mm step thickness (under the gate) shape factor decreases due to turbulence and lack of the liquid because of solid-liquid segregation.

Table 2. The graphite shape-factor before and after thixo-forming

Thickness (mm)	Shape factor	Thickness (mm)	Shape factor
30	0.82	3	0.95
25	0.84	1	0.94
15	0.85	Wedge	0.97
10	0.90	As-cast	0.97
5	0.91	As-reheat	0.98

Liquid fraction vs. different step thicknesses is shown in Fig. 10. Liquid fraction increases with decreasing thickness. This phenomenon can be due to difference between liquid and solid flow velocity. The probability of moving the large particle to thinner steps is low, therefore the large particle stay in thicker steps while it is easier to pass the liquid through different thicknesses.

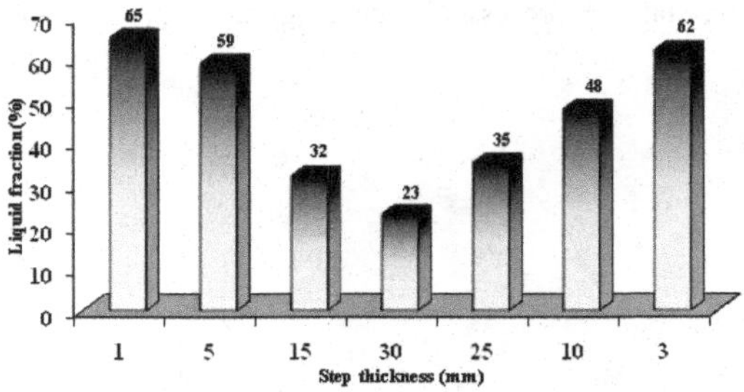

Fig.10. liquid fraction changes with changing in thickness

Conclusions

1. Thixo-forming of ductile iron in thick and thin sections (less than 1 mm) was successfully developed.
2. Die temperature and injection velocity are the most effective parameters to thixo-forming particularly for high melting point materials such as ductile cast iron.
3. Ductile iron shows good fluidity in semi-solid state.
4. Shape factor of graphite shows a minor change (less than 15%) in thixo-forming of ductile iron.
5. Liquid fraction increases with decreasing thickness. This phenomenon can be due to difference between liquid and solid flow velocity.

References

1. C. Labrecque, M. Gagné, "Optimizing the Mechanical Properties of Thin-Wall Ductile Iron Castings", AFS Transactions, American Foundry Society, Schaumburg, IL USA, 2005.
2. Ductile Iron Databook for Design Engineers, Rio Tinto Iron and Titanium, Montréal, 1999.
3. Chengchang Ji, Shigen Zhu, "Study of a new type ductile iron for rolling: Composition design (1)", Materials Science and Engineering A, 2006, Vol. 419, pp. 318–325.
4. J. Achary, "Tensile properties of ADI under thermomechanical treatment", Journal of Material Engineering and Performance, 2000, Vol. 9, No. 1.
5. Nuchthana Poolthong, Peiqi Qui, Hiroyuki Nomura, "Primary particle movement and change of property of cast iron by centrifugal effect in semi-solid processing", Science and Technology of Advanced Materials , 2003, Vol. 4, pp. 481–489.
6. Masayuki Tsuchiya, Hiroaki Ueno, Isamu Takagi, "research of semi solid casting of iron", JSAE Review, 2003, Vol. 24, pp. 205–214.
7. Z. Fan: Int. Mater. Rev., 2002, 47, 49-85.
8. F. Pahlevani, M. Nili-Ahmadabadi, "Development of semi-solid ductile cast iron", International Journal of Cast Metals Research, 2004, Vol. 17, No. 3, pp. 157-161.
9. B. Heidarian, M. Nili-Ahmadabadi, H. Mehrara, "The Role of Alloy Design in Semi-Solid Casting of Ductile Iron Processed by Inclined Cooling Plate", Submitted for publication.
10. S. Ashouri and M. Nili-Ahmadabadi: Solid State Phenomena, 2006, 116-117, 201-204.
11. P.K. Seo, C.G. Kang, Materials processing Technology, 2005, pp. 402- 409.

SHAPE CASTING:
3rd International Symposium
2009

Modeling

Session Chair:
Christof Heisser

Shape Casting: The 3rd International Symposium
Edited by: John Campbell, Paul N. Crepeau, and Murat Tiryakioğlu
TMS (The Minerals, Metals & Materials Society), 2009

PREDICTION OF COLUMNAR TO EQUIAXED TRANSITION IN ALLOY CASTINGS WITH CONVECTIVE HEAT TRANSFER AND EQUIAXED GRAIN TRANSPORTATION

Wajira U. Mirihanage[1], Shaun McFadden[1], David J. Browne[1]

[1]School of Electrical, Electronics and Mechanical Engineering,
University College Dublin, Belfield, Dublin 4, Ireland.

Keywords: CET, Equiaxed nucleation, Grain transport

Abstract

A macroscopic, non-equilibrium model of the Columnar to Equiaxed Transition (CET) in alloy shape casting is presented. Convective heat transfer in the liquid metal and equiaxed grain transportation by fluid flow is included in the model. Nucleation from mould walls is used as the mechanism for columnar grain initiation. Nucleation from inoculants in under-cooled liquid-ahead of the columnar front is considered for equiaxed grain formation. The front tracking model computes the advancement of the columnar front while the average growth of the equiaxed grain envelopes is simultaneously simulated. Latent heat release is incorporated in the model. The columnar mush and the coherent equiaxed dendrites are treated as porous media for convective flow. When equiaxed fraction is sufficient, no further advancement of the columnar front is permitted and the CET position is determined. CET is simulated for solidification of an aluminum-silicon alloy along with predictions of average equiaxed grain sizes.

Introduction

Columnar and equiaxed grains are normally contained in the macrostructure of alloy castings. When both types co-exist; a distinguishable transition is called the Columnar-to-Equiaxed Transition (CET) is often visible. Over the years, much research attention has been given to analytical, experimental and computer modeling of CET phenomena. CET models have been presented at the various scales [1-8]. The analytical and experimental work of CET has been recently reviewed by Spittle[9]. McFadden and Browne [1,2] presented a Front Tracking (FT) model to predict CET in alloy castings. This FT model provides promising macrostructure CET predictions with reasonable use of computer time and memory [10]. However, this CET model is limited to handle the scenario with pure thermal diffusion only and no consideration of natural thermal convection is included. The FT model for columnar growth, considering thermal convection is already published [11]. Apart from some small fragmentation, the bulk of the columnar mushy zone is stationary. But unlike the columnar zone, equiaxed grains are not stationary in the equiaxed mush as a result of convective flow effects. Also in the equiaxed case, the existing FT model [1,2] tracks each equiaxed grain separately. Hence, it is very difficult to simulate moving grains using FT and it will demand considerable computational power to perform such a simulation, especially when the casting is relatively large (as is the case for most industrial shape castings). Therefore, to give a CET prediction in the presence of thermal convection, an average equiaxed model is integrated to the existing columnar FT to give a novel FT model of CET. In the new model it is not necessary to track crystallographic or microscopic details for each individual dendrite, thus enabling a considerable reduction in computational overhead. As an initial development, solutal interactions are not considered. A version of this new variant of the model, but with thermal diffusion only, was already presented with experimental validation [12]. Further, attempts have been made (and are presented) to include physical identifiable sources of nuclei to form equiaxed grains. Two nucleation mechanisms are discussed: extrinsic and intrinsic mechanisms. Extrinsic nucleation particles are seeds that are introduced on purpose (e.g. inoculation using TiB_2). Intrinsic nucleation occurs as new crystal embryos are developed over time as solidification progresses (e.g., via fragmentation of the columnar front) [13].

The Mathematical Model

Initially temperature and convective flow velocities in each control volume are determined by solving conservation equations in the time domain. The transient state conservation equations govern the transport of energy, momentum and mass. Variables are expressed in 2D with the assumption of an incompressible liquid alloy,

$$\frac{\partial T}{\partial t} + \frac{\partial (uT)}{\partial x} + \frac{\partial (vT)}{\partial y} = \frac{k}{\rho C_p}\left[\frac{\partial^2 T}{\partial x^2} + \frac{\partial^2 T}{\partial y^2}\right] + E \tag{1}$$

$$\frac{\partial u}{\partial t} + \frac{\partial u^2}{\partial x} + \frac{\partial (uv)}{\partial y} = -\frac{1}{\rho}\frac{\partial p}{\partial x} + \frac{\mu}{\rho}\left[\frac{\partial^2 u}{\partial x^2} + \frac{\partial^2 u}{\partial y^2}\right] + (T - T_{ref})\beta g_x + S_x \tag{2}$$

$$\frac{\partial v}{\partial t} + \frac{\partial (uv)}{\partial x} + \frac{\partial v^2}{\partial y} = -\frac{1}{\rho}\frac{\partial p}{\partial p} + \frac{\mu}{\rho}\left[\frac{\partial^2 v}{\partial x^2} + \frac{\partial^2 v}{\partial y^2}\right] + (T - T_{ref})\beta g_y + S_y \tag{3}$$

$$\frac{\partial u}{\partial x} + \frac{\partial v}{\partial y} = 0 \tag{4}$$

Where, $u, v, T, t, p, \mu, \rho, C_p, k$ and β are velocity in x direction, velocity in y direction, temperature, time, pressure, dynamic viscosity, density, specific heat, thermal conductivity and thermal expansion coefficient respectively. T_{ref} is the reference temperature. S and E are source terms while g is an external force (gravity) acting on x and y directions. For this case g_x is zero.

Heterogeneous nucleation at the mould wall is considered for the columnar grains. So at a given nucleation undercooling, columnar grains start to grow from the domain boundaries. For equiaxed grains, free growth from inoculant particles is considered where a seed's diameter, d, determines the initiation nucleation undercooling , ΔT_i. The relationship between these parameters are given by [14] as,

$$\Delta T_i = \frac{4\gamma}{\Delta S_v d} \tag{5}$$

Where, ΔS_V is the volumetric entropy of fusion and γ is the solid-liquid interfacial energy. Purposely added inoculants are normally present in most industrial castings. The model assumes the presence of constant and uniformly distributed inoculants throughout the solidification period in molten metal. According to Quested and Greer [14] the size distribution for the diameter, d, of commercial inoculants in industrial aluminum castings follows a log-normal distribution, which is mathematically represent as,

$$f(d) = \frac{1}{\sigma d \sqrt{2\pi}} \exp\left(-\frac{(\ln d - \ln \varphi)^2}{2\sigma^2} \right) \tag{6}$$

Where, σ and φ are relevant to the standard deviation and the geometric mean. In the scenario where no purposely added inoculants are present, a simple time-dependant nucleation particle formation (addition to the molten melt) is taken in to the account. These creations of new seed embryos are possible due to columnar tip fragmentation [15]. In this scenario the total number of new seeds n_{tot} added to the melt is assumed as function of time and mathematically represented as,

$$n_{tot} = C_{fr} \int_0^t f(t)\, dt \tag{7}$$

$$f(t) = 0 \qquad \forall \qquad t < t_{start}$$

$$f(t) = (t - t_{start})^m \qquad \forall \qquad t \geq t_{start}$$

Where, C_{fr} and m are constants.

The dendrite tip velocity, v_t, is calculated according to Burden and Hunt [16] dendrite tip kinetics adapted for relatively low gradients. Therefore, dendrite tip velocity is given by,

$$v_t = C\, \Delta T_t^{\,2} \tag{8}$$

Where, C is the dendrite growth coefficient. Columnar growth is simulated using the FT. The columnar front tracking algorithm used here is described in detail in literature [17].

Equiaxed grains are assumed to be spherical in shape. Growth of this spherical mushy envelope is computed using the growth law for dendrite tips (equation 8). Equiaxed growth becomes complicated due to the new nucleation events over time; hence, in each time step average grain size and total number of equiaxed grains are computed. The total equiaxed volume in each control volume (CV) is given by,

$$V' = n\, V'_{old} + (n' - n)\, V_{new} \tag{9}$$

Here, V is volume and n is the number of nucleation seeds. Primes represent the values at the current time step and no primes represent the values at the old time step. The volume, V, with subscripts *new* and *old* indicates volumes of newly nucleated equiaxed grains and volume of previously nucleated equiaxed grains, respectively.
To consider grain impingement, the sum of all envelope volumes is taken as the extended volume $\phi_{ext.}$. Net volume fraction of the equiaxed mushy envelops ϕ, is calculated through the Avrami equation [18] .

$$\phi = 1 - \exp[-\phi_{ext}] \quad (10)$$

The transportation of nucleated equiaxed grains with melt flow is performed by considering the system as a slurry. Hence it is assumed that solid crystals are flowing with the equal velocity to the liquid in the molten metal. As a result of melt flow, number of equiaxed grains and grain volumes are continuously changing in each CV due to addition to the effects of local nucleation and growth. Therefore, the resultant number of equiaxed grains, N^F, and volumes, V^F, at the end of the time step in each CV is given by,

$$N^F = N + \Delta t.\frac{1}{\Delta x \Delta y}\sum_{sides} \pm N_i |u_i| \Delta z. \quad (11)$$

$$V^F = V + \Delta t.\frac{1}{\Delta x \Delta y}\sum_{sides} \pm V_i |u_i| \Delta z \quad (12)$$

Where, *N* and *V* are number and volume of equiaxed grains after the local progress alone, *u* is the flow velocity from the side and ± represent the appropriate direction of the flow. The numerical parameters Δt, Δx, Δy represent small time step, CV sizes in the *x*-direction and *y*-direction respectively. The parameter Δz is the respective CV interface from Δx or Δy.

Latent heat effects are incorporated by considering changes in the grain volume and local solid fraction [17]. The source term *E* in the energy equation accounts for the latent heat effects. These changes include the latent heat release during growth as well as absorption of latent heat during re-melting of the solidified grains when transported to the superheated liquid regions.

$$E = \left[g_s \frac{\partial V}{\partial t} + V \frac{\partial g_s}{\partial t} \right] \frac{L}{C_p} \quad (13)$$

Here, g_s is local solid fraction. Local solid fraction is calculated using a simple linear solid fraction relationship when in the freezing range [17].

$$g_s = \begin{cases} 0 & \forall \quad T \geq T_L \\ \dfrac{T_L - T}{T_L - T_S} & \forall \quad T_S < T < T_L \\ 1 & \forall \quad T \leq T_S \end{cases} \quad (14)$$

At low equiaxed solid volume fractions, the equiaxed region is modeled as a slurry. For slurry, viscosity is a function of the fraction of the solid volume, f_s; and hence modified viscosity, μ, has been obtained using the Thomas's empirical approximation cited in [19] and is given as,

$$\mu = \mu_0 \left[1 + 2.5 f_s + 10.05 f_s^2 + A \exp(B f_s) \right] \quad (15)$$

Where *A* and *B* are constants and μ_0 is the viscosity of the pure liquid. The columnar zone is treated as a porous medium [11,20]. At the higher equiaxed solid volume fractions, which exceeded the coherency level, f_{coh}, the equiaxed region is also treated as a porous media similar to the columnar zone. In porous media, the phase interaction forces are proportional to the artificial liquid velocity and source terms S_x and S_y are given by [21],

$$S_x = -\frac{\mu_0}{\rho} K \bar{u} \tag{16}$$

$$S_y = -\frac{\mu_0}{\rho} K \bar{v} \tag{17}$$

Where, K is a component of the permeability tensor – a physical property of a porous medium. It is a common practice to simplify these equations by the assumption that the mush is isotropic [19,22]. Therefore, using the permeability tensor K defined by Blake-Carman-Kozeny model using morphological constant K_O ,

$$K = K_O \frac{f_s^{\,2}}{(1-f_s)^3} \tag{18}$$

The evolution of equiaxed and columnar solid fractions in each CV are separately calculated by average equiaxed calculation and columnar front tracking, respectively. The columnar front is free to progress until it meets a control volume that is fully occupied by the equiaxed grains. At this point, no further growth of the columnar front is possible and a CET is predicted [12].

Simulation Results and Discussion

Model simulations are presented for Al-7%wt.Si. Aluminum and silicon have similar densities so solutal convection can be ignored; only thermal convection may be considered. We consider a rectangular mould casting of a size of *0.2m X 0.2m* (height-'*y*' *direction* and width-'*x*' *direction*) with the assumption that the front and back faces are adiabatic. This brings zero heat flux along the perpendicular direction ('*z*'); which converts the system in to 2D. Initial pouring temperature is set to *900K*. Initial imposed flow after pouring is assumed to be zero. The mould temperature is set at *825K*. The thermo-physical properties of the alloy and other parameters used for simulations are ; $L = 400\ kJkg^{-1}$, $k = 100 Wm^{-1}K^{-1}$, $\rho = 2600\ kgm^{-3}$, $C_p = 1000\ kgK^{-1}$, $\mu_0 = 1.3X10^{-3}\ kgm^{-1}s^{-1}$, $\beta = 7.1X10^{-4}K^{-1}$, $C = 5.0\ X\ 10^{-6}K^2m^{-1}s$, $K_0 = 4.92X10^7\ m^{-1}$, $T_l = 891K$, $T_s = 850K$, $f_{coh} = 0.6$ and $g = 9.81\ ms^{-2}$.

The number of nucleation seeds is set at *6000* and the size distribution follows a log-normal distribution. The mean diameter of seeds is *0.085μm* with a standard deviation *1.5.* This resembles the purposely-added inoculants. Figure 1.a shows the predicted CET from the model simulation and the equiaxed volume fraction variation throughout the casting. Figure 1.b present the predicted equiaxed grain size map. The average grain size of the casting is *3.9 mm*.

The model simulation is repeated to simulate the same casting, but without purposely-added nucleation seeds (inoculants). Intrinsic nuclei formation is simulated by using the simple relationship in equation (7); and no seed formation is allowed in the first *30* seconds of the process. Values of $C_{fr} = 9.15X10^{-5}$ and $m = 3$ are used. The same mean diameter and standard deviation of the previous extrinsic inoculants scenario is used this intrinsic case. Similar to the previous simulation with inoculants, figure 2.a shows the predicted CET position the details of the equiaxed volume fraction variation in the casting. Figure 2.b presents the predicted equiaxed grain size distribution. Average grain size in the equiaxed zone is *6.7mm.*

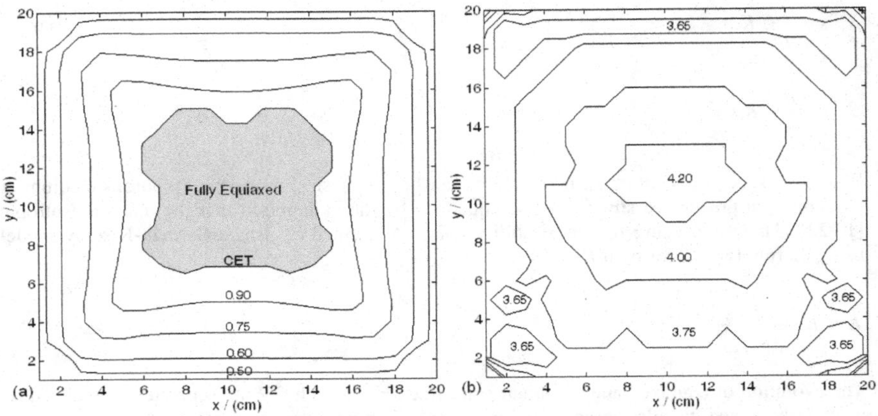

Figure 1. Results for casting with inoculants; (a) CET and Equiaxed volume fraction contours (b) Equiaxed grain diameter (mm) contour graph

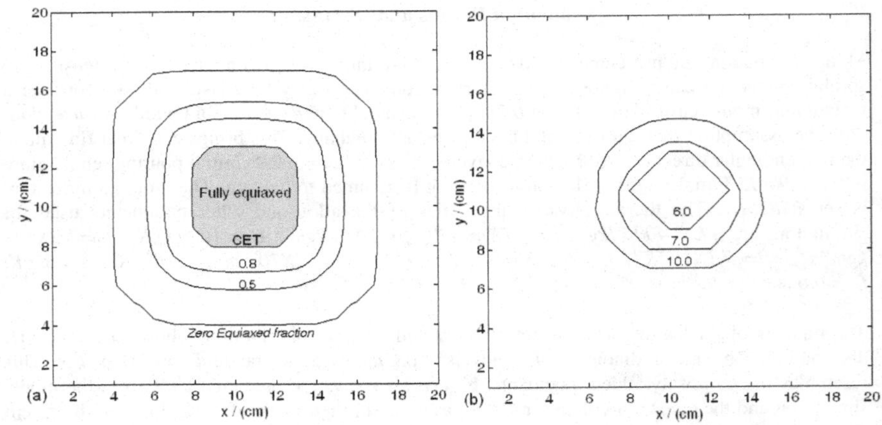

Figure2. Results for casting without initial inoculants and time dependant seed generation; (a) CET and Equiaxed volume fraction contours (b) Equiaxed grain diameter (mm)

A larger equiaxed zone is observed in figure 1 and CET is also observed to occur earlier in the inoculated melt than the non-inoculated one. Relatively high equiaxed volume fraction is present in the columnar region of the inoculated melt. According to these columnar and equiaxed volume fractions, it is more like a mixed columnar-equiaxed region. In a practical sense, these equiaxed grains close to the mould wall could appear elongated in shape, due to the presence of fairly high thermal gradient. Therefore, the equiaxed grains could appear quite columnar in real macrographs. Hence, equiaxed grain fraction could be lower in reality. With slightly larger inoculant particles or increased number of inoculants, the model would predict near completely equiaxed results in model simulations (as is the case in many industrial aluminum castings). The main aim of this contribution is to demonstrate CET prediction capabilities and, therefore, small inoculant particle sizes were used. Also, the numbers of inoculants was selected to maintain the final equiaxed grain size in the millimeter range.

In the simulation of inoculated solidification, the total number of equiaxed grains appearing is approximately 3300. This is nearly half of the total number of seed particles used for the simulation. The remaining smaller nucleation particles are unable to nucleate at the available undercooling prior to the occupation of space by growth. Apart from this, some amount (not specifically calculated) of nucleated equiaxed grains disappear due to transport into the superheated liquid, by fluid flow. The instantaneous dissolving of such grains is assumed in this contribution. In reality these crystal embryos could survive up to a certain time period and could lead to grain formation again, as discussed in ref. [23].

Conclusion

A 2D CET model is presented and a case with a Al-7%wt.Si alloy casting is simulated. Thermal convection is considered in the model and the resultant flow effects and equiaxed grain transportation effects are included. Simulations were performed for extrinsic and intrinsic nucleation particles. Extrinsic nucleation particles were introduced on purpose (e.g. inoculation). Intrinsic nucleation occurs as new crystals are developed over time (e.g., fragmentation of the columnar front). Equiaxed grain sizes were also predicted from the simulations. This model is limited to the growth scenarios where solutal diffusion lengths are reasonably smaller than the equiaxed grain sizes [24] and where solutal convection is weak. The sedimentation of equiaxed crystals has yet to be incorporated in to the model.

Acknowledgments

The authors wish to acknowledge the support of European Space Agency (ESA) under the PRODEX funding. This work is part of the ESA-MAP (Microgravity Applications Project) project CETSOL.

References

1. McFadden S., "*A Front tracking Model for Predict Columnar To Equiaxed Transition in Alloy Solidification*" (PhD Thesis, University College Dublin-Ireland,2007)
2. McFadden S., Browne D.J., *Appl. Math Modell.*(2008), doi:10.1016/j.apm. 2008.01.027
3. Hunt J.D., *Mater. Sci. Eng.*, 1984, vol 65, pp 73-83
4. Martorano M.A., Beckermann C., Gandin Ch-A., *Metall. Mater. Trans. A,* 2003, vol 34A, pp 1657-74
5. Rappaz M., Gandin Ch-A., *Acta Metall.* 1989, 23, pp.1777

6. Wang C.Y., Beckermann C., *Metall. Mater. Trans. A* , 1996, vol 27A, pp2755-64

7. Jacot A., Maijer D., Cockcroft S., *Metall. Mater. Trans. A*, 2000, 31A, pp2059-68

8. Flood S.C., Hunt J.D., *J. Cryst. Growth*, 1987, 82, pp 552-560

9. Spittle J.A., *Inter. Materials Reviews*, 2006, vol 51; no: 4, pp 247-269

10. S. McFadden , L. Sturz, H. Jung, N. Mangelinck-Noël, H. Nguyen-Thi, G. Zimmermann, B. Billia, D.J. Browne, D. Voss and D. Jarvis: J. Jpn Soc. Microgravity Appl., 2008, vol. 25, pp. 489-94

11. Banaszek J., Browne D.J., 2005, *Mater. Trans.*, Vol. 46 no.6, pp 1378-87

12. Mirihanage W.U., McFadden S., Browne D.J., "*A New Macroscopic Model for Prediction of Columnar to Equiaxed Transition using Front Tracking*" (Paper presented at 5th International Conference on Solidification and Gravity, Miskolc-Hungary, 3rd Sep. 2008), p.63

13. R.H. Mathiesen, L. Arnberg, P. Bleuet, and A. Somogyi: Metall. Mater. Trans. A, 2006, vol. 37A, pp. 2515-2525

14. Quested T.E., Greer A.L., Acta Mater., 2005, 53, pp 4643-53

15. Jackson K.A., Hunt J.D., Uhlmamm D.R., Seward T.P., *Trans.of Metall. Soc.of AIME*, 1966, vol. 236, pp 140 -158

16. Burden M.H., Hunt J.D., *J. Cryst. Growth*, 1974, 22, pp 109-116

17. Browne D.J., Hunt J.D., *Num. Heat. Trans. B*, 2004, 45, pp 395-419

18. Christian J.W., Theory *of Transformations in Metals and Alloys*, Pergamon, 1975, p 17

19. Qin R.S., Fan Z., 2001, *Matter. Sci. Technol.* 17, pp 1149-52

20. Banaszek J, McFadden S., Browne D.J., Sturz L., Zimmermann G., *Metal. Mater. Trans. A..*, 2007, Vol. 38A, pp 1476-84

21. Brent A.D., Voller V.R.,. Reid, K.J, 1988, *Num. Heat Trans.* 13, pp 297-318

22. Ilegbusi O.J., Mat M.D., *Mater. Sci. Eng.* 1998, A247, pp 135-141

23. Han Q., Hellawell A., *Metall. Mater. Trans. B*, 1997,Vol. 28B, pp 169-173

24. McFadden S., Browne D.J., *Scripta Mater.*, 2006, 55, pp 847-850

Shape Casting: The 3rd International Symposium
Edited by: John Campbell, Paul N. Crepeau, and Murat Tiryakioğlu
TMS (The Minerals, Metals & Materials Society), 2009

PREDICTION OF DEFORMATION AND HOT TEAR FORMATION USING A VISCOPLASTIC MODEL WITH DAMAGE

M.G. Pokorny, C.A. Monroe, C. Beckermann

Dept. Mechanical and Industrial Engineering, University of Iowa, Iowa City, Iowa 52242, USA

Keywords: Casting, Simulation, Stress

Abstract

A viscoplastic deformation model considering material damage is used to predict hot tear formation in a permanent mold magnesium alloy casting. The simulation model calculates the solid deformation and ductile damage. The viscoplastic constitutive theory accounts for temperature dependent properties and includes creep and isotropic hardening. Ductile damage theory is used to find mechanically induced voiding, and hot tears are expected in regions of extensive damage. Previously performed experiments are used to validate the simulation predictions. The test casting consists of a long horizontal bar connected to a vertical sprue on one side and an anchoring flange on the other. Hot tears are observed at the junction between the horizontal bar and the vertical sprue. The hot tearing severity is manipulated by adjusting the initial mold temperature. The simulation results are in good agreement with the experimental observations, both in terms of location and severity of the hot tears.

Introduction

The automotive industry is showing greater interest in magnesium alloys as they have a high strength to weight ratio when compared to steel or aluminum alloys. Because of their low density, incorporating magnesium alloys into the design of new vehicles decreases weight and increases fuel efficiency. This is especially important in helping to reduce carbon emissions that contribute to global climate change. However, some magnesium alloys show a high susceptibility to hot tearing, especially if cast in a permanent mold.

Hot tears are irreversible cracks that form in the semi-solid stage, called the mushy zone, during casting [1]. Hot tears develop as a result of thermal and mechanical strains due to the contraction of the casting and geometric constraints of the mold. In the mushy zone, porosity can form during late stages of solidification due to shrinkage. With sufficient deformation, this porosity may act as a nucleation site for hot tears. Hot tears can also form in the absence of porosity, but the lack of feeding flow is a necessary condition for a tear to remain open. As reviewed by Monroe and Beckermann [2], numerous attempts have been made in the past to understand the effect of various casting variables on hot tear formation and to develop criteria for predicting hot tears in castings. However, a truly predictive and reliable hot tear model is not yet available.

In the present study, a newly developed viscoplastic model that calculates deformation and damage is used to predict hot tears in a AZ91D magnesium alloy steel mold casting. The model is implemented in a general purpose casting simulation code. The model predictions are compared with the experimental results of Bichler *et al.* [1].

Description of Experiments by Bichler *et al.* [1]

The experiments by Bichler *et al.* [1] explored the hot tearing susceptibility of an AZ91D magnesium alloy test casting in a permanent steel mold. The composition of the AZ91D magnesium alloy used in the experiments was 8.61% aluminum, 0.6% zinc, 0.23% manganese, 0.017% silicon, 0.003% copper, 0.0038% iron, 0.0014% nickel, 0.0012% beryllium, and balance magnesium. The 20 mm thick casting consisted of a 260 mm (10.2 in) long horizontal bar connected to a vertical sprue and a flange, as shown in Figure 1. The figure also indicates thermocouple locations (TC1 to TC3), which recorded temperatures during casting at a rate of 7 readings/second. Temperatures of five separate castings poured at 720°C (1,328°F) were recorded using these thermocouples.

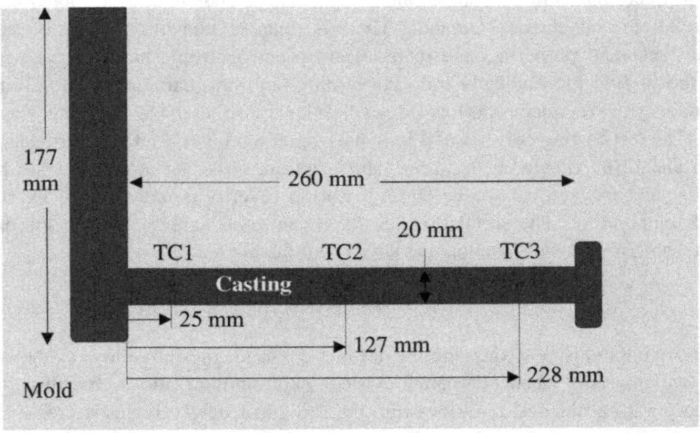

Figure 1. Casting and mold geometry with thermocouple locations [1].

During solidification, the contraction of the horizontal bar is restrained by both the sprue and the flange. This restraint can cause hot tears to form. The hot tears always occurred at the junction of the sprue and the horizontal bar [1]. The induced hot tearing was varied by changing the initial temperature of the steel mold. For a pouring temperature of 700°C (1,292°F), test castings were made at seven different initial mold temperatures ranging from 140°C (284°F) to 380°C (716°F). A semi-quantitative method was adopted to represent the severity of the hot tears observed in the test castings. This measure was called the hot tearing susceptibility index (HSI) [1]. The HSI was found to increase with decreasing initial mold temperature. In this paper, the thermocouple results and the hot tearing tendencies are compared to simulations.

Model

The filling, solidification and stress simulations were performed using a modified version of MAGMAsoft [3]. The simulations require inputs such as the three-dimensional geometry, mold temperature, pour temperature, thermo-physical properties, mechanical properties, mold-metal interfacial heat transfer coefficient, and others. Using these conditions and material properties, the temperature variations in both the casting and mold are predicted. The temperatures are then used in subsequent deformation calculations. Material deformation depends on the temperature results, since deformation is driven by density changes during casting solidification and further cooling within a rigid mold.

As part of the current study, a newly developed temperature-dependent viscoplastic deformation model was implemented in a module of MAGMAsoft called MAGMAstress. The details of this model are rather complex, and only a brief overview is presented here. The model considers the solid and liquid phases in the mushy zone along with damage induced porosity. The deformation of the mushy zone is a function of the solid fraction, which is a unique feature of the present model. Conventional models, such as the von Mises plasticity model, do not account for plastic volume change and no damage can be predicted. The model used in the current research includes the effect of plastic volume changes due to tensile (or compressive) strains. Assuming small strain theory, the total solid strain is decomposed into the elastic, thermal, and viscoplastic components. The elastic strain is governed by Hooke's law. The thermal strain is given by the linear thermal expansion coefficient, which is calculated from the density. In the mushy zone, thermal strains are assumed to be present only for temperatures below the eutectic start temperature. The viscoplastic or creep strains are determined by the flow condition. The flow condition limits the maximum stress the material can hold by keeping the equivalent stress below the yield stress. The equivalent stress depends on the solid fraction according to Cocks model [4]. In the limit where the solid fraction is unity, Cocks model returns to the von Mises solution. The yield stress function is a power law model including both isotropic hardening and creep.

Damage due to solid deformation is porosity created by volumetric plastic strain. The volume fraction damage (porosity) is found by integrating over time the volumetric part of the viscoplastic strain rate. The integration is started when the shrinkage porosity calculations indicate that the feeding flow is cut off. It should be noted, that the predicted damage is only an indicator of where hot tears may form in a casting; it does not predict the exact shape or size of the final hot tear. The potential for hot tearing increases with an increase in magnitude of this damage indicator. Damage provides an estimate of the volume that the crack occupies.

Additional details of the model, including the governing equations, can be found in reference [5]. The thermo-physical properties and the solidification path for the AZ91D magnesium alloy were calculated using the software JMatPro [6]. Reference [5] also contains a description of the method used to estimate the mechanical properties of the AZ91D magnesium alloy as a function of temperature.

Results

Temperature Predictions

The temperature measurements were used to determine the mold-metal interfacial heat transfer coefficient and to confirm the accuracy of the solidification and heat transfer simulations. The heat transfer coefficient was obtained using a trial-and-error procedure in which the predicted temperatures were matched with the experimental thermocouple data. The resulting heat transfer coefficient variation with temperature is shown in Figure 2(a). Above the liquidus temperature, a heat transfer coefficient of 6000 W/m^{2}K was found to result in good agreement between measured and predicted temperatures. Through the solidification range, the heat transfer coefficient was varied with the solid fraction to a final value of 1000 W/m^{2}K at 100% solid. From 100% solid to room temperature, the heat transfer coefficient was decreased cubically to 100 W/m^{2}K. The decrease in the heat transfer coefficient with temperature reflects the formation of an air gap between the casting and the mold during cooling.

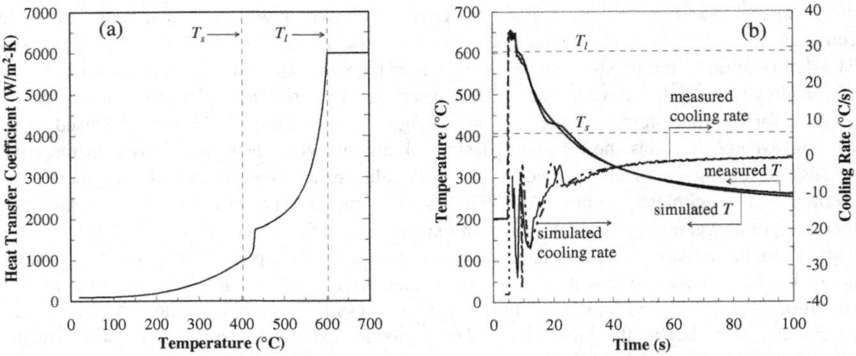

Figure 2. Interfacial heat transfer coefficient as a function of temperature (a); measured [1] and simulated thermocouple results for TC1 and an initial mold temperature of 202°C (b).

Figure 2(b) shows an example of a measured and predicted temperature comparison. Temperature versus time curves are shown for a thermocouple indicated as TC1 in Figure 1. The measured and predicted temperatures can be seen to be in generally good agreement. Similar agreement was obtained for all experiments in which temperatures were measured. More insight can be obtained by examining the measured and predicted cooling rate curves that are also shown in Figure 2(b). The cooling rate curves were obtained using a five point moving average of the time derivative of the temperature data. The first major peak in the measured cooling rate curve indicates nucleation of the primary Mg-rich solid. The peak corresponds to a temperature of 598°C (1,108°F) (average value from all temperature measurements). This nucleation temperature is 5°C (9°F) below the liquidus temperature of 603°C (1,117°F) predicted by JMatPro. The difference may be attributed to the presence of some nucleation undercooling; in fact, a temperature recalescence can be observed in the measurements. Nucleation is not modeled by JMatPro. A second major peak in the cooling rate curve indicates the start of eutectic solidification. The measured eutectic start temperature is equal to 421°C (790°F) (average value from all temperature measurements). This value is 10°C (18°F) lower than the eutectic start temperature predicted by JMatPro. The difference may be attributed to inaccuracies in JMatPro, in particular the Scheil analysis used to model solidification and the neglect of eutectic undercooling. The final 100% solid temperature cannot be inferred from the measured and predicted cooling rate curves due to the absence of any discernible peak. Despite the inaccuracies in the JMatPro data, the agreement between the measured and predicted temperatures was still deemed satisfactory for the present purposes.

Deformation and Hot Tear Predictions

Model predictions are compared to experimental results for three mold temperatures, 140°C (284°F), 260°C (500°F) and 380°C (716°F), and a pouring temperature of 700°C (1,292°F). Figure 3 shows the calculated final damage field and distortion of the casting magnified by 20 times for the three different initial mold temperatures. The simulations were terminated at 350 s after pouring; at this time the casting was fully solid and at a temperature close to the initial mold temperature. In Figure 3, the solid black line represents the original non-deformed casting. It can be seen that the sprue undergoes free contraction along its height. The magnitude of the contraction decreases with increasing initial mold temperature, since the thermal strain is less for a smaller temperature interval between the end of solidification and the final casting temperature (which is approximately given by the initial mold temperature). Several contact points can be observed along the mold-metal interface, which provide the necessary restraint for deformation and hot tears to occur. The most important contact points are seen at the junction of the sprue and the bar and at the flange end. Although the entire bar experiences contraction and restraint, the deformation causes the most significant damage on the sprue side of the bar. As expected, more damage (and distortion) is predicted to occur as the initial mold temperature is decreased.

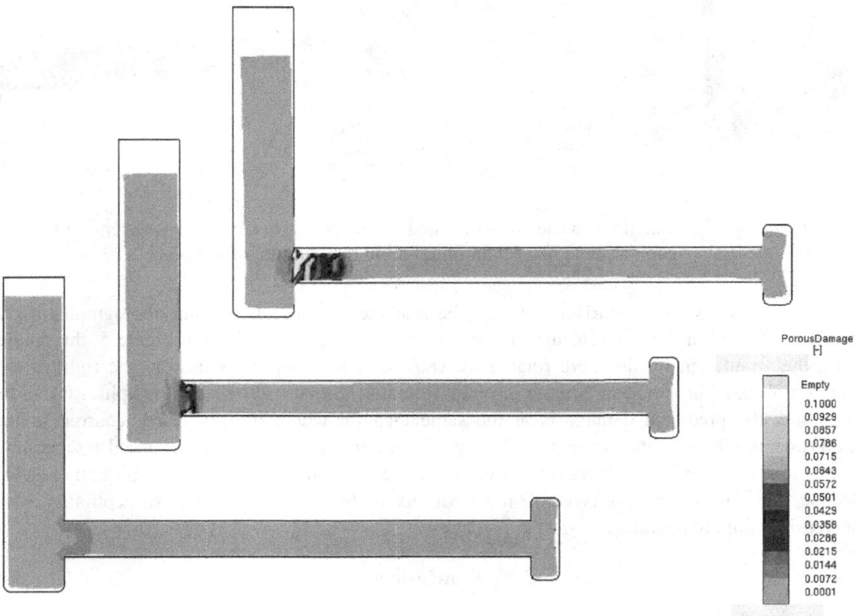

Figure 3. Calculated distortion, magnified 20 times, and damage field for initial mold temperatures of 140°C, 260°C and 380°C (from top to bottom).

Figure 4 shows the calculated von Mises stress and plastic effective strain at the end of the simulation for an initial mold temperature of 260°C (500°F). The von Mises stress is largest in the bar and almost zero in the sprue, as can be seen in Figure 4(a). In Figure 4(b), the plastic effective strain can be seen to be large in the bar with the largest values near the corners at the flange end. Hence, a hot tear criterion based on a von Mises model and effective strain alone would predict hot tear formation at the flange end where the stress and effective plastic strain are highest. However, in the experiments the hot tears did not form at the flange end, but at the junction with the sprue. Therefore, a prediction based on the von Mises stress and effective plastic strains would be inadequate for this casting.

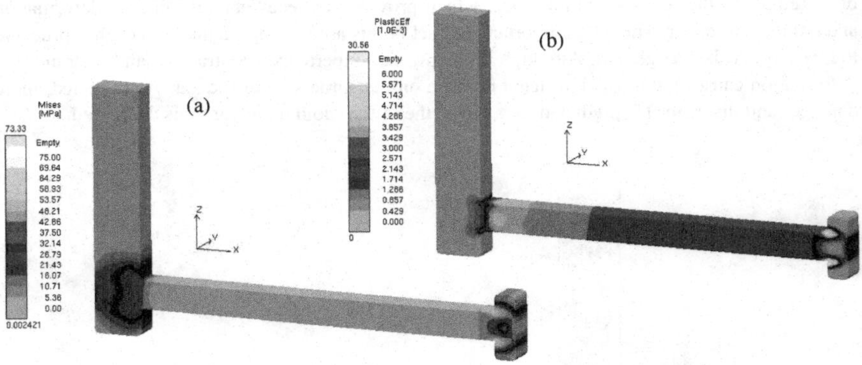

Figure 4. Calculated von Mises stress (a) and plastic effective strain (b) at the end of the simulation, 350s, for an initial mold temperature of 260°C.

Figure 5 shows a comparison between the predicted damage fields and photographs of the hot tears formed in the experiments for the three mold temperatures [1]. In Figure 5, the graphs with the simulation results were rotated by 180° so that the sprue is now to the right of the horizontal bar. This rotation is done in order to match the view in the photographs. It can be seen that the predicted damage is at the same location where the hot tears occurred in the experiments. The strong increase in the calculated damage with decreasing mold temperature corresponds well with the increasing severity of the hot tears seen in the experimental results. Hence, the simulation results confirm the observed decrease in hot tear susceptibility with increasing mold temperature.

Conclusions

A viscoplastic deformation model was used to predict hot tears in an AZ91D magnesium alloy permanent mold casting. The model calculates deformation and material damage. Preliminary estimates of temperature and strain-rate dependent mechanical properties were

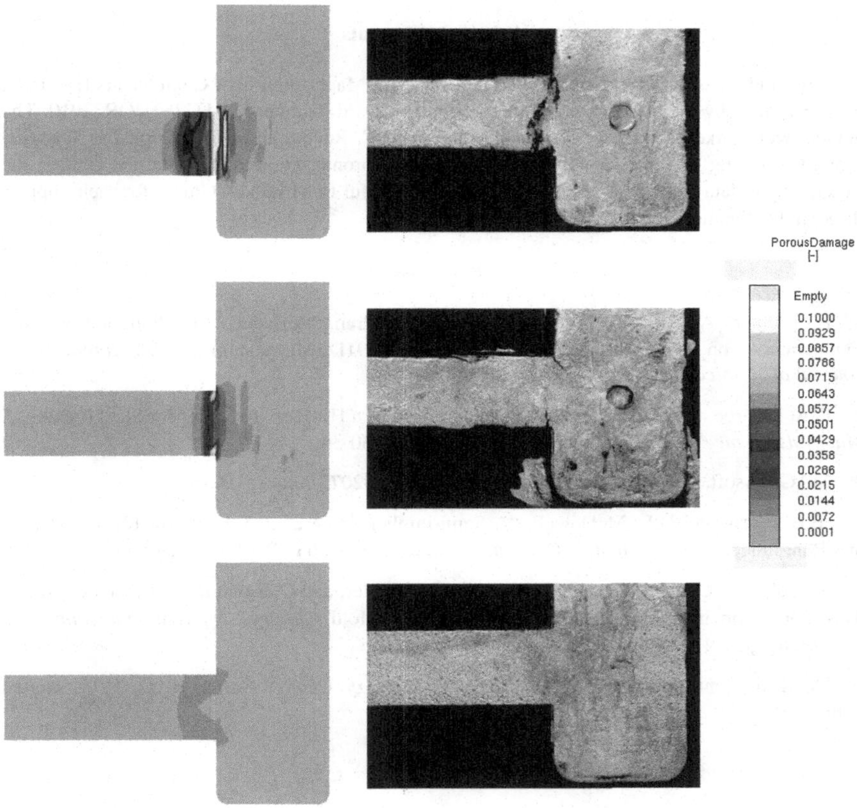

Figure 5. Simulation results showing calculated damage (left side) and the corresponding experimental results (right side) [1] for initial mold temperatures of 140°C, 260°C and 380°C (from top to bottom).

obtained from stress-strain data found in the literature [5]. Simulations were performed of experimental test castings of Bichler *et al.* [1]. The predicted damage from the simulations was found to be in good agreement with the hot tears observed in the experiments, both in terms of location and severity. The simulation results corroborate that the hot tears form most likely at the junction between the horizontal bar and the vertical sprue. The simulation results also confirm that hot tear susceptibility decreases with increasing mold temperature. These results indicate that the damage calculated using the viscoplastic deformation model is a reasonable predictor of hot tearing. Future work will include the measurement of more accurate mechanical properties. In addition, it will be desirable to compare the predicted stresses and strains with direct experimental measurements.

Acknowledgements

This study was performed under the High Integrity Magnesium Cast Components (HIMAC) Project funded by the Department of Energy under Award Number DE-FC26-02OR22910. The authors would like to thank Dr. L. Bichler and Prof. C. Ravindran of the Centre for Near-net-shape Processing of Materials, Ryerson University, Toronto, Ontario, Canada, for making the experimental data available. The authors are also grateful to MAGMA GmbH for their support through the donation of software, time and information.

References

1. L. Bichler, A. Elsayed, K. Lee, and C. Ravindran, "Influence of Mold and Pouring Temperatures on Hot Tearing Susceptibility of AZ91D Magnesium Alloy," *International Journal of Metalcasting*, 2 (1) (2008), 43-56.

2. C.A. Monroe and C. Beckermann, "Development of a Hot Tear Indicator for Steel Castings," *Materials Science and Engineering A*, 413-414 (2005), 30-36.

3. MAGMAsoft, MAGMA GmbH, Kackerstrasse 11, 52072 Aachen, Germany.

4. E.B. Marin and D.L. McDowell, "A Semi-implicit Integration Scheme for Rate-dependent and Rate-independent Plasticity," *Computers and Structures*, 63 (3) (1997), 579-600.

5. M. Pokorny, C.A. Monroe, C. Beckermann, L. Bichler, and C. Ravindran, "Prediction of Hot Tear Formation in a Magnesium Alloy Permanent Mold Casting," *International Journal of Metalcasting*, 2 (4) (2008), 41-53.

6. JMatPro, Sente Software Ltd., Surrey Technology Centre, Surrey GU2 7YG, United Kingdom.

Shape Casting: The 3rd International Symposium
Edited by: John Campbell, Paul N. Crepeau, and Murat Tiryakioğlu
TMS (The Minerals, Metals & Materials Society), 2009

A DIFFUSING RUNNER FOR GRAVITY CASTING

Fu-Yuan Hsu and Huey-Jiuan Lin

Department of Materials Science and Engineering, National United University,
No.1 Lein-Da, Kung-Ching Li, MiaoLi, 36003 Taiwan, R.O.C.

Keywords: diffuser, runner system, aluminium gravity casting, critical gating velocity.

Abstract

In gravity casting, the quality of an aluminium alloy casting relies on, among other things, the design of the runner system in which the gate velocity into the mould cavity should be controlled to stay under a critical velocity (close to 0.5 m/s). In this study a diffuser was proposed to reduce the velocity of liquid metal to below this critical value, while the flow rate remained almost unchanged. Flow separation and dead-zones in the diffuser design were avoided. A computational modeling package and real casting experiment (water analogy method) were employed for exploring and verifying the new design. The efficiency of the diffuser was quantified by the measurement of coefficient of discharge, Cd.

Introduction

Aluminum alloys, as the liquid aluminium contacts the surrounding atmosphere containing the oxygen, have a tendency to form an insoluble oxide-film on the surface. With velocities higher than 0.5 m/s, "fountaining" with a mushroom shape was observed, which generates severe rates of oxides entrainment [1]. Also, no surface defect was found when an aluminum alloy (with 7% Mg) was cast in a runner system with a careful control of gate velocity (0.46 m/s) [2].

High reliability aluminum alloy (i.e. Al-7Si-Mg) castings were obtained by bottom gating a sand mould at an initial gate velocity less than or equal to 0.5 m/s [3]. Consequently, the liquid aluminum has a critical ingate velocity of less than 0.5 m/s [4]. Thus, one of the desirable functions of runners or gates is to decelerate the liquid metal to a speed below this critical ingate velocity.

As a large fraction of the dynamic pressure (or kinetic head) can convert into static pressure (or potential head) in the form of a pressure rise, a reduction of velocity can be achieved. This may be accomplished by decelerating the flow in a diffusing passage (or diffuser).

In an attempt to reduce the velocity of the metal flow by pressure recovery, many authors used diffusers in their experiments [5-7]. But most of them did not report satisfactory results afterwards. This is probably because of a sudden increase in the cross sectional area of geometry in their design. As a result, the liquid metal will break away from the wall and becomes a free stream. This unstrained stream will damage the flow quality as it is incorporated or mixed with the surrounding atmosphere (i.e. air or mould gases).

Therefore, the current study focused on developing a diffusing runner, which could reduce the velocity of flow under the critical velocity of 0.5 m/s. Also, there is no flow separation from the wall of diffusing runner.

Method

Diffusing runner

Diffuser data are most commonly presented in terms of a pressure recovery coefficient, C_p, defined as the ratio of static pressure rise to inlet dynamic pressure,

$$C_p \equiv \frac{p_2 - p_1}{\frac{1}{2}\rho \cdot \overline{V}_1^2} \qquad \text{Equation 1}$$

The above equation, where $\overline{V}_1$ is the average inlet velocity, (p_2 - p_1) is the pressure difference between the outlet and inlet, and ρ is density of liquid, can be obtained from the Bernoulli equation.

In order to achieve the energy head of flow transforming from velocity to pressure heads, the pressure recovery should be derived from liquid metal flow into a diffusing geometry. By creating various heights of a baffle within the diffusing runner (Figure 1 (c)), the pressure head could be recovered and the flow velocity could be reduced.

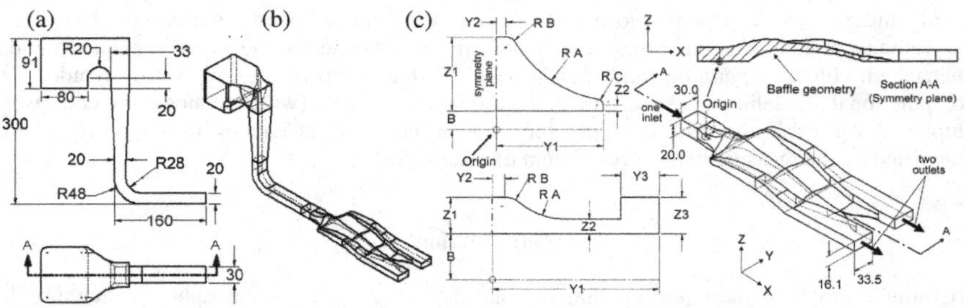

Figure 1 The runner systems for the experiments: (a) the system without the diffuser, (b) the system with the diffuser, and (c) the diffuser.

Figure 1(c) showed the schematics of diffusing runner for this experiment. In this diffuser, one inlet and two outlet runners were comprised. The baffle height B changed along the diffuser as shown in the Table 1. Moreover, the profile of y-z sections altered along x direction. The area of two outlet runners expanded from zero to 539.4 (i.e., 16.1×33.5) mm^2. But the inlet region of 600 (i.e., 20.0×30.0) mm^2 reduced along the x direction until zero. The area of outlet runners is 1.8 times of that of inlet runner.

Table 1 The dimensions for various cross-sections of the diffuser shown in Figure 1. (Unit: mm)

x	B	Y1	Z1	Y2	Z2	Y3	Z3	RA	RB	RC
25.5	0	18.0	23.0	6.7	6.2			30.0	6.0	6.0
39.4	0	26.3	26.5	4.0	2.0			84.9	5.0	3.0
45.3	3.6	34.0	24.8	3.0	2.0			40.0	3.5	2.0
61.73	12.7	51.7	18.3	1.6	5.0	6.0	7.0	40.0	2.0	
81.2	15.6	54.0	14.6	3.0	5.0	10.5	11.0	35.0	5.0	
145.7	12.0	54.0	8.0	4.5	4.0	17.5	18.0	47.5	29.2	
197.8	0	54.0	6.8	3.0	4.0	22.0	16.1	30.0	30.0	

Water analogy method

The water analogy method was used in two experiments for the runner systems without and with the diffuser (Figure 1(a) and Figure 1(b)). In the runner system without the diffuser, there is a pouring basin, sprue, a bend-shape runner, and detailed dimensions are shown in Figure 1(a). The diffuser assembled at the end of the previous system with total height of 300mm was demonstrated in Figure 1(b).

Transparent plastic, Polymethylmethacrylate (PMMA), was used as mould material for machining the shape of the runner. Liquid water was poured into the pouring basin, where a plastic stopper was placed at the entrance of the sprue. As the basin was filled completely, the stopper was abruptly lifted clear from the sprue entrance, permitting the start of the filling process. Additional water was poured continuously into the basin to maintain a constant head height in the basin during the filling of the runner system. For clear observations, red colored ink was added into the water. The filling sequences of the dyed water were recorded at the speed of 25 frames per second by a camera.

Coefficient of discharge

The efficiency of the diffuser can be quantified by the coefficient of discharge. Due to energy losses through the wall friction in the diffuser, the actual discharge will be slightly less than the one of the original. These energy losses are therefore accounted for by introducing a coefficient of discharge Cd,

$$Cd = \frac{Actual\ \ flow\ \ rate}{Original\ \ flow\ \ rate} \qquad \text{Equation 2}$$

The original flow rate of the diffuser can be calculated from the runner system without the diffuser (Figure 1(a)). Then, the actual flow rate can be derived from the runner system with the diffuser (Figure 1(b)).

The flow rate could be obtained from the product of the cross-sectional area of flow and its velocity. If the flow was completely filling a runner cross-sectional area, the flow is equal to that runner geometry. Therefore, the actual velocity could be determined by the trajectory method as shown in the following.

The trajectory method

When the flow exits the end of the horizontal channel, its trajectory has two components, horizontal and vertical. If the air resistance during the fall is neglected, the launch velocity V_o (or the velocity at the end of that channel) is therefore,

$$V_o = D_h \sqrt{\frac{g}{2D_v}} \qquad \text{Equation 3}$$

where D_h and D_v are the distance of horizontal and vertical components respectively, g is gravitational acceleration.

The actual flow rate issuing from the end of the channel is:

Actual flow rate = $A_c \cdot V_o$ Equation 4

where A_c is the cross-sectional area of the launch flow stream.

Computational modeling

The purpose of computational modeling is to understand the filling sequences in the diffuser. The flow transformation from kinetic to pressure heads then could be studied from the modeling results.

A computational fluid dynamics (CFD) code, Flow-3D™, has been used for these investigations. Since the flow phenomena in running systems were mainly considered here, one fluid (i.e. liquid metal) with sharp interface tracking (i.e. free surface boundary tracking) of a VOF algorithm was employed. Isothermal conditions were assumed since the change of viscosity of the liquid could be ignored as approximately only 1 per cent of heat loss was expected [8].

Result

Two parts of experiment results are showed in the followings. There are computational modeling and water analogy.

<u>Computational modeling</u>
From the modeling results, the velocity contours at various time frames in isometric view were illustrated in Figure 2.

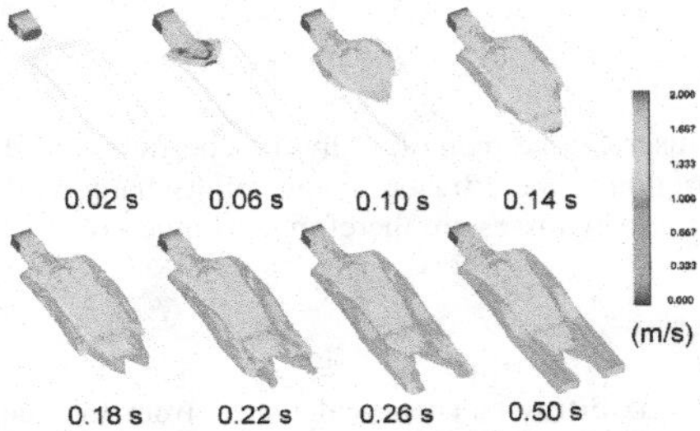

Figure 2 Filling profiles of the diffuser at various time frames in an isometric view.

Since the initial inlet velocity of 2.0 m/s was fixed, the velocity contour at the runner entrance was in red color before impacting the baffle geometry within the diffuser. At 0.06s the high velocity liquid aluminium contacted the baffle and started to reduce velocity. At this time the flow commenced to spread laterally across the baffle geometry of the diffuser (c.f. Figure 1). From the time frame of 0.10s to 0.18s, the spreading flow filled the two outlet regions. Until 0.5s the flow filled the diffuser sequentially without large empty regions remaining.

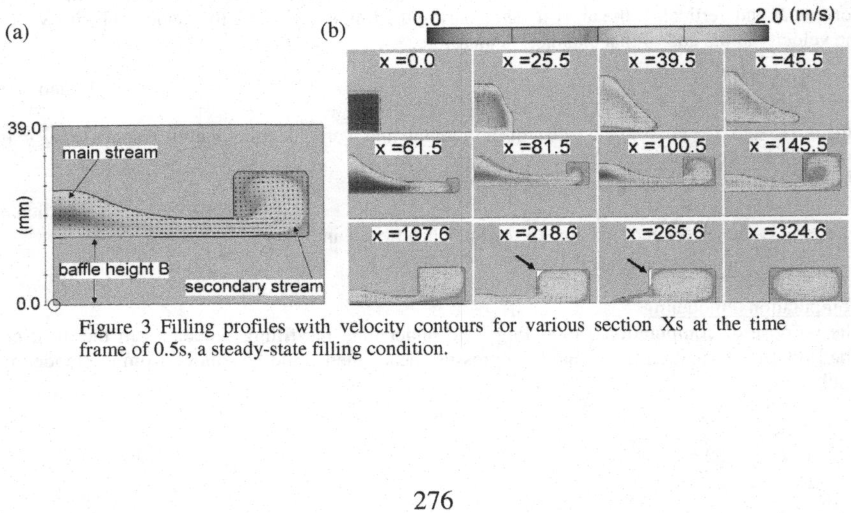

Figure 3 Filling profiles with velocity contours for various section Xs at the time frame of 0.5s, a steady-state filling condition.

Regarding the magnitude of the velocity, it decreased from 2.0 m/s (i.e., red color) in the inlet runner to approximate 1.0 m/s (i.e., yellow and green color) in the outlet runners.
Figure 3 (a) illustrates the velocity magnitude and vectors of the y-z cross-section at the x coordinate of 109.5mm. In this section the length of vector in the main stream is short (or reduced to a point). It means that the direction of flow in this region is normalized to the section. Moreover, in the secondary stream region the vector length is long and the lateral direction of the flow is transformed. For the velocity magnitude, the high velocity (i.e., close to red color) in the mainstream was depicted while the low velocity (i.e., yellow and green color) in the secondary stream was shown.
When liquid aluminum filled the diffuser in a steady-state condition (e.g., a long computational time, 0.5s in Figure 3 (b)), the flow in the diffuser would not change a much. Therefore, the filling conditions in various y-z cross sections could be demonstrated.
In the region of X sections from 218.6 to 265.6 (mm) (c.f., Figure 3 (b)), there is small empty region at the upper-left corner of the outlet runner. In the diffuser, other than in this region, the liquid aluminum filled the runner completely.

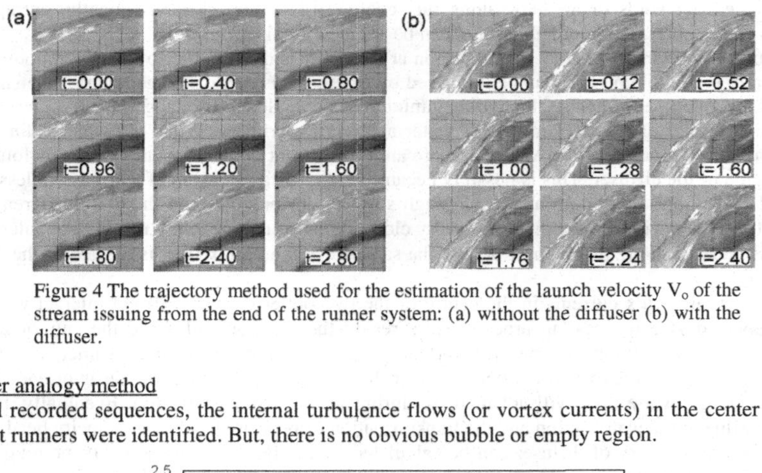

Figure 4 The trajectory method used for the estimation of the launch velocity V_o of the stream issuing from the end of the runner system: (a) without the diffuser (b) with the diffuser.

Water analogy method

In all recorded sequences, the internal turbulence flows (or vortex currents) in the center of the outlet runners were identified. But, there is no obvious bubble or empty region.

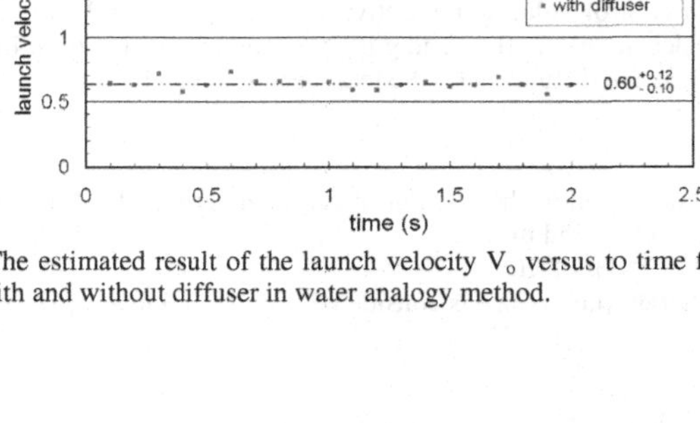

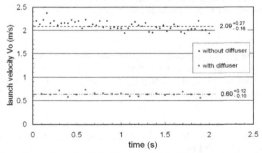

Figure 5 The estimated result of the launch velocity V_o versus to time for the runner systems with and without diffuser in water analogy method.

Figure 4 shows the trajectory stream issuing from the end of the runner systems without and with the diffuser. The estimated result of the launch velocity V_o versus the time for these two runner systems is revealed in Figure 5. Before the diffuser was added in the system, the velocity of 2.09 (m/s) was estimated. After the diffuser is included, a velocity of 0.60 (m/s) was measured.
Considering the flow stream completely filled the runner system, the flow rates of 1.25×10^{-3} (i.e., $2.09\times600\times10^{-6}$) and 0.65×10^{-3} (i.e., $0.60\times2\times539.4\times10^{-6}$) m^3/s were estimated across the inlet and outlet runners in the diffuser, respectively. For the diffuser the coefficient of discharge Cd is therefore 0.52.

Discussion

In this diffuser system with a total head height of 300mm, the average velocity of 0.6 m/s, which is close to the critical gate velocity for aluminium alloy casting, is achieved. If a gate velocity of less than 0.5 m/s is preferred, an additional diffuser could be connected to the end of this system. Assembling of diffusers is an opportunity to control the gate velocity within the required value.
The filling sequences in the diffuser, as shown in Figure 2, confirm that filling in one-pass without leaving voids or air gaps along this geometry is achieved. Consequently, entrapping oxides and bubbles into liquid aluminum can be avoided in this diffuser.
In Figure 3 (b), there is a small empty region at the upper and left corner of the X-sections from 218.6 mm to 265.6mm. This can be improved by filleting this part of the geometry. The smaller size of this region ensures that the liquid aluminum can completely fill the geometry.
But this result is different from that of water analogy experiment. When the water filling was tested in the transparent PMMA mould, there are no apparent bubbles or empty regions found in this region of the diffuser. This is possibly because of the surface tension of water being less than that of real liquid aluminium alloy. Although surface tensions between them are different, the water filling behaviour in the diffuser is very close to that of the modeling result. The inlet flow spreads laterally along the diffuser and, at the same time, the vortex flow is created in the outlet runner.
These internal vortex currents in the centre of the two outlet runners will not interact with the atmosphere during the casting process. As a result, the diffuser will avoid the entrapment of oxide films and air bubbles, when the vortexes form. By reducing the area of inlet runner, the flow is forced to fill across to the outlet runner. If enough area of outlet runner is provided, the kinetic energy of the flow will not increase during the flow transformation. Eventually, it will decrease due to the wall friction and the transformation of pressure head from velocity head.
The pressure recovery of diffuser can be calculated by the Bernoulli equation. If the total head loss of the flow in the diffuser is denoted as h_L, the equation can be obtained as follows.

$$\frac{p_1}{\rho g}+\frac{u_1^2}{2g}+z_1=\frac{p_2}{\rho g}+\frac{u_2^2}{2g}+z_2+h_L \qquad \text{Equation 5}$$

where $\mathbf{p_1}$,$\mathbf{u_1}$, **and** $\mathbf{z_1}$ are pressure, average velocity, and elevation height of a considered flow stream entering into a control volume, respectively, while $\mathbf{p_2}$,$\mathbf{u_2}$, **and** $\mathbf{z_2}$ are those of the flow leaving from. ρ is the density of the flow and g is gravitational acceleration. In abrupt changes in area or geometry, such as the diffuser here, the total head loss can be expressed as:

$$h_L = K_L\frac{u_2^2}{2g} \qquad \text{Equation 6}$$

Where K_L is the loss coefficient of the diffuser, it can be determined by coefficient of discharge, Cd, of the diffuser (i.e., 0.52); and that is:

$$K_L = (1/Cd^2 - 1) = (1/0.52^2-1) = 2.70 \qquad \text{Equation 7}$$

If the same elevation height (i.e., $z_1=z_2$) is considered, Equation 5 may be rewritten as:

$$\frac{u_1^2 - u_2^2}{2g} = \frac{p_2 - p_1}{\rho g} + K_L \frac{u_2^2}{2g}$$ Equation 8

By replacing the values derived from the result of water experiment, Equation 8 then can be calculated as:

$$\frac{2.09^2 - 0.60^2}{2\times 9.81} = \frac{p_2 - p_1}{\rho g} + 2.70\frac{0.60^2}{2\times 9.81}$$

$$0.204 = \frac{p_2 - p_1}{\rho g} + 0.050$$

Therefore, pressure head differences between entering and leaving the diffuser is:

$$\frac{p_2 - p_1}{\rho g} = 0.154 \text{ (m)}$$ Equation 9

If the density of liquid aluminium, 2430 kg/m^3, is used, the pressure difference then can be decided.

$p_2 - p_1 = 3671$ (Pa) Equation 10

If the minimum entering velocity is considered 1.93 (i.e., 2.09-0.16, as shown in Figure 5) m/s, the pressure difference can be recalculated as 2896 Pa. This value is very close to the pressure difference between the flow impacting on the baffle geometry, 2986 (i.e., 4436-1145) Pa, as shown in Figure 6.

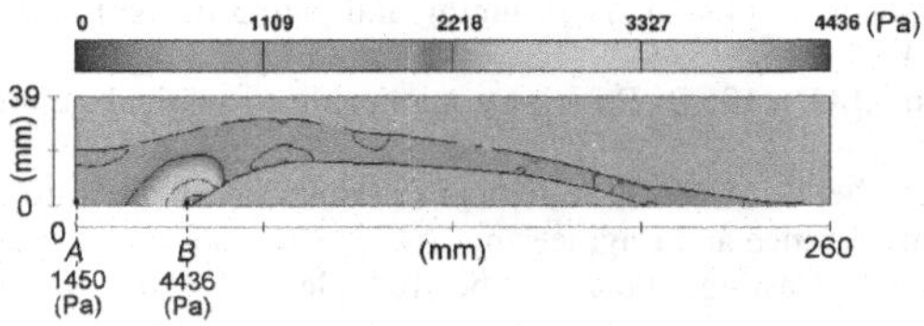

Figure 6 The pressure contour in the X-Z section on the symmetry plane of the diffuser. The pressures at two points read from the modeling result.

Thus, the pressure recovery from this impact at the beginning of the baffle geometry (i.e., the point B in Figure 6) can be verified by this calculation. Afterwards, the function of the remaining part of the diffuser geometry is holding the transformed flow without creating the detrimental oxide films and bubbles. But the size of this part should be kept as small as possible, since a large area of geometry will increase the area of wall frictions during the filling, resulting in big energy head losses and small coefficient of discharge for the whole geometry. Hence, the role of the baffle geometry is to recover the pressure head from the kinetic head as much as possible, while that of the rest part of the diffuser is to constrain the flow completely and to minimize the head losses. Furthermore, by replacing the values derived previously, the pressure recovery coefficient (Equation 1), C_p, for the diffuser in this study, can be estimated as:

$$C_p = \frac{p_2 - p_1}{\frac{1}{2}\rho \cdot \overline{V}_1^2} = \frac{3671}{\frac{1}{2}\times 2430 \times 2.09^2} = 0.69$$

The values of C_p, K_L, and Cd, are useful for future application when an accurate prediction of flow rate in a runner system is needed.

Conclusion

1. In this diffuser system with a total head height of 300mm, the gate velocity of 0.6 m/s is obtained.
2. Kinetic energy is efficiently transformed to pressure head via this diffuser. The pressure recovery coefficient, C_p, of 0.69 is achieved, as the loss coefficient, K_L, of 2.7 or the coefficient of discharge, Cd, of 0.52 is estimated.
3. Considering the functions in individual geometry, the diffuser comprises two parts. One is to maximize the pressure recovery in the baffle geometry and the other is to constrain liquid aluminium without generation of oxide films and bubbles, and to maintain its size as small as possible.
4. Vortex currents, produced in the core of the outlet runner, will not damage the free surface of liquid aluminium. An optimizing shape of outlet runner can be accomplished by tailoring the geometry to meet the actual profile of flow. The harmful oxide films and bubbles can be prevented in this diffuser.

Acknowledgements

F-Y Hsu acknowledges the help of Yau-Ming Yang and Mei-Yu Yang, and the sponsorship of National Science Council in Taiwan (R.O.C.), with project no. of NSC 97-2218-E-239 -002.

References

1. Runyoro J., Campbell J., (1992), The running and gating of light alloys, The Foundryman, Apr., pp.117-124.
2. Grube K.R., Kura J.G., (1955), Principles applicable to vertical gating, Trans. AFS, v.63, pp.35-48.
3. Green N.R., Campbell J., (1993), Statistical distributions of fracture strengths of cast Al-7Si-Mg, Materials Science and Engineering, A173, 261-266.
4. Campbell J., 2004, Castings Practice the 10 rules of castings, Elsevier Butterworth-Heinemann.
5. Swift R.E., Jackson J.H., Eastwood L.W., (1949), A study of the principles of gating, Trans. AFS, v.57, pp.76-88
6. Robertson J.T. and Hardy R.C. ,(1946), Trans AFS, vol. 54, p732
7. Sirrell B., Campbell J., (1997), Mechanism of Filtration in Reduction of Casting Defects due to Surface Turbulence during Mold Filling, Trans. AFS, pp.645-654.
8. Richins D.S., Wetmore W.O., (1952), Hydraulics applied to molten aluminum, Trans. ASME, July, pp.725-732.

Shape Casting: The 3rd International Symposium
Edited by: John Campbell, Paul N. Crepeau, and Murat Tiryakioğlu
TMS (The Minerals, Metals & Materials Society), 2009

PROCESS MODELLING AND MICROSTRUCTURE PREDICTION IN GRAVITY DIE ALUMINUM CASTINGS WITH SAND CORES

Rosario Squatrito, Ivan Todaro, Luca Tomesani[1]

[1] DIEM - Department of Mechanical, Nuclear, Aviation, and Metallurgical Engineering - University of Bologna, Viale Risorgimento 2, Bologna, Italy, 40136

Keywords: Gravity Die Casting, A356, Numerical Simulation, SDAS, Shrinkage Porosity

Abstract

In this work a full numerical analysis has been carried out on the gravity die casting process of a 8 cylinders A356 aluminum engine head, which currently is in production. The complete mould system consisted of seven steel parts, three cast iron supports with heating devices, and seven high-complexity sand cores.
The fully assembled FEM model consisted of $11.8*10^6$ elements. Filling and solidification analysis were done by carefully replicating all the available process monitoring data, such as alloy composition, casting temperature, filling strategy, die temperatures and gradients, as well as cooling times. Unknown simulation parameters, such as boundary HTC at various interfaces were fine tuned within acceptable ranges, in order to fit experimental observations.
Microstructure characteristics, such as SDAS, were evaluated by using existing relationships with cooling times and then correlated with the experimental ones at 640 different measuring points, covering four different sections of the head. Correlation factors between predicted and measured values were evaluated for each section.
The porosity content was also evaluated via the shrinkage porosity module of the adopted numerical code and then verified with 640 corresponding experimental values. This also led to correlation factors being evaluated between predicted and experimental numbers.

Introduction

Main competitors ruling the automotive sector are fighting their struggle by focusing efforts on improving the power-to-weight ratio by increasing the structural performances and quality standards of the manufactured components. This can be achieved by optimizing the production technologies and by having a more and more precise and consistent knowledge about microstructure and defects formation. This leads to the possibility to reduce safety factors in the design process and to reduce the overall weight of the casting.
During the last years, foundries producing car components are ever more frequently adopting process simulation tools. These can forecast the final mechanical characteristics of the castings by correlating the particular production process to the microstructure evolution and to the insurgence of defects within the products.
The physical and mathematical models used by these software tools have been widely improved and many validation tests showed their good reliability. However, their user-friendly interface often leads to underestimate the real complexity of the casting technology: fluid dynamics play a role in filling [1], as well as heat exchange, phase changes, and micro structure evolution during solidification.
Thus, the task of reproducing all the real physical conditions of a production process in a simulation can be a really complex one. Not only owing to the geometrical complexity of a casting, but also because the precise operating conditions of the casting devices are often

difficult to evaluate and control. Moreover, the wrong evaluation of the boundary conditions can heavily affect simulation reliability. In this perspective an accurate set-up through direct measuring campaigns is absolutely necessary.
In this paper we propose the analysis performed on a V8 4.2 liter cylinder head obtained by gravity die casting at Ferrari S.P.A.. The gravity die casting process for aluminum engine heads is by far one of the most complex of the casting industry: the high geometrical complexity (the complete mould system consists of many parts: mould tools, supports, sand cores) is added to the demand of the highest mechanical properties, lowest levels of porosity, together with the need for a careful control of the tool's temperatures throughout the production cycle.
The aim of the work is, first, to assess the technical feasibility of such analysis, trying to replicate all the monitored data by tuning the numerous boundary conditions of the thermal problem; secondly, to understand the robustness of existing algorithms for microstructure evaluation such as SDAS and porosity, by which the final mechanical properties can be predicted.
The numerical analysis of this research has been carried out using the commercial software for casting simulation, Procast V.2008 by ESI Software [2].
The numerical model consisted of $11.8*10^6$ elements. The filling analysis was validated with data coming from the casting device: casting time, metal head evolution in the basin, time for not-moving metal on the casting top. The fully coupled solidification analysis was carried out by adjusting unknown simulation parameters such as boundary HTC at various interfaces, within acceptable ranges [3], in order to fit experimental observations.
Microstructure characteristics, such as SDAS, were evaluated by using existing relationships with cooling times and then correlated with the experimental ones at 640 different measuring points, covering four different sections of the head. Correlation factors between predicted and measured values were evaluated for each section.
The porosity content was also evaluated via the shrinkage porosity module of the adopted numerical code and then verified with 640 correspondent experimental values. This also led to correlation factors to be evaluated between predicted and experimental numbers.

Process Description

From the foundryman's point of view, the production of this casting is a challenge, due to the great geometrical complexity coming from functional features required. Moreover, the casting has to be completely defects free on some key surfaces and volumes and has to exhibit excellent mechanical characteristics once in use, also at very high service temperatures (combustion chamber walls reach temperatures up to 300°C).
Those needs, together with the relatively high number of pieces to be produced, determined the gravity die casting process to be selected for this aim. This technology permits:

- Good dimensional and surface tolerance
- Relatively tranquil filling, in order to reduce inclusions and oxid-film content
- High heat rates, in order to have a fine microstructure

The mould consists of seven parts (four of them sliding) made of tool steel, three cast iron supports and seven sand cores. In particular, the mould consists of (Figure1):

- One "bottom tool": it is a not moving part. It is placed on a cast iron platform and has two different cooling circuits: a longitudinal one to control and limit thermal deformation of the mould, the other one to remove heat from the valve plate
- Two "side tools" reproducing the left and right parts of casting: side tools are linked at two cast iron supports equipped with heater devices.
- Two "rear and front tools" sliding inside two respective "towers". The front tower holds the pouring basin.

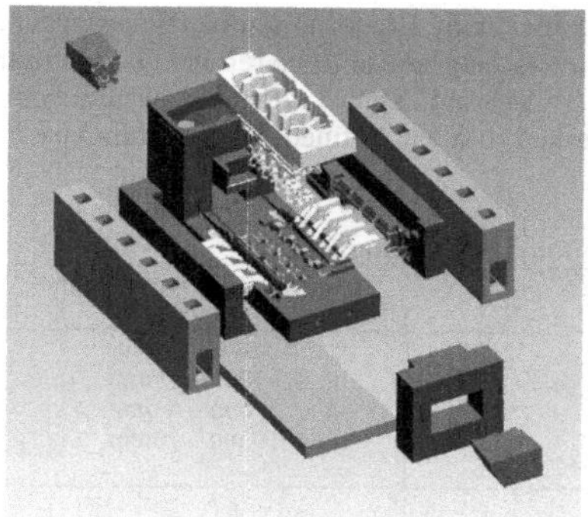

Figure 1. Exploded view of the casting system

Sand cores, some of which are previously assembled in one pack, are put into the mould first by an operator, when the tool system is open, and then, with the assistance of a robot, when the mould pack is closed. The system is completed by filters added at each entrance of the two side-filling arms.
The alloy is strictly derived from A356, refined with Ti an B and modified with Sr. The molten metal can be poured only if the temperatures checked on the bottom and side tools are satisfied.
Molten metal is kept at the desired temperature in a low pressure furnace near the casting device; from there, it is drawn off by imposing a pressure slope. This slope is determined, together with an accurate dimensioning of the whole filling system, to allow the basin to be quickly filled. In this way the basin and the sprue can be kept full, avoiding air entrapment during this phase.
The solidification of the casting is controlled by monitoring some key temperatures of the mould. Those temperatures operate the thermoregulation system, consisting of heaters in the side tools and coolers in the bottom tool.
After solidification, the moving parts of the device are opened, the casting is extracted and marked with a code, which univocally links it to process parameters recorded during its production.

Numerical analysis

The whole simulation analysis has been carried out in different steps, during which a comparison between numerical results and monitored data has been acquired. Those steps were:

1. Simulating the average working condition of the casting device in terms of temperature distribution. This target has been accomplished through thermal cycling simulations: the temperature distribution at the end of one cooling cycle has been used as initial condition for the following one until a steady state was reached for the initial condition of the process;
2. Filling simulation, adopting the temperature distribution of step 1 as initial condition;
3. Solidification simulation, running in the conditions obtained at the end of step 2

The evolution of microstructure and shrinkage porosity is heavily affected by the cooling conditions of the casting. For this reason, all the components of the mould have been included in the calculation domain of the thermal analysis.
The geometrical model has been meshed using tetrahedral elements. To have the best sensitivity in the heat flux calculations, boundary surfaces of the casting were meshed using a dense grid.

The model complexity, together with a lack of symmetry surfaces, required the whole casting device to be meshed. Moreover, the grid had to be refined enough in the thinnest zones (4-5 mm) of the casting (i.e., cooling, intake and exhaust ducts). The risers and the external surfaces of the casting device could be meshed in a less refined way. A summary of component dimensions is given in Table 1.

Table 1. Finite Element model description

Part	Dimension of elements (min-max)	Number of elements
Casting	1mm -10 mm	2951137
Mould	1 mm – 10 mm	6088088
Sand cores	3 mm – 5 mm	2285533
Heaters and Platform	10mm – 20mm	528560
TOTAL		11853318

To avoid interpolation in calculating heat fluxes between two adjacent parts of the calculation domain (e.g., between the casting and the side tool), the nodes of the border surfaces were set to be coincident. This means that, given one node of the border surface, a "twin" node with the same spatial coordinates is created, the first one being assigned to one part of the domain (e.g. the casting), the second one being assigned to the other part of the domain (e.g. the side tool), with a particular HTC imposed between the coincident nodes.

This choice led, on one hand, to more accurate result, on the other hand, to the amount of border nodes being doubled. Thus, it increased computational time.

A measurement campaign was carried out using six thermocouples placed in determinate points of the mould. This allowed us to collect precise information about temperature values and trends, to evaluate the reliability of the Heat Transfer Coefficient (HTC) used, and to allow the validation of the thermal cycling simulation. Thermocouple positioning was made as follows:

- TC1, TC2, TC3 in the right side tool, respectively at 10 mm, 30mm and 50 mm from the interface with the casting; this allowed the acquisition of thermal gradients in a zone of the tool side close to the casting. The temperature evolution curves were compared with the numerical ones, thus allowing us to adjust and validate HTC values throughout the consecutive analysis.
- TC4, TC5 respectively in the right (intake) and left (exhaust) side tools, at 15 mm and 10 mm, respectively, from the interface between the side tool and the cast iron heater: this made it possible to determine the right HTC boundary condition of the outer surface of the side tool, which was influenced by the presence of the heater.
- TC6 in the bottom tool, near the cooling circuit: this allowed the tuning of the HTC value in this zone.

Temperatures were acquired via a 16 channel multiplexer with 0.5 Hz acquisition frequency. The overall distribution of temperature was also refined by means of thermal images of the outer surfaces of the tool system, taken at both the open and closed positions.

Results and discussions

In the first stage of the study, the evaluation of the results obtained from the cycling simulation was compared with the measured temperatures as function of time.

The comparison of trends for the predicted and measured curves shows the good reliability of simulation results in predicting correct thermal conditions in the mould components. In the three nodes of the side tool mesh corresponding to thermocouples TC1,TC2 and TC3 (Figure 2a), it can be observed that the thermal flow in the mould are correctly predicted and minimum and maximum temperatures have a good match. The slopes of the curve are comparable during mold closing time (heating) and opening time (cooling).

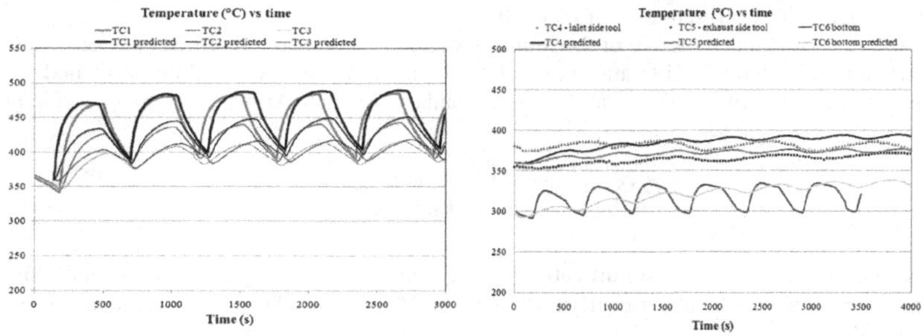

Figure 2: Predicted and experimental temperatures at: a) TC1, TC2, TC3 locations; b) TC4, TC5, TC6 locations.

In Figure 2b predicted and experimental values are compared at the TC4, TC5 and TC6 locations. Concerning the TC4 and TC5 values, that were taken in proximity of the burners, they show a smooth and stable development with time, with simulated values in very close agreement with measured ones.

A somewhat different behavior can be seen at the TC6 location, where the two curves evidently show overlapping trends, but with different thermal flow during each production cycle. These differences can be explained by the difficulties for a correct tip positioning of a very long drilled hole for the thermocouple: few degrees of deviations could then lead to some millimeters of deviation in the thermocouple location.

In the second stage of the study, relationships between numerical and measured results have been investigated based on the metallurgical analysis of the final product, where average SDAS and porosity distribution were evaluated.

Four sections were extracted from one cylinder head cast in average processing conditions. Each section was divided in subsections. Each subsection was further subdivided in cells by a regular grid obtaining 150-180 specimens per section (Figure 3).

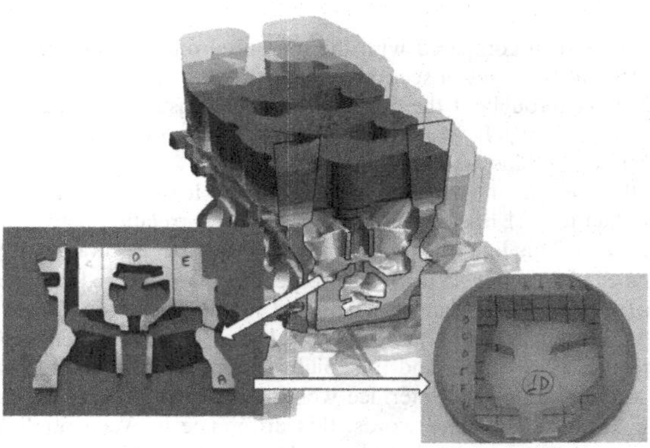

Figure 3. Description of specimen extraction

To allow the comparison between experimental and numerical data, an algorithm was written to extract numerical data from the mesh at the locations defined by the "physical" cells. In particular, all the SDAS and porosity values of the nodes within a cell were averaged in order to have a unique value representative of that cell.

Here, the results obtained on two of the four investigated section are presented (sections 1 and 4). Numerical results about solidification rate calculated in all nodes of the FEM casting model were processed in order to obtain the local SDAS (Secondary Dendrite Arm Spacing) by the rule [4]:

$$SDAS = \left[\left(166 \cdot \frac{\Gamma D \ln(c_{eut} / c_0)}{m(1-k)(c_0 - c_{eut})} \right) \cdot t_f \right]^{1/3}$$

Where t_f is the average solidification time, calculated between Liquidus and eutectic temperature, D is the diffusion coefficient in liquid, Γ is Gibbs-Thomson coefficient, m is the Liquidus slope.

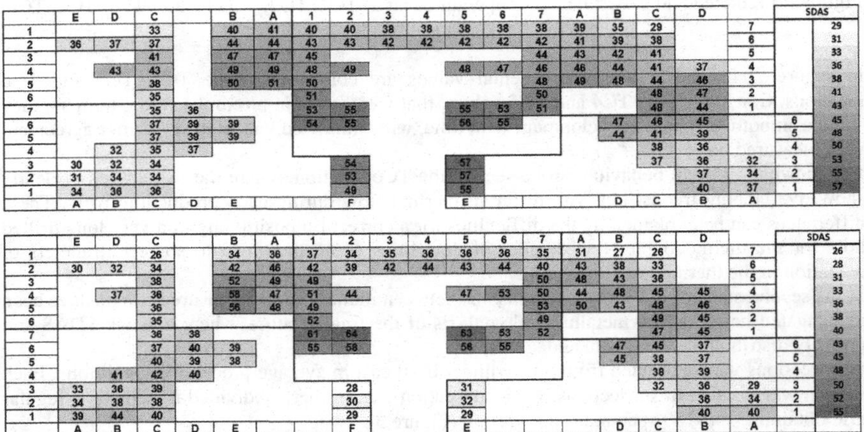

Figure 4. a) Predicted SDAS on section 1; b) Measured SDAS on section 1

Numerical results were then compared with the measured ones. In Figure 4a-b two calculated and measured SDAS matrix values of section 1 are reported.

The overall comparison throughout the section between measured and calculated SDAS value distribution shows a good match. Moreover, all trends along each matrix row suggest a good accuracy in boundary conditions assumption.

Figure 5a) shows the correlation for all cells of section 1. It shows the general correlation line with a lower slope than the 1:1 line, representing the ideal correlation between experimental and predicted SDAS values. This line mainly depends on the presence of a small group of points at very high predicted SDAS (about 55 μm) with respect to the measured ones (around 30 μm). A linear correlation factor R=0.605 can be calculated in this condition. These points are those located in two thin zones surrounded by the central sand core. It is believed that the thermal modulus of these zones is so small, and the solidification times so short, that the global HTC assumed to exist at the casting-sand interface would not fully represent the actual heat transfer condition in these locations. In other words, this difference between predicted and measured SDAS should be avoided by considering a time-dependent HTC [3]. If those points were not considered, the linear correlation factor would become R=0.862. In the authors' opinion, this R value is more representative of the whole simulation capabilities, and, moreover, it is about the same value obtained in all the other investigated sections. As an example, results of section 4 (Figure 5b) show a linear correlation factor R= 0.957.

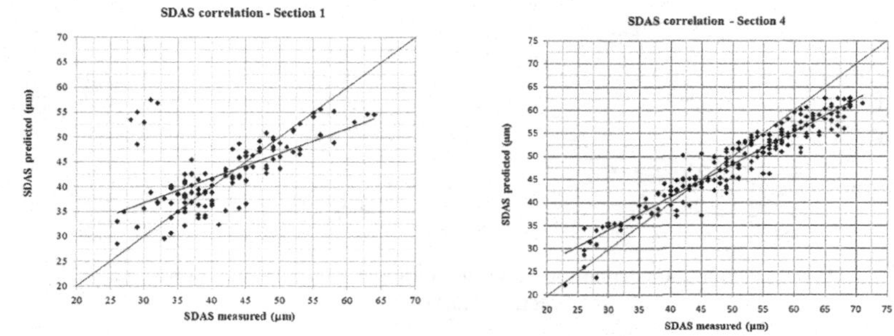

Figure 5. General correlation between predicted and measured SDAS at:
a) section 1; b) section 4.

Correlation charts for section 1 and section 4 generally show that SDAS values are slightly over-predicted in low SDAS zones and under-predicted in high SDAS zones.

The shrinkage porosity model of the adopted commercial code [5] was used to evaluate the porosity distribution at all the cell locations already shown. The same procedure adopted for the SDAS comparison was then used to correlate predicted and measured porosities (Figure 6a-b).

By the comparison between calculated and measured values of shrinkage porosity (Figure 7a-b), a high dispersion of data in all investigated zones was observed. It is, however, important to note that the maximum porosity level throughout the sections is very low (below 1%), hence very far from standard quality criteria for scrap determination in industrial foundry processes.

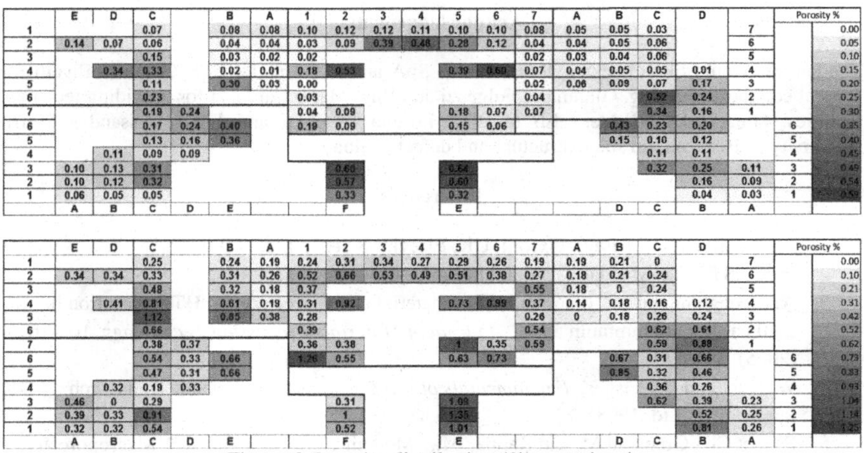

Figure 6. Porosity distribution (%) – section 1
a) predicted; b) measured.

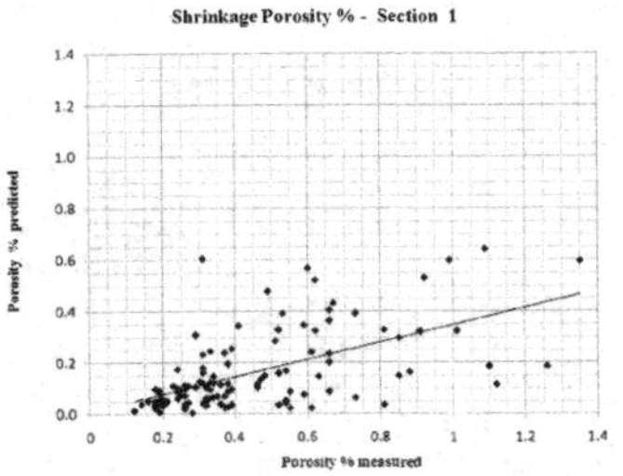

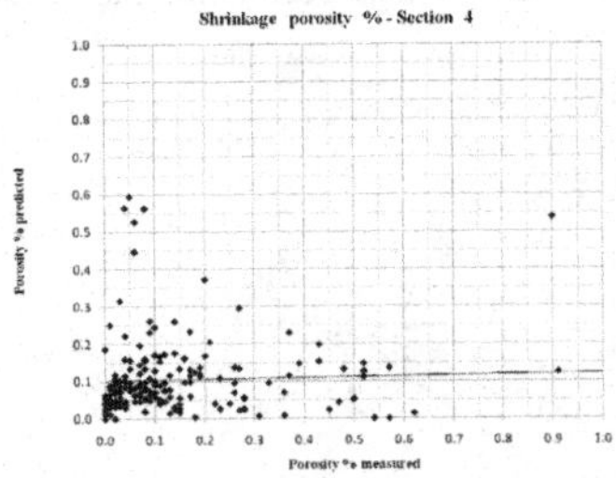

Figure 7. General correlation between predicted and measured porosity at: a) section 1; b) section 4.

Conclusions

A full numerical analysis has been carried out on the gravity die casting process of an 8 cylinder A356 aluminum engine head in current production. Both filling and solidification analysis were done by carefully replicating all the available process monitoring data.
Microstructure characteristics, such as SDAS, were evaluated by using existing relationships with cooling times and correlated with the experimental ones at 640 different points. Correlation factors between predicted and measured values were evaluated for each section and found to be globally in the range of 0.85-0.96.
The porosity content was also evaluated via the shrinkage porosity module of the adopted numerical code and then verified with the same procedure of the SDAS. There, correlation factors were found to be in the range of 0.04-0.58, although the maximum measured porosity was 1% and 1.6% respectively, thus below the sensitivity limit stated by the software producer.

Acknowledgements

The authors would like to acknowledge Ferrari SpA and in particular Eng. Gianluca Pivetti for general collaboration, Eng. Gianmaria Fulgenzi and Eng. Mario Camozzi for providing technical facilities. Great acknowledges also to Prof. Lorella Ceschini and Eng. Alessandro Morri, University of Bologna, for microstructure and defect evaluation.

References

1. John Campbell, *Casting*, (Oxford, UK, Elsevier Science Ltd, 2003)
2. ProCAST User Manual v. 2008, 2008, 479-480
3. A. Meneghini and L. Tomesani, "Chill material and size effects on HTC evolution in sand casting of aluminum alloy", *Journal of Materials Processing Technology*, 162-163 (2005), 534-539
4. W. Kurz and D.J.Fisher, *Fundamentals of solidification* (Switzerland, Trans Tech Publications Ltd, 1998)
5. Pequet Ch., Gremaud M. and Rappaz M., "Modeling of Microporosity, Macroporosity, and Pipe-Shrikage Formation during the Solidification of Alloys Using a Mushy_zone Refinement Method: Applications Aluminium Alloys", *Metallurgical and material transactions*, 33A (2002), 2095-2106.

Shape Casting: The 3rd International Symposium
Edited by: John Campbell, Paul N. Crepeau, and Murat Tiryakioğlu
TMS (The Minerals, Metals & Materials Society), 2009

Autonomous Optimization in Casting Process Simulation

Christof Heisser

MAGMA Foundry Technologies, Inc.
10 N. Martingale Road, Suite 425
Schaumburg, IL 60173

Keywords: Casting Simulation, Solidification Simulation, Autonomous Optimization

Abstract

Computer processing speed has changed dramatically in the last 10 years. Traditional casting process modeling is based on "what-if" scenarios, requiring the user to make a decision, implement changes and start a simulation. To gain the most advantage out of computer advances, casting process modeling is now combined with genetic autonomous optimization tools, which produce a simulation based on a range of parameters rather than specific design points. Besides the optimization of designs, process parameters, as well as mechanical properties can be optimized too. The production of high quality castings depends on many factors. As state-of-the-art simulation tools consider many of these factors, multi-objective autonomous optimization opens a whole new level of accuracy. Input parameters and thermophysical properties can be optimized to characterize the specific melt of a foundry or production line. This paper will show the background of genetic optimization tools and examples of optimized castings utilizing this technology.

Introduction

The development of computer processing speed over the last decade has huge implications on the use of casting process simulation for castings. First and foremost, simulation is much more usable as it provides answers much faster. Secondly, it leads to a paradigm shift: Traditional casting process simulation is based on "What-If" scenarios, demanding an operator to make a decision, implement changes and start another "run." The user has to wait for the completion of a simulation, sometimes over night, before he can check the results the next morning. Today, if the operator starts a simulation just before he leaves for home, the computer might finish the simulation in less than one hour and sits idle for the next 14 hours! The solution to this dilemma is to combine the casting process simulation tool with an "Autonomous Optimization Tool." The latter already have been used for many years in the design community to optimize designs.

BACKGROUND

Casting process simulation has aided foundry operations for the last 20 years. However, the current market demands leave less and less time to find a working casting and rigging design, and process window, as well as leaving literally no room for mistakes or extended trial runs. Autonomous casting process optimization tools were recently made available to the casting design and foundry community to address these challenges. The autonomous casting process optimization tool used in the following examples is based on a generic algorithm. Initially a group of designs is created and simulations are run for all members of this group. After

evaluating all results of this generation, the genetic algorithm decides which of the previously calculated results show promise to lead to a better, finally optimal, design. The algorithm then decides for each design if it should be eliminated, modified, combined with one of the already calculated designs, combined with a new, not yet calculated design or kept unchanged. Thereby the algorithm creates a next generation of designs. This process is repeated until no improvement can be found
anymore, or the operator decides to end the optimization process (**Figure 1**).

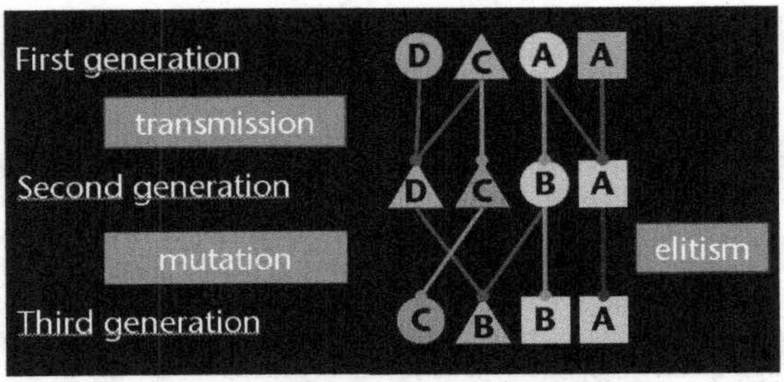

Figure 1: Principle of genetic optimization algorithm.

AUTONOMOUS OPTIMIZATION OF DUCTILE IRON CASTINGS

In the first example a 3-step process was used to improve the yield of an existing high production, safety-critical automotive casting run on a vertically parted mold. This was a defect free part currently in production, but the foundry strove to find the optimum gating and riser design.

Step 1: Optimize the "outside" (cold) riser. As with many castings, this one could be divided into several independent areas, so it made sense to first optimize the outside riser (**Figure 2**). Furthermore, as the gating layout assured very similar filling and thereby temperature conditions in each of the castings, the initial optimization could be conducted on just one casting. This procedure is commonly used at the beginning of complex optimizations to reduce the number of potential designs, as well as the size of each simulation run, primarily to save time. The optimization program requires inputs about objectives like "minimize shrinkage" and "minimize riser volume" (improve yield), as well as the allowable range to use for each dimension (constrains), i.e. it would be bad if the riser would overlap the casting or other risers above or below it. A fixed shape for the riser neck was used, as the contact patch on the casting could not be changed.

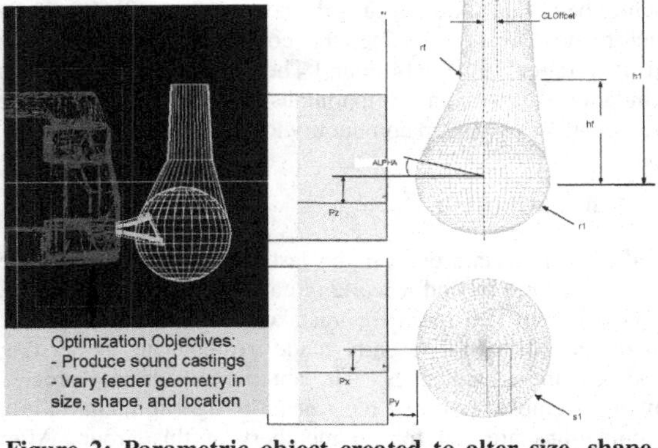

Figure 2: Parametric object created to alter size, shape, and location of outside riser.

The optimization tool creates many different designs and decides after every generation of designs, which to try next (see background). The decision is based on a genetic algorithm that selects and modifies previously successful designs or combines previous designs to find a better one. The program creates a report comparing all results, but depicting the best ones (**Figure 3**).

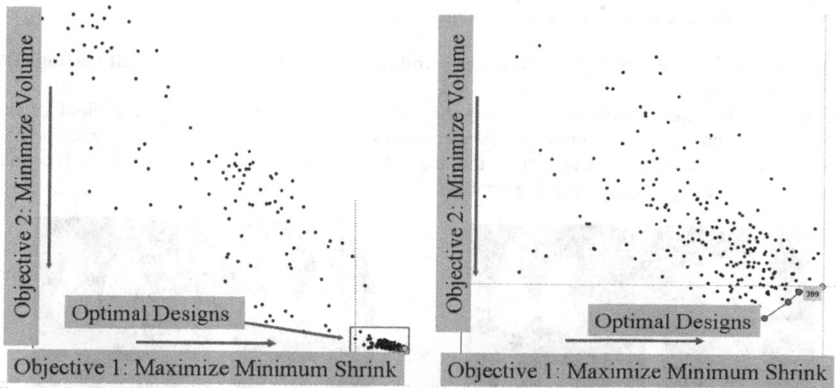

Figure 3: Each dot represents one design. In this case, the best ones are found in the righthand lower corner.

"Design 309" showed the lowest riser volume and created a sound casting. **Figure 4** shows this optimum riser shape in its optimum location on the outside of each casting.

Step 2: Optimize the "inside" riser. As the inside riser has different temperature conditions, than the outside one, a separate optimization run was used to optimize the inside riser. Using the experience from the first optimization step it was decided to start the optimization of the inside riser with a design close to the optimium riser design found for the outside riser. However, as changes in the design of the inside riser could modify the filling profile of the entire casting and thereby potentially change the temperature of the melt for the outside riser, the outside riser was still allowed to change, but in a narrower window. Evaluation areas were used to assure that these critical areas of the castings would be defect free (**Figure 4**).

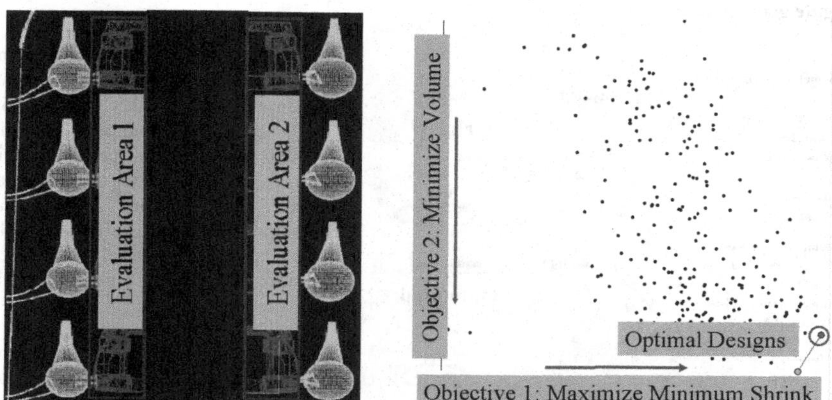

Figure 4: Evaluation areas focus optimization on critical areas.

The ability to handle these multi-objective optimizations is the key to the application of the genetic algorithm in casting process optimizations.

Step 3: Optimize yield of cup, down runner and ingates while assuring that all castings fill in the same time.

The final multi-objective optimization put the most emphasis on improving the yield of the gating system, but under the limitation that all castings have to be filled in the same time (as they all have the same sized risers), as well as that the critical areas have to stay defect free. The final ingate and riser design is shown in **Figure 5**.

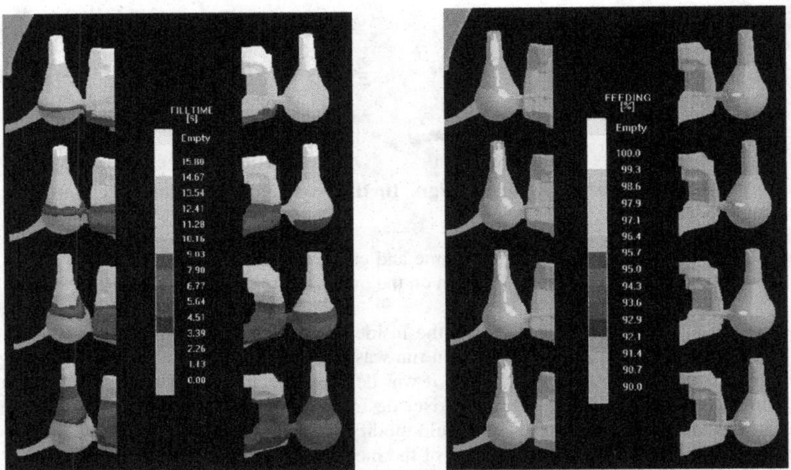

Figure 5: Figures show very similar filling times and defect free (translucent) castings (midsection of castings are blocked for confidentiality). Also note the slight differences in ingate geometries.

volumes and masses:

material	volume [l]	mass [kg]
part		Original
+ machining allowance		
= raw cast		
+ gating	0.69	4.88
+ feeders	1.51	10.71
= casting	3.88	27.49

volumes and masses:

material	volume [l]	mass [kg]
part		Optimized
+ machining allowance		
= raw cast		
+ gating	0.32	2.28
+ feeders	1.25	8.85
= casting	3.22	22.85

Figure 6: Optimized gating and riser geometry saves a total of almost 5kg in pouring weight.

The final design improves the yield of the entire mold by nearly 5kg without reducing casting quality (**Figure 6**). The use of autonomous casting process optimization provided this optimal gating and riser design in far shorter time than conventional simulations or casting trials would have (the traditional approach provided the starting condition for this optimization).

The second example shows the application of autonomous casting process simulation on a printing machine lever. The challenge was not

Figure 7: Original gating and riser layout.

only to improve the yield, but also to assure a minimum tensile strength of 660MPa in the thick section of the casting. Due to that requirement, the foundry did not place risers on that critical area, but chose side risers (**Figure 7**).

This arrangement limited the amount of castings that could be placed in each mold. A top riser would allow for more castings to be placed on the pattern. Therefore, the optimization setup included the option to place 3, 4, or 5 castings next to each other. Depending on the size of each exothermic sleeve and its location, the number of castings per mold could change as well.

The first optimization run, which focused on one casting to find the optimum riser size and location, produced a sound casting with the desired tensile strength. The second run, however, showed that the distance between the castings has a big impact on the tensile strength as well (**Figure 8**).

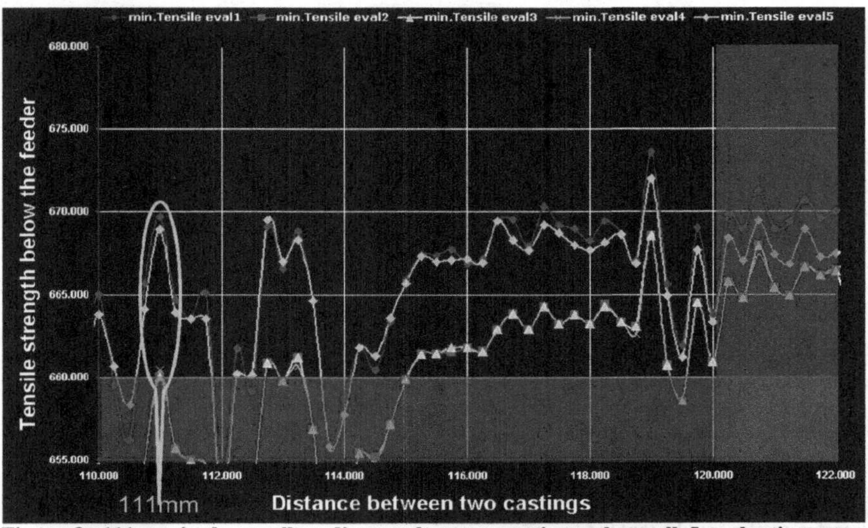

Figure 8: 111mm is the smallest distance between castings where all 5 evaluation areas reached 660MPa.

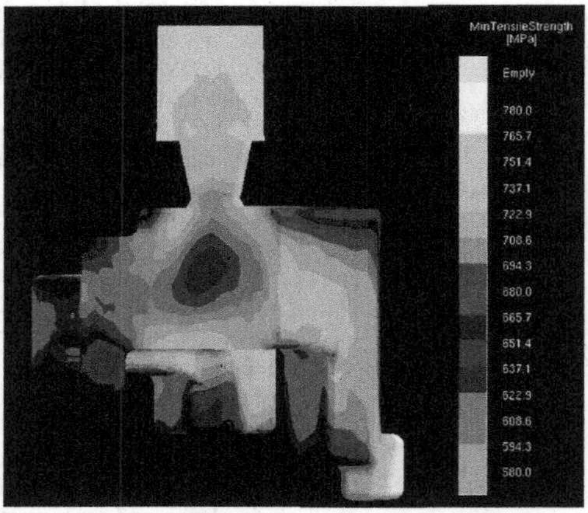

Figure 10: Tensile strength distribution in optimum design.

Figure 10 shows the tensile strength of the parts with the optimum riser design and minimal distance between castings. This optimization led to a cost reduction of approx. US$6.00 per casting - a 35% improvement.

INVERSE OPTIMIZATION

Another area where autonomous casting process simulation can be used is "inverse optimization", i.e. to match thermocouple curves with simulated curves to calculate heat transfer coefficients or thermo-physical properties. **Figure 11** shows a conventional thermal analysis cup used in foundries as a quality control tool. Virtual thermocouples placed in the same location as the actual one, as well as throughout the cup allow the evaluation of the solidification process in detail.

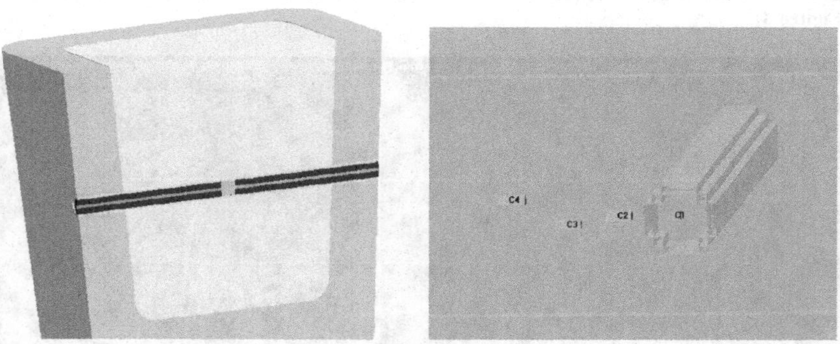

Figure 11: Test cup and placement of virtual thermocouples.

As the conditions found in a test cup are not anywhere near the conditions castings experience in the mold, specific boundary conditions, different from the ones used in regular casting process simulations, have to be derived. The autonomous optimization tool used here can read in measured curves and compare them to simulated ones. Two of the biggest factors on the cooling behavior are the heat transfer between the melt and the sand wall of the test cup, as well as the one between the melt surface and the air on the open top surface. As the outside can be simulated as a cooling medium and as the rate of the heat loss to that medium can be controlled by a heat transfer coefficient, the tool has to

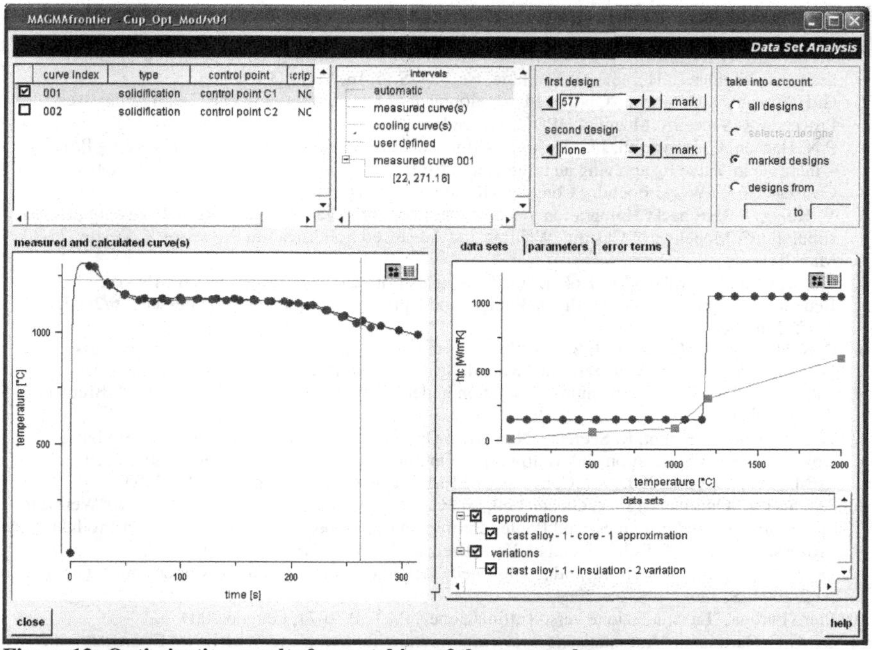

Figure 12: Optimization results for matching of thermocouple curves.

vary both heat transfer coefficients to match the measured with the calculated curves. **Figure 12** shows a good match between the measure (red) and calculated (blue) curve using two independent temperature dependent heat transfer coefficients between casting and sand wall (red) and melt and air (green). This example shows that autonomous optimization tools can be used to fine-tune the boundary conditions in casting process simulation tools to match measured and simulated values.

SUMMARY

Three examples were used to show the capability of an autonomous casting process optimization tool to improve the yield of ductile iron castings considering casting quality, as well as microstructures and mechanical properties. An inverse optimization example was chosen to show the capability of these tools to match measured and calculated thermocouple curves.

ACKNOWLEDGMENTS

The authors would like to thank Heidelberger Druckmaschinen AG, Heidelberg, Germany and ThyssenKrupp Waupaca, Waupaca, USA for the permission to use their castings in this paper.

REFERENCES

1. M. Saillard, G.Hartmann, A. Egner-Walter,"Nouvelles possibilités de conception et d'optimisation des pièces moulées", Hommes et Fonderie, Nr. 324, S.12-16, May 2002
2. G. Hartmann, J. C. Sturm, "Optimized development of Die Castings and Die Casting Processes" Proceedings Vincenza Meeting, HPDC, September 2002
3. P.N. Hansen, G. Hartmann, J.C. Sturm, "Optimized Development for Castings and Casting Processes – Increase in Value by applying an integrated CAE Chain for the Development of Automotive Castings", 65th World Foundry Congress, Kyongju, Korea October 2002
4. V. Kokot, P. Bernbeck, "Integration and application of optimization algorithms with casting process simulation", Modeling of Casting, Welding and Advanced Solidification Processes X, Destin, 25-30 May 2003
5. G Hartmann, P Bernbeck, V. Kokot, "Gießereien als Entwicklungspartner der OEM's - die Bedeutung computergestützter Entwicklungs- und Optimierungswerkzeuge", Giesserei, 6/2003, S. 44-55,Jun 2003
6. P. N. Hansen, G. Hartmann, J. Sturm, "Optimierte Entwicklung für Gußteile und Gießprozesse – Wertsteigerung durch die integrierte Anwendung einer CAE Prozeßkette für Gußteile aus der Automobilindustrie", Symposium „Simulation in der Produkt- und Prozessentwicklung", Bremen, 5.-7. November 2003
7. G. Hartmann, V. Kokot, R. Seefeld, "Numerical Optimization of Casting Processes-Leveraging Coupled Process Simulation and Multi-Object Optimization to the Manufacturing Level", 21st CAD-FEM User´s Meeting 2003, Int. Congress on FEM-Technology, November 12-14, 2003
8. J.C. Sturm, "Optimierung von Gießtechnik und Gussteilen. Gießtechnische Simulation als Werkzeug zur automatischen Fertigungs- und Bauteiloptimierung", Symposium "Simulation in der Produkt- und Prozessentwicklung", 5.-7. November 2003, Bremen, S. 39-46, November 2003
9. Götz C. Hartmann, Rudolf Seefeldt, "Die zweite Generation von Simulationswerkzeugen", Giesserei, Nr.2/2004, S. 38-42, Februar 2004
10. Piero Parona, "La simulazione verso l'ottimazione", P&T, P.70-71, Giugno 2004
11. Dr. G.C. Hartmann, "Automatische Speiseroptimierung für komplexe Modelle im Eisenguss – die zweite Generation der Gießsimulation",Giesserei Rundschau, H. 7/8, S.145, Juli/August 2004
12. J.C. Sturm, "Optimization – Integration – Casting Property Prediction", Proceedings of World Foundry Congress, 5.-9. September 2004, Istanbul, Turkey, September 2004
13. Dr. A. Egner-Walter, H. Dannbauer, "Integration lokaler Bauteileigenschaften gegossener Fahrwerkssteile in die Betriebsfestigkeitsberechnung", In: VDI-Gesellschaft Fahrzeug- und Verkehrstechnik, VDI-Berichte Nr. 1846: Berechnung und Simulation im Fahrzeugbau, Tagung Würzburg 29. – 30.9.2004, September 2004
14. G. Hartmann, G. Busch, I. Ivica, "Autonome rechnerische Speiseroptimierung an komplexen Modellplatten"; Giesserei 91, Nr. 10, S. 22 – 30, Oktober 2004
15. MAGMAfrontier Documentation, Basics of Design Optimization, 2005.
16. modeFRONTIER Documentation, Why do Design Optimization? & User Manual
17. David E. Goldberg, Genetic Algorithms in Search, Optimization and Machine Learning; Addison-Wesley, 1989.
18. David E. Goldberg et al., Genetic algorithms: a bibliography (IlliGAL Report n. 97011), University of Illinois at Urbana-Champaign, Illinois Genetic Algorithm Laboratory, Urbana, USA, 1997.

ADDITIONAL RESOURCES

Please contact:
MAGMA Foundry Technologies, Inc.
847-969-1001
info@magmasoft.com
www.magmasoft.com

Shape Casting: The 3rd International Symposium
Edited by: John Campbell, Paul N. Crepeau, and Murat Tiryakioğlu
TMS (The Minerals, Metals & Materials Society), 2009

MODELING THE FORMATION OF POROSITY DURING LOW PRESSURE DIE CASTING (LPDC) OF ALUMINUM ALLOY A356

Ehsan Khajeh[1], Xinmei Shi[2], Daan M. Maijer[1]

[1] Department of Materials Engineering, The University of British Columbia, 6350 Stores Road, Vancouver, BC, Canada V6T 1Z4
[2] Department of Electrical and Computer Engineering, The University of British Columbia, 2332 Main Mall, Vancouver, BC, Canada V6T 1Z4

Keywords: Microporosity, Macroporosity, Permeability, Casting, LPDC, A356

Abstract

In this study, the formation of macro and micro porosity during Low Pressure Die Casting (LPDC) of Aluminum alloy A356 has been modeled based on the solution of the heat conduction, continuity and Darcy equations. In this model a two-stage approach has been used to predict the formation of porosity. In the first stage, the evolution of porosity has been determined based on the partitioning of hydrogen between solid, liquid and pore, whereas, in the second stage, when the calculated pressures reach the vapor pressure of the liquid, the evolution of porosity is calculated based on the volume of isolated liquid. By studying the role of permeability as the critical factor in determining the local pressures, the feed-ability of the mushy zone has been assessed. In order to validate the current method, the model has been applied to examine the evolution of pores in a trial casting.

Introduction

The formation of porosity in cast Al alloys is detrimental to the mechanical properties and surface quality of cast components. The formation of these defects is governed by the complex interaction of phenomena occurring during solidification including the liquid/solid hydrogen solubility, volumetric expansion due to pore formation, volumetric shrinkage due to phase change, solute partitioning, and resistance of dendritic network to flow which leads to a pressure drop. Although progress has been made in modeling the formation of porosity, issues such as the dendritic network permeability, the critical super-saturation for pore nucleation (linked to curvature effects), and hydrogen partitioning mechanisms, require further investigation [1,2].
The first methodology to predict microporosity formation was presented by Kubo and Pehlke [3] in which they considered the effects of both hydrogen segregation and solidification shrinkage. In this early work, however, only interdendritic feeding was considered and mass feeding was neglected. Their proposed methodology has been the basis for other studies such as that of Zhu et al. [1]. Since the accurate computation of pressure is essential for microporosity predictions, various approaches have been adopted to solve for the pressure field within the mushy zone ranging from straightforward analytical expressions for steady state solidification [4] to more realistic transient solutions based on Darcy's law and the continuity equation [3]. Macroporosity, on the other hand, can be calculated from a simplified shrinkage model based on the solution of the energy equation and evaluating the potential volumetric shrinkage of each isolated liquid region [5]. There have been other more comprehensive methods reported employing the solution of the full Navier-Stokes and energy equations [6] which can be coupled with conventional

methods to track the free surface development [7], but due to convergence problems and long computation times they are less favorable. Most researchers study the formation of either macroporosity or microporosity independently. The first to consider both was Pequet et al. [8]. Their model was based on the solution of Darcy's equation and the micro-segregation of hydrogen in the mushy zone, following by identification of the various liquid regions that appear during solidification (open regions, partially closed regions and closed regions). In their model, they used the traditional Carman-Kozeny expression for the permeability calculation.

Any model describing the flow of interdendritic liquid with the Darcy equation is very sensitive to the permeability since it affects the pressure calculation. There is still ambiguity at the present time whether the Carman-Kozeny or any other analytical relation is applicable over the complete range of solid fractions and geometry variations [9]. In this study, the formation of micro and macro porosity has been modeled based on the solution of the heat conduction, continuity and Darcy equations. In order to validate the current method, the model has been applied to examine the evolution of porosity in aluminum alloy A356 castings produced via LPDC process. By studying the role of permeability, as the critical factor in determining the local pressures, the feed-ability of the mushy zone has been assessed [10].

Experimental Procedure

A series of unmodified aluminum alloy A356 experimental castings were produced by LPDC. A ½ section of a die half, shown in Figure 1, reveals the casting geometry and the cooling channel configuration. Figure 1 shows a schematic illustration of die and cooling configuration. This configuration was designed to produce varying amounts of porosity depending on the cooling intensity. The LPDC process is a cyclic process beginning with the pressurization of the furnace. The excess pressure in the furnace forces the liquid metal up into the die cavity where it is cooled and solidified by heat transfer to the die. After completion of solidification, the die is opened and the casting is ejected. After a short delay, the die is closed and the next cycle begins. The operational conditions employed during the plant trial are summarized in Table I for the castings discussed in this study. To provide thermal data to tune and validate the model, 8 thermocouples were placed at different locations within die. Castings from selected cycles were cut and the corresponding sections polished to 1μm for image analysis to measure the porosity distribution.

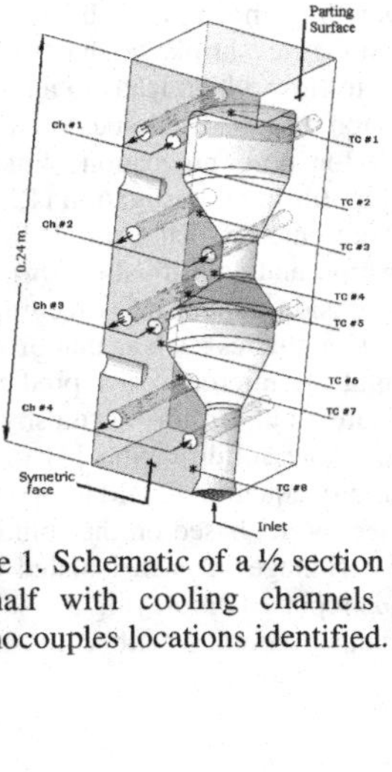

Figure 1. Schematic of a ½ section of a die half with cooling channels and thermocouples locations identified.

Table I. Experimental parameters.

Trial No.	Cycle No.	Parameters				
		Die close time (s)		Die open time (s)	Cooling channel (Ch #1)	
		Pressure On	Pressure Off		Time (s)	Flow rate (L/min)
1	13	32	88	60	-	-
2	15	32	88	45	60	600
3	16	32	88	45	120	600

Model Development

Model description

The numerical model consists of a conduction-only thermal model (neglects advection) which describes the heat transfer between the casting and die during solidification and a fluid flow model describing the compensatory flow due to the solidification. Beginning with the flow model, it has been assumed that the interdendritic flow is a creep flow which can be described by the Darcy equation [8]:

$$v = -\frac{K}{\mu}\nabla p_l \tag{1}$$

where K is the mushy zone permeability tensor, μ is the dynamic viscosity of interdendritic liquid, p_l is the liquid dynamic pressure (Pa), and v is the superficial velocity vector. The velocity may be calculated from the mass conservation equation [2,8]:

$$\nabla \cdot (\rho_l v) + \frac{\partial \bar{\rho}}{\partial t} - \rho_l \frac{\partial f_{gp}}{\partial t} = 0 \tag{2}$$

where ρ_l is the interdendritic liquid density, $\bar{\rho}$ is the average density of the solid-liquid mixture which is a function of temperature calculated from thermal model, f_{gp} is the fraction of gas porosity, and t is the time. The last term on the LHS in (2) incorporates the evolution of microporosity. Assuming the complete diffusion of hydrogen in the solid and liquid and locally closed regions for hydrogen redistribution as in [8], the hydrogen balance before the formation of porosity can be written as:

$$\rho_l[H]_0 = [H]_l(k_H \rho_s f_s + \rho_l f_l) \tag{3}$$

where $[H]_0$ is the initial hydrogen content in computational domain, $[H]_l$ is the hydrogen content in liquid, and k_H is the partition coefficient of hydrogen. As solidification proceeds, the hydrogen content in the liquid increases due to partitioning and may eventually exceed the solubility limit. At the critical super-saturation, a pore nucleates [1]. After the formation of gas porosity, equation (3) is re-written as:

$$\rho_l[H]_0 = ([H]_{l-e} + \Delta[H]_{s-s})(k_H \rho_s f_s + \rho_l f_l) + \alpha \frac{f_{gp} P_{gp}}{T} \tag{4}$$

where p_{gp} is the pressure in the porosity, α is a conversion factor, $\Delta[H]_{s-s}$ is the required super-saturation for nucleation, and $[H]_{l-e}$ is the solubility limit of hydrogen in liquid aluminum which is determined using Siverts' law [1]:

$$[H]_{l-e} = S_l\sqrt{p_{gp}} = S_l\sqrt{p_a + p_h + p_l + p_\sigma} \tag{5}$$

In equation (5), S_l is the equilibrium constant. The pressure in the pore (p_{gp}) is calculated as the sum of the imposed inlet pressure, p_a, the metalostatic pressure, p_h, the dynamic pressure, p_l, and the over pressure due to the capillary effects, p_σ, which is calculated assuming a spherically shaped pore surrounded by secondary dendrite arms.

$$p_\sigma = \frac{2}{r_p} = \frac{4}{f_l d_2} \tag{6}$$

where r_p is the pore radius and d_2 is the secondary arm spacing which may calculated from experimentally derived data as in [1]:

$$d_2 = \frac{158.49}{R^{0.34}}$$

(7)

where R is the cooling rate calculated at an undercooling of *10 K.* Using equations (3) and (5), the hydrogen content of the liquid phase, before and after the formation of a pore can be written as:

$$[H]_l^* = \min([H]_{l-e} + \Delta[H]_{s-s}, [H]_l) \quad (8)$$

By rearranging (4) and including (8),

$$f_{gp}^* = \frac{[H]_0 \rho_l - [H]_l^* (k_H \rho_s f_s - \rho_l f_l)}{(\frac{\alpha . p_{gp}}{T} - [H]_l^* \rho_l)} \quad (9)$$

f_{gp}^* returns zero when $[H]_l^*=[H]_l$ and it becomes (4) when $[H]_l^*=[H]_{l-e}+\Delta[H]_{s-s}$. It is assumed that gas porosity will only form when the total pressure is above the vapor pressure of liquid. By defining a step function for the total pressure as:

$$\chi(x) = \begin{cases} 1 & x > 0 \\ 0 & x < 0 \end{cases} \quad (10)$$

The total gas porosity, f_{gp}^T in each computational cell is calculated as:

$$f_{gp}^T = \int_0^{t_f} \frac{\partial f_{gp}^*}{\partial t} \cdot \chi(P_{total} - P_{vapor}) dt \quad (11)$$

where t_f is the solidification time. Below the vapor pressure of at least one of the alloy components, the vapor is the stable phase and a pore containing the vapor of that substance will form. Therefore, total corresponding shrinkage porosity, f_{sp}^T for each cell is calculated as:

$$f_{sp}^T = \beta . [1 + \int_0^{t_f} \frac{\partial f_l}{\partial t} \cdot \chi(P_{total} - P_{vapor}) dt] \quad (12)$$

where β is the volume shrinkage ratio of the alloy. The total porosity is calculated as $f_p^T = f_{gp}^T + f_{sp}^T$. Above a critical liquid fraction, f_l^{cr}, the solid network is not coherent and porosity formed above this value will form a non-dispersed pore. In this case, the volume of porosity is determined by integrating the total porosity, and assigning it to the corresponding region.

<u>Solution methodology</u>

A 3D transient heat transfer model of the process used to produce the experimental castings (refer to Figure 1 and Table I) was developed with the commercial Finite Element (FE) software, ABAQUS. The model includes geometry corresponding to ¼ sections of the casting and die. The heat transfer coefficients at die/casting, die/cooling channels, and die/environment interfaces were specified as functions of temperature as in [11]. The ambient temperature of air in the cooling channels and around the die was assumed to be 100°C. The liquid metal was assumed to be initially at a uniform temperature of 700°C at the start of each cycle. An initial temperature of 400°C was assumed for the die in the first cycle. The end-cycle temperature distribution in the die was set as the initial temperature for each subsequent cycle. By comparing the experimental and simulated cooling curves at 8 locations in the die, the heat transfer coefficient at the interfaces were adjusted. At temperatures above liquidus, better contact between the casting and die results in higher heat transfer coefficients while after formation of a solid shell, poorer contact results in smaller values.

The initial hydrogen content was assumed to be 0.15 cc/100g and the required super-saturation of hydrogen for nucleation was assumed to be 0.05 cc/100g. The FE thermal solutions at each time step was interpolated to the Finite Volume (FV) mesh to facilitate the porosity calculation. To calculate the pressure field, the Poisson's equation has been solved at each time step using an iterative method to reach an acceptable residual:

$$\nabla \cdot (\rho_l D_p \nabla p_l) + S_p = 0 \tag{13}$$

where D_p is defined as the pressure diffusion coefficient and S_p is the source term. For the mushy zone, D_p and S_p can be calculated from (1) and (2). In (1), the permeability is determined using Kozeny-Carman expression:

$$K = \frac{(1-f_s)^3}{k_C S_v^2 f_s^2} \tag{14}$$

where S_v is the solid/liquid interfacial area per unit volume of solid (in this study the structure was equiaxed and the S_v equates to d_2 [9,10]) and k_C is the Kozeny-Carman constant. While k_C is often assumed to be equal to *5*, the value of *1* has been shown to be more reasonable [9,10]. For liquid regions, since the pressure drop is much smaller than that in the mushy zone, a relatively high value of diffusion coefficient can be chosen. Also the source term was assumed to be zero.

Results and discussion

Figure 2 shows the comparison between predicted and measured cooling curves for 3 of the 8 thermocouples locations placed within die. As seen, by tuning the heat transfer coefficient at interfaces, an acceptable match has been obtained.

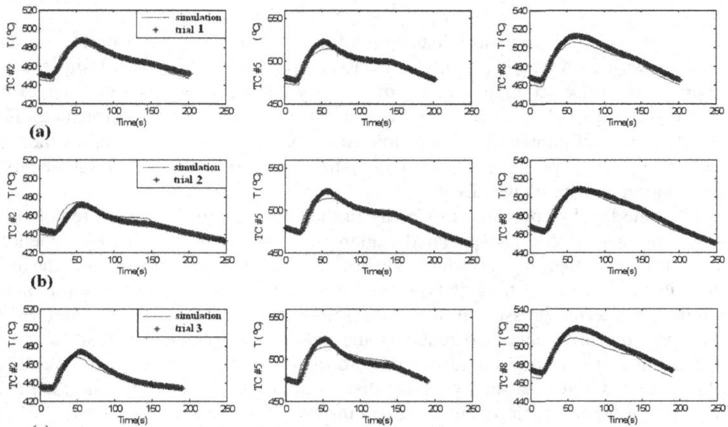

Figure 2. Comparison between predicted and measured cooling curves for three locations (TC# 2, TC #5 and TC #8 as in figure 1) in the die. a) Trial 1, b)Trial 2, and c) Trial 3.

One of the

difficulties in modeling of solidification, particularly for alloys with wide freezing ranges, is the uncertainty of permeability over the complete range of solid fractions. Since incorrectly predicting of the pressure drop affects the prediction of mushy zone feeding, over or under estimation of porosity occurs with improper use of permeability expressions. Using non-isotropic expression for the permeability is essential in models involving columnar dendritic solidification and is critical in the final stages of solidification. In equiaxed structures, the one-length scale (*Sv*) isotropic expression given by the Kozeny-Carman is satisfactory but *Sv* and k_c must be properly defined [10]. Figure 3 shows the calculated pressure field in a 1D domain without considering the formation of microporosity. The pressure drop is quite sensitive to the variation of permeability caused by changing k_c.

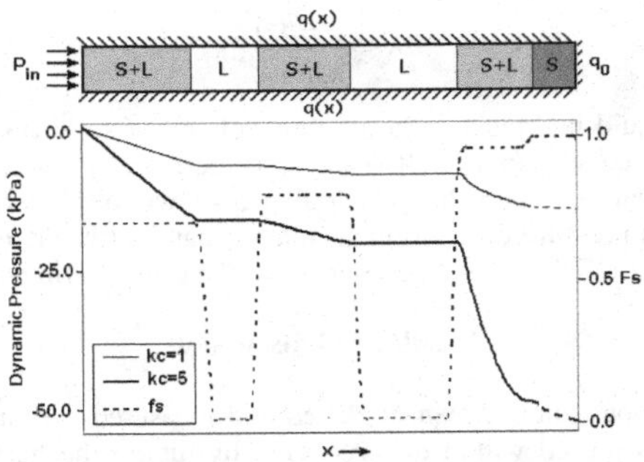

Figure 3. Dynamic pressure field in a 1D test case with the Kozeny-Carman expression for permeability with k_c=1 and k_c=5.

Figure 4 shows the predicted porosity distribution for Trial 1 conditions using two permeability conditions along with a schematic plot of the measured porosity distribution. Using the Kozeny-Carman expression with $k_c=5$, higher values of porosity are predicted due to the larger pressure drop. As discussed in [9,10], $k_c=1$ is a more reasonable value for equiaxed structures. Figures 5 and 6 show the predicted porosity for $k_c=1$ along with the experimental results for the two other casting trial conditions. The majority of the porosity observed in these castings occurs due to liquid encapsulation and shrinkage effects.

Even with k_c=1, the level of porosity, especially in the upper volume, is over-predicted for each trail casting. The size of the encapsulated region and the extent of porosity depend on the coherency of the solidified region which blocks off the incoming liquid and the extent of solidification in the overall casting. If liquid encapsulation occurs when the solid shell of the casting is thin, the internal pressure in the casting may cause the solid shell to deform thereby relieving the pressure difference and reducing the amount of shrinkage porosity. The exterior surface of the trial castings shows evidence of deformation and may account for the differences between the predicted and measured porosity distributions. An iso-contour line identifying the shape of the critical liquid fraction that delineates the boundary between coherent and distributed porosity based on the temperature at which the solid network was assumed to be coherent and un-deformable is also shown in Figures 4 – 6. In this region, macroporosity was expected but not observed in the trial castings. Instead, the deformation of the casting's surfaces were likely large enough to eliminate macro-shrinkage. Although, permeability is an important parameter for modeling porosity formation, the conditions required for gas pore nucleation such as initial pore size and shape and the critical super-saturation condition requires further investigation. In the current study, more than 70% of the porosity was predicted to form below the vapor pressure of the liquid (i.e. shrinkage porosity). For directional solidification conditions where conditions favor gas porosity, the nucleation and growth mechanism of these pores is as important as permeability.

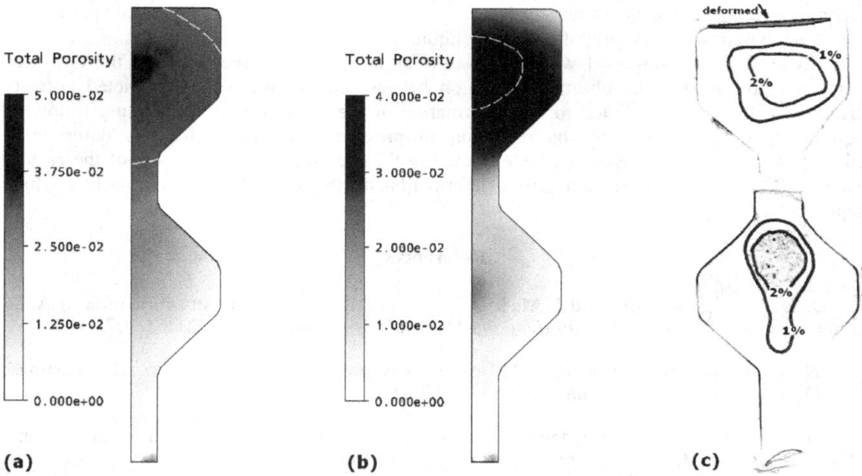

Figure 4. Predicted porosity distribution for Trial 1 with (a) k_c=5 and (b) k_c=1, and (c) measured porosity distribution. Dashed line is the boundary between the porosity formed below and above f_l^{cr}=0.6.

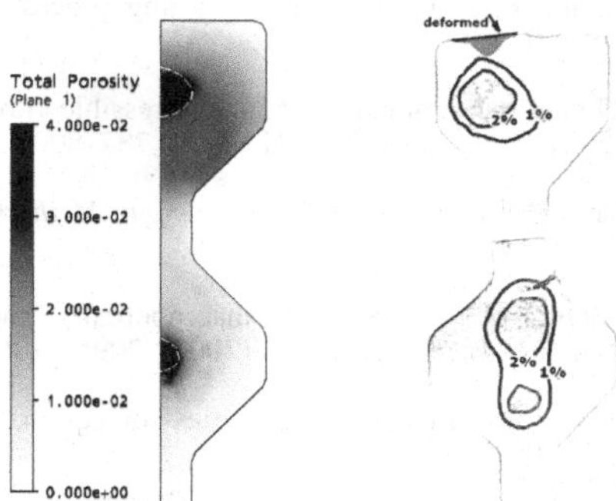

Figure 5. (a) Predicted porosity distribution for Trial 2 with k_c=1 and (b) measured porosity distribution. Dashed line is the boundary between the porosity formed below and above f_l^{cr}=0.6.

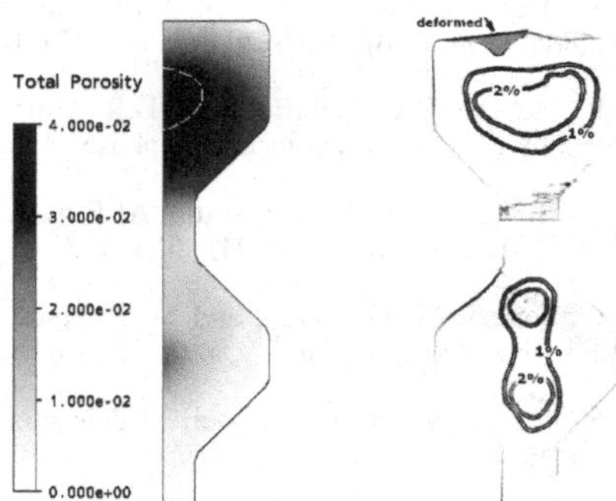

Figure 6. (a) Predicted porosity distribution for Trial 3 with k_c=1 and (b) measured porosity distribution. Dashed line is the boundary between the porosity formed below and above f_l^{cr}=0.6.

Summary and Conclusion

A model was developed and tested to predict the distribution of porosity in a series unmodified aluminum alloy A356 casting produced via LPDC. A two-stage approach has been used to describe porosity formation. In the first stage, porosity is formed by accounting for hydrogen partitioning in the solid/liquid and assuming a critical super-saturation for nucleation. In the second stage, when the pressure is below the vapor pressure of the liquid metal, porosity is calculated based on the fraction of remaining liquid.

Porosity predictions using k_c=1 were shown to be in much better agreement with the measured porosity distributions. The observed deviation between the measured and predicted porosity distribution has been attributed to the deformation of the solid shell of the casting following liquid encapsulation. Further, the reduction in pressure associated with this deformation eliminated the macro-shrinkage cavity expected in these castings. A comparison of the casting shape with the die indicates substantial deformation on the top surface of the casting which supports this analysis.

References

1. J.D. Zhu, S.L. Cockcroft, and D.M. Maijer, "Modeling of microporosity formation in A356 Aluminum alloy casting", *Metallurgical and Materials Transaction A*, 37 (2006), 1075-1085.

2. A.S. Sabau, and S. Viswanathan, "Microporosity prediction in Aluminum alloy casting", *Metallurgical and Materials Transaction B*, 33 (2002), 243-255.

3. K. Kubo, and R.D. Pehlke, "Mathematical modeling of porosity formation in solidification", *Metallurgical and Materials Transaction B*, 16 (1985), 359-366.

4. T. S. Piwonka, M. C. Flemings, "Pore formation in solidification", *AFS Transaction*, 236 (1966), 1157-1165

5. R. Tavakoli, and P. Davami, "Automatic optimal feeder design in steel casting process", *Comput. Methods Appl. Mech. Engrg.*, 197 (2008), 921–932.

6. E. Mcbride, J.C. Heinrich, and D.R. Poirier, "Numerical simulation of incompressible flow driven by density variations during phase change", *J. Numer. Meth. Fluid*, 31 (1999) 787–800.

7. M. Bellet, and V.D. Fachinotti, "ALE method for solidification modeling", *Comput. Methods Appl. Mech. Engrg.*, 193 (2004) 4355–4381

8. C. Pequet, M. Gremaud, and M. Rappaz, "Modeling of microporosity, macroporosity, and pipe-shrinkage formation ...", *Metallurgical and Materials Transaction A*, 33(2002), 2095-2106.

9. S.G.R. Brown et al., "Numerical determination of liquid flow peremeabilities for equiaxed dendritic structures", *Acta Materialia*, 50 (2002), 1559–1569.

10. E. Khajeh, and D.M. Maijer, "Numerical modeling of liquid flow permeability on 3D microtomographic geometry of Al-Cu alloys", (Present paper at 138th TMS Annual Meeting, San Francisco, 15th February 2009)

11. B. Zhang, D.M. Maijer, and S.L. Cockcroft, "Development of a 3-D thermal model of the low-pressure die-cast (LPDC) process of A356 aluminum alloy wheels", *Materials Science and Engineering A*, 464 (2007), 295-305.

Shape Casting: The 3rd International Symposium
Edited by: John Campbell, Paul N. Crepeau, and Murat Tiryakioğlu
TMS (The Minerals, Metals & Materials Society), 2009

PREDICTING RESIDUAL STRESSES CASUED BY HEAT TREATING CAST ALUMINUM ALLOY COMPONENTS

Chang-Kai Wu, Makhlouf M. Makhlouf
Advanced Casting Research Center
Worcester Polytechnic Institute (WPI); 100 Institute Rd., Worcester, MA 01609, USA

Keywords: Heat treatment, Modeling, Distortion, Residual Stresses

Abstract

The objective of this work is to develop and verify a mathematical model that enables the prediction of residual stresses caused by heat treating cast aluminum alloy components. The model uses the commercially available software (ABAQUS) and an extensive database that is developed specifically for aluminum alloy under consideration (in this case, A356). The database includes the mechanical, physical, and thermal properties of the alloy all as functions of temperature. The database is obtained through a search of the open literature and measurements made on A356 alloy specimens using an Instron tensile testing machine. In addition, boundary conditions – in the form of heat transfer coefficients for each of the heat treatment steps - are obtained from measurements performed with a special quenching system. The database and boundary conditions are used in the software to predict the residual stresses that develop in a commercial A356 cast component that is subjected to a standard commercial heat treating cycle. In order to verify the accuracy of the model predictions, the predicted residual stresses are compared to residual stresses measured using x-ray diffraction.

Introduction

The mechanical properties of aluminum alloy castings can be greatly improved by a precipitation hardening heat treatment. Typically, this heat treatment consists of three steps: (1) solutionizing, (2) quenching, and (3) aging; and is performed by first heating the casting to and maintaining it at a temperature that is a few degrees lower than the solidus temperature of the alloy in order to form a single-phase solid solution. Then rapidly quenching the casting in a cold (or warm) fluid in order to form a supersaturated non-equilibrium solid solution; and finally, reheating the casting to the aging temperature where nucleation and growth of the strengthening precipitate(s) can occur [1]. Obviously, these processing steps involve significant thermal changes that may be different from location to location in the casting.

Since most of the quality assurances criteria that cast components have to meet include prescribed minimum mechanical properties and compliance with dimensional tolerances, it is necessary for casters to be able to accurately predict these changes in order to take appropriate measures to insure the production of parts that meet the required specifications. Several software packages that are capable of predicting the heat treatment response of wrought steels are available commercially [2, 3], but none of them have been shown to be able to accurately predict the response of cast aluminum alloy components. In this project, a finite element-based model capable of predicting the response of cast aluminum alloy components to heat treatment is being developed. The model is based on the finite element solver ABAQUS and requires an extensive database, which includes temperature-dependent mechanical, physical, and thermal properties of the casting alloy. This database has been assembled for A356 alloy and used in the model in order to predict the residual stresses that develop in a typical cast A356 component upon heat treatment. In addition, the residual stresses in a commercially produced and heat treated part that was manufactured from A356 alloy were measured and the predictions of the model were found to be in good agreement with the measured values.

Creating the Database

Aluminum casting alloy A356 was used to develop and demonstrate the procedure for obtaining the necessary database and modeling the response of aluminum alloy cast components to heat treatment. The data includes mechanical and physical properties, and heat transfer coefficients for various process steps as functions of temperature. Other required thermal and physical properties, such as density, specific heat, etc., were obtained from JMatPro Software[1]. The methodology developed in modeling A356 alloy castings can be extrapolated to other aluminum alloy castings.

The quenching heat transfer coefficient

The apparatus shown in Figure 1 was used to measure the heat transfer coefficient during quenching. The quenching heat transfer coefficient is used by the thermal module in ABAQUS to compute the heat that is transferred out of the part during quenching. A small cylindrical probe (9.5 mm in diameter and 38mm long) shown schematically in Figure 2, was machined from a A356 alloy casting. A hole was drilled down to the geometrical center of this probe and a k-type thermocouple was inserted for measuring the time-temperature profile. Graphite powder was packed into the hole before the thermocouple was inserted in order to ensure intimate contact between the probe and the thermocouple. The probe was heated to the solutionizing temperature and held at that temperature for 12 hours in order to ensure homogenization. Subsequently, the probe was quenched in water that was maintained at room temperature. While quenching, the temperature of the probe was acquired as a function of time using a fast data acquisition system at a scan rate of 1000 scans/sec. A quench tank with two liters of water was used and the probe was immersed completely in the water. The temperature of the water before and after quenching remained constant at 25°C (77°F).

A heat balance applied to the probe results in Equation (1), which was used to calculate the heat transfer coefficient at the surface of the probe.

$$h = -\frac{\rho V C_p}{A_s\left(T_s - T_f\right)}\frac{dT}{dt} \tag{1}$$

In Equation (1), h is the heat transfer coefficient at the surface of the probe, ρ, V, C_p, and A_s are the density, volume, specific heat, and surface area of the probe, respectively. T_s is the temperature at the surface of the probe, which, due to the geometry of the probe, is approximately equal to the measured temperature at the center of the probe. T_f is the bulk temperature of the quenching medium. The derivative of temperature with respect to time in Equation (1) is calculated from the measured temperature vs. time data [4].

---

[1] Developed and marketed by Sente Software Ltd., Surrey Technology Centre, 40 Occam Road, GU2 7YG, United Kingdom.

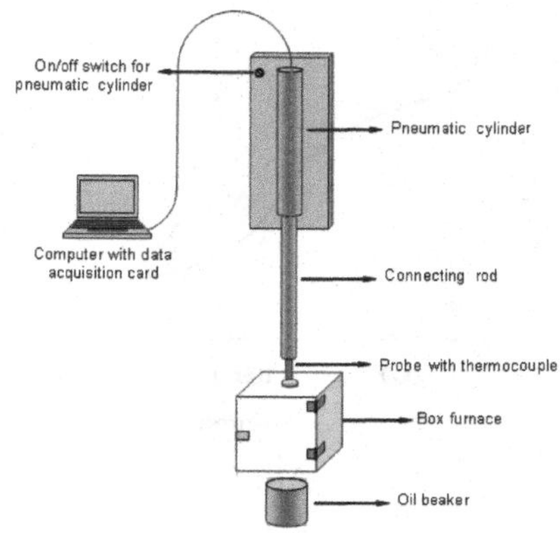

Figure 1. The quenching system used for measuring the heat transfer coefficient.

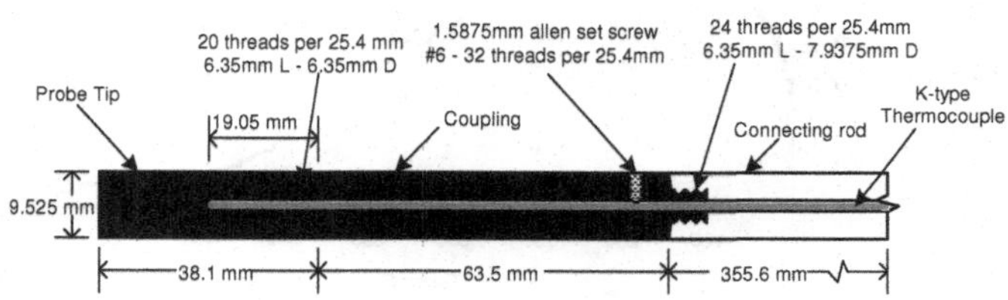

Figure 2. The quench probe-coupling-connecting rod assembly.

Measuring surface roughness

In order to guarantee a similar surface micro-profile for the quenching probe and the modeled component, surface roughness measurements were performed on both of them. Measurements were made by a UBM Laser Microscope[2] and showed that the superficial roughness is 0.501 μm Sa (mean superficial micro-profile amplitude) for the modeled component and 0.398 μm Sa for the quenching probe. Muojekwu, et al. [5] have shown that such small difference in surface roughness have negligible effect on the magnitude of the quenching heat transfer coefficient.

Determining the quenching velocity

The quenching heat transfer coefficient was measured for quenching at 3 different velocities, namely 1,000, 1,100, and 1,200 mm/s. Figure 3 shows the calculated cooling rate vs. temperature, and the quenching heat transfer coefficient for the different quenching velocities.

---

[2] Manufactured by Solarius Development Inc., 550 Weddell Drive, Suite 3, Sunnyvale, CA 94089.

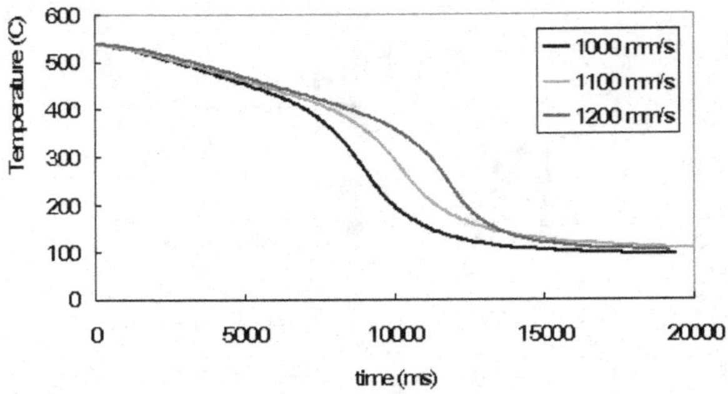

(a)

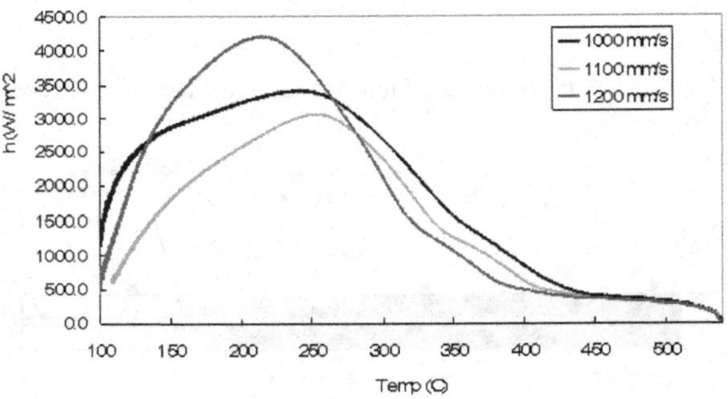

(b)

Figure 3. (a) Temperature vs. time and (b) quenching heat transfer coefficient vs. temperature.

Measuring the mechanical properties

An Instron universal testing machine[3] was used for measuring the room temperature mechanical properties of A356 alloy. The elastic modulus, yield stress, and plastic strain of the alloy were calculated from these measurements. Two types of specimens were tested: (1) specimens that were solutionized at 538°C and then rapidly quenched in room temperature water, and (2) specimens that were solutionized at 538°C and then furnace cooled to room temperature.

The resulting stress-strain curves are shown in Figure 4. Water quenched tensile bars show higher ultimate tensile stress and yield stress, and lower Elastic modulus. For the time being, the necessary high temperature tensile properties are obtained from the open literature [6]. Work is on-going to measure these properties by means of a Gleeble machine.

---

[3] Instron Worldwide Headquarters, 825 University Ave, Norwood, MA 02062-2643, USA

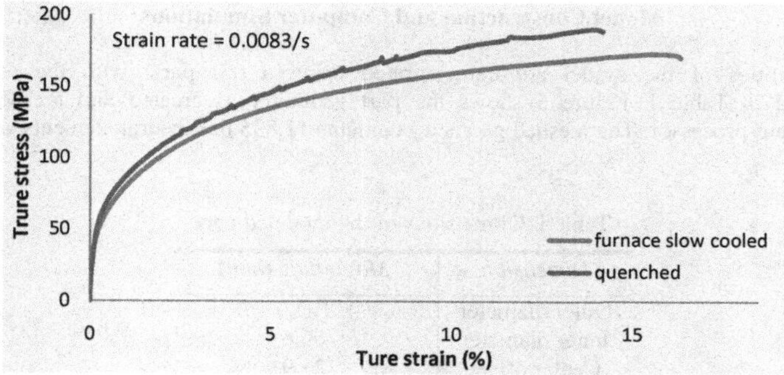

(a)

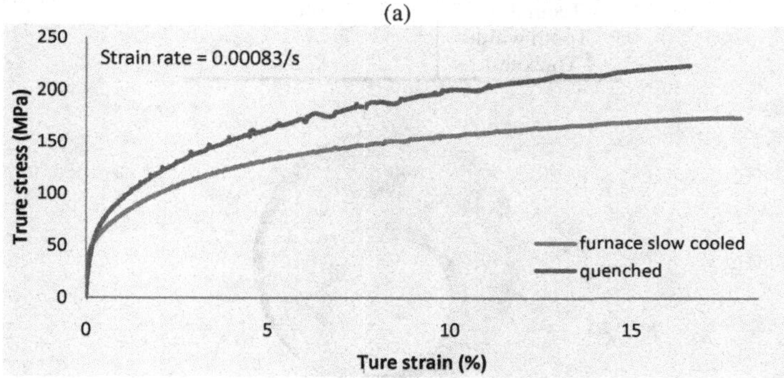

(b)

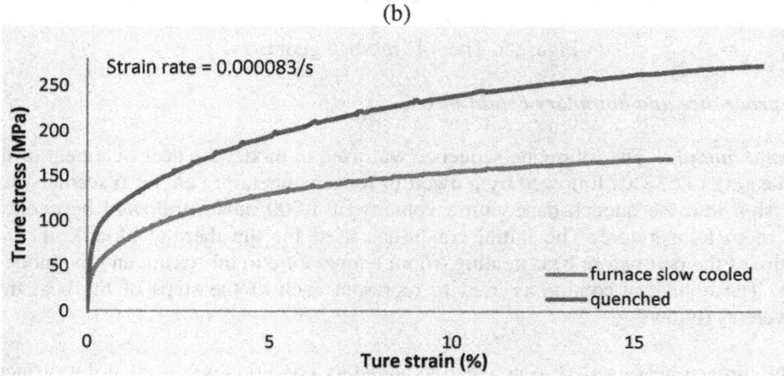

(c)

Figure 4. True stress-strain curves (a) at strain rate = 0.0083/s, (b) at strain rate = 0.00083/s, and (c) at strain rate = 0.000083/s.

Model Construction and Computer Simulations

The capabilities of the model are demonstrated using a test part[4] with the dimensions summarized in Table I. Figure 5 shows the part geometry as created and meshed by the ABAQUS pre-processor. The meshed geometry contains 11,835 hexahedral elements and 14,510 nodes.

Table I. Dimensions of the modeled part.

Dimension	*Magnitude (mm)*
Outer diameter	122.9
Inner diameter	66.6
Center offset	23.9
Teeth deep	6.2
Tooth width	7
Thickness	17.5

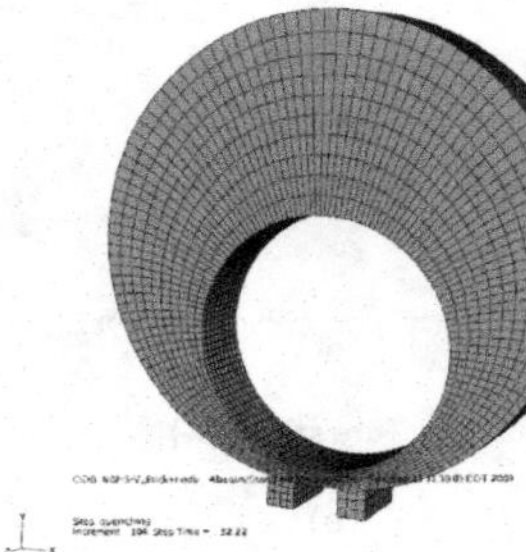

Figure 5. The 3-D meshed geometry.

Solution procedure and boundary conditions

The thermal module –The following sequence was used to model the heat treatment of the part: Furnace heating to 538°C, followed by a dwell in room temperature air for 6 seconds, followed by immersion into the quench tank with a velocity of 1,000 mm/s, followed by quenching in water to room temperature. The initial conditions used for the thermal module included the temperature of the part before heat-treating (room temperature in this case), and the mode of heat treatment. The boundary conditions used to represent each of the steps of the heat treatment process were as follows:

- For the furnace-heating step: A convective boundary condition was used at the surface of the part by providing the measured heat transfer coefficient for heating the part in the furnace up to the homogenization temperature of 538°C.

---

[4] Courtesy of Montupet S.A., 60180 Nogent Sur Oise, France.

- For the dwell step: A convective boundary condition was used at the surface of the part by providing the air heat transfer coefficient (200 W/m^2). The ambient temperature was room temperature.

- For the immersion step: The direction and velocity of immersing the part into the quench tank were defined. This step is important in order to capture the temperature gradient along the immersion length of the part. In this demonstration, the part was immersed along (1) its length and (2) its thickness with a velocity of 1,000 mm/s, and the process time for this step was 0.1284 seconds for the vertical quenching and 0.0015 seconds for the horizontal quenching.

- For the quenching step: A convective boundary condition at the surface of the part was used by providing the measured heat transfer coefficient for quenching the part in water from the homogenization temperature down to room temperature.

The stress module – This module uses mainly the time-temperature history of the part, which is generated by the thermal module in order to calculate the residual stresses. For initial condition, the stress at all the nodes was set to zero. If a known initial stress state existed, the appropriate values could be used. Nodal constraints are required in order to prevent rigid body displacement and rotation. This requirement applies to all the process steps, and is defined only once in the model input file. Referring to the 3-D geometry in Figure 5, 3 nodes at the center of the top face were constrained from moving.

Simulation Results and Comparison to Model Predictions

In order to verify the accuracy of the model, the model predicted residual stresses were compared to measurements performed on a component of configuration and dimensions similar to those used in the model. The standard x-ray diffraction method for measuring residual stresses in metallic components was used. In this method, line shifts due to a uniform strain in the component are measured and then the stresses in the component are determined by a calculation involving the elastic constants of the material [7-9]. Measurements were made in an x-ray diffractometer equipped with a stress analysis module[5]. The residual stresses were measured at the inner face of the hole in the thinnest section since this location is expected to have the highest magnitude of residual stress [10]. Figure 6 shows a comparison between the measured and model predicted magnitude of residual stress in the part. It is clear that there is very good agreement between the measured and the model predicted residual stresses and that quenching the part vertically creates more residual stresses in the part than quenching it horizontally.

[5] Model X'Pert Pro Diffractometer manufactured by PANalytical, Inc., Natick, MA, USA.

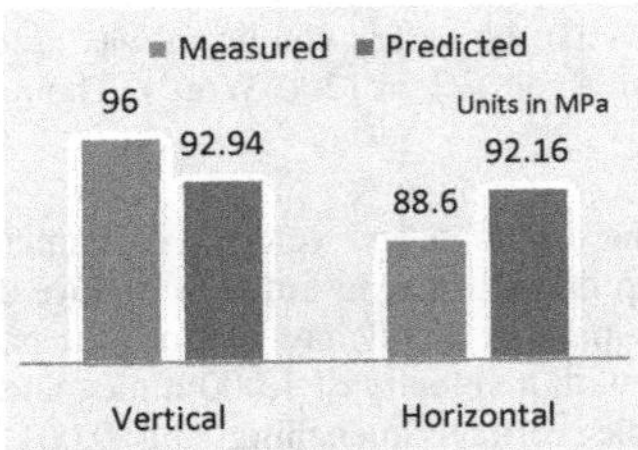

Figure 6. Comparison between measured and predicted residual stresses generated in the part by (a) quenching it vertically, and (b) quenching it horizontally in the quenching tank.

Summary and Conclusions

A model has been developed using the ABAQUS finite element software to predict the response of cast aluminum alloy components to heat treatment. The necessary database for the model is currently being generated. However, by using data from the open literature in place of the "yet to be generated" high temperature mechanical properties of the alloy, it was possible to simulate the response of a typical cast component to solutionizing and quenching. The model predictions were verified by measurements made on heat treated commercially cast parts, and the model predicted residual stresses were found to be in very good agreement with residual stresses measured by the x-ray diffraction method.

References

1. D. Emad et al., *Optimal Heat Treatment of A356.2 Alloy*. Light Metals, (The Minerals, Metals & Materials Society, 2003): 983.
2. D.J. Bammann, M.L. Chiesa, and G.C. Johnson, "Modeling Large Deformation and Failure in Manufacturing Processes," *Proceedings of the Nineteenth International Congress on Theoretical and Applied Mechanics*, 1996, 359-376.
3. V. Warke et al., "Modeling the Heat Treatment of Powder Metallurgy Steels & Particulate Materials," *International Conf. On Powder Metallurgy & Particulate Materials*, 2004, 39-53.
4. M. Maniruzzaman et al., "CHTE Quench Probe System – A NEW Quenchant Characterization System", *Proceedings of the Fifth International Conference on Frontiers of Design and Manufacturing* (ICFDM 2002), 1 (2002), 619-625.
5. C.A. Muojekwu, I.V. Samarasekera and J.K. Brimacombe, "Casting-chill interface heat transfer during solidification of an aluminum alloy," *Metall. Mater. Trans*, 26 (B), (1995), 361-382.
6. C.M. Estey et al., "Constitutive behavior of A356 during the quenching operation", *Materials Science and Engineering,* 383 (A), (2004), 245-251.
7. P.S. Prevey, "A method of Determining Elastic Constants in Selected Crystallographic direction for X-Ray diffraction residual stress measurement," *Advances in X-Ray Analysis*, 20 (1977), 345-354.
8. B.D. Cullity and S.R. Stock, "Elements of X-Ray Diffraction", *3rd edition, Princeton Hall Publications*, 2004, 435-468.
9. "Residual Stress Measurement by X-Ray Diffraction", *SAE Standard*, J784a, August 1971.
10. V. Guley, "Residual stress and retained austenite X-Ray diffraction measurements on ball bearings" (Job report D030914J. PANalytical Co., 2003).

Shape Casting: The 3rd International Symposium
Edited by: John Campbell, Paul N. Crepeau, and Murat Tiryakioğlu
TMS (The Minerals, Metals & Materials Society), 2009

INTERNET-BASED CASTING CAE SYSTEM

Tao JING, Baicheng LIU

Key Laboratory for Advanced Materials Processing Technology, Ministry of Education,

Department of Mechanical Engineering, Tsinghua University, Beijing 100084, China

Keywords: numerical simulation, casting, CAE, Internet

Abstract

This paper presented the development of a Web-based Casting CAE System. The premise of this work is that the numerical simulation of solidification process has become a valuable tool for evaluating and optimizing casting pattern and rigging design to dramatically reduce the cost of reaching a satisfactory design that could be very expensive and time-consuming if a trial-and-error approach is used. The major competitive advantage of a Web-based casting CAE system, compared to a stand alone system, lies in its capability to share the limited resources through Internet within quite a lot of foundries, especially those medium and small scale foundries which may have difficulties to set up their own system alone. Java-based multi-tier architecture, CORBA-based distributed computing technology and Java Applet, Swing and Java Severlet technology are used for the implementation of the system.

Introduction

Application Service Provider (ASP) provides online application and management service, it obtains income and benefits from the rentals for these service from the customers. ASP manages, maintains and upgrades these service frequently and distributes and improves the software, hardware, network and professional technology efficiently to provide the customers better service. It is believed that ASP has the potential to bring about a completely different ways of service.

Simulation of solidification process plays an important role in bridging casting CAD and CAM, and attracts more and more attention in foundry. To establish such a simulation system, a significant amount of capital expenditure is required for software, hardware and maintenance, which hindered the application of the technology of simulation of solidification process in medium or small enterprises.

Based on ASP philosophy this paper put forwards a web-based solution for casting CAE system. The primary objective of this research is to explore the way of combining the simulation technology with Internet to provide lower costs and security, efficiency, and high quality services for foundry enterprises.

The frame of web-based casting CAE system

The clients make the CAD file using standard data format, then login the web sever, upload the file and fill out any necessary process parameters based on the manufacturing condition for computing. Figure 1 shows the web user interface of system. Both the CAD file and process parameters will be transferred to the web server. Once the computing request is submitted, the Web sever will queue this task and select the available sub-sever to do the computation, and the clients can inquiry the computing process at any time, download the result after the computing is finished. Figure 2 depicts an example output of post-processing module. Then the clients can

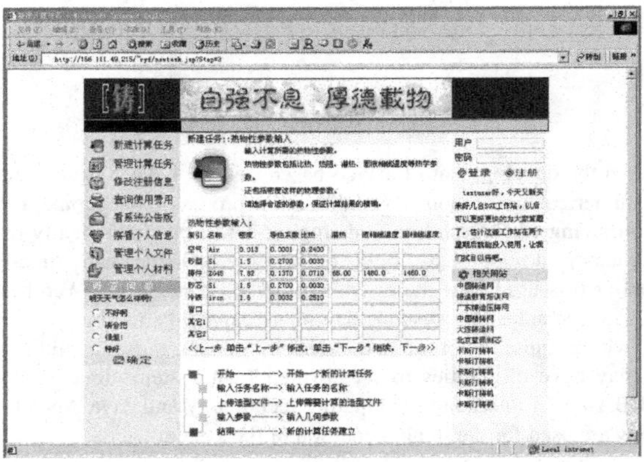

Fig. 1 Web user interface of system

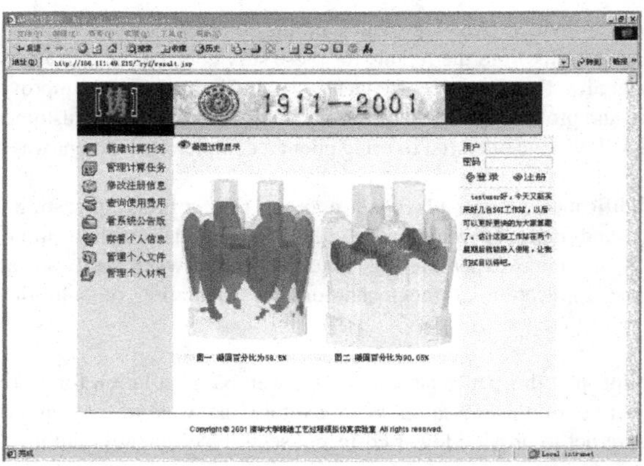

Fig. 2 An example output of post-processing module

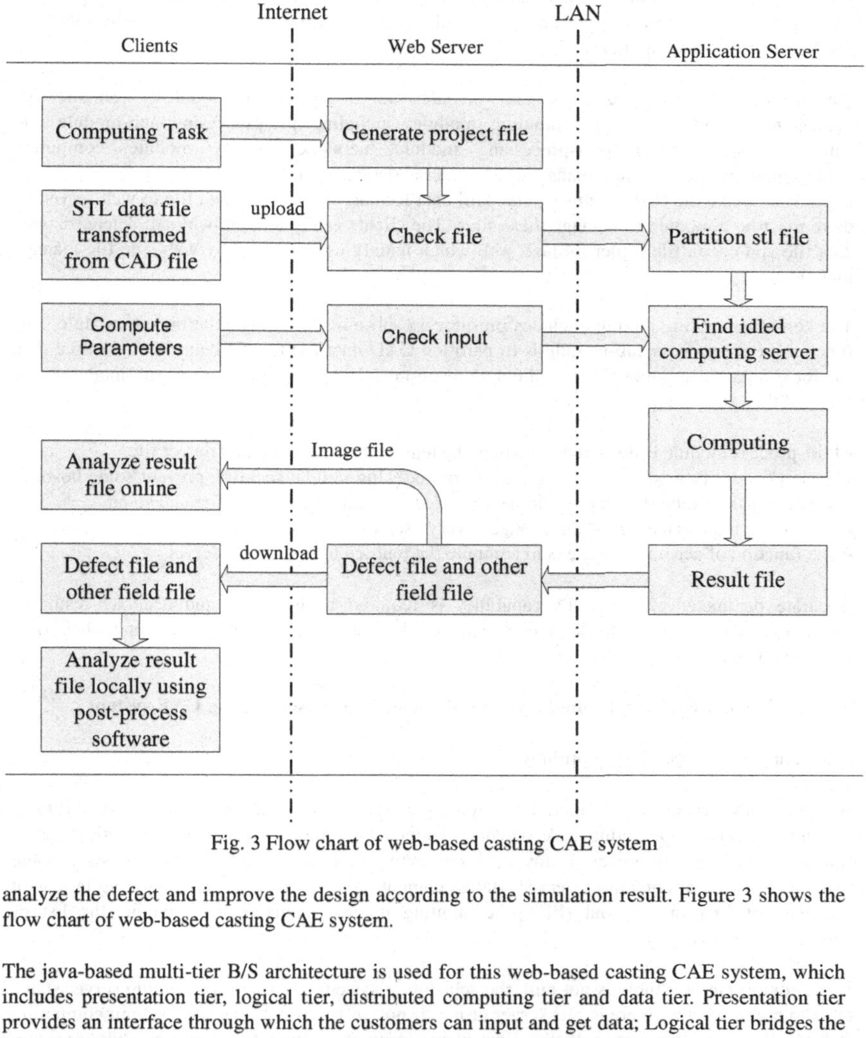

Fig. 3 Flow chart of web-based casting CAE system

analyze the defect and improve the design according to the simulation result. Figure 3 shows the flow chart of web-based casting CAE system.

The java-based multi-tier B/S architecture is used for this web-based casting CAE system, which includes presentation tier, logical tier, distributed computing tier and data tier. Presentation tier provides an interface through which the customers can input and get data; Logical tier bridges the presentation tier, the distributed computing tier and the data tier. It responds to the user request, scheduling tasks, takes the data from the data tier and transfers the necessary data to computing server. Distributed computing tier, which is in charge of the numerical simulation, is the core tier of the system. It is called by the logical tier, returns the results to the logical tier and saves some important data in the data tier. Data tier determines and maintains the integrity and safety of the data, and it responds to the request of the logical tier, accesses the data and saves the results from the distributed computing tier.

The web-based casting CAE system includes seven functional modules: web-side file management module, kernel computing module (including pre-processing sub-module and simulation sub-module), post-processing module, network security module, computing management module, account management, material database module.
The clients make the CAD file using standard data format, then upload data file to web server, so there must be a module to manage these files. The clients can upload, download, delete, rename data file and create file folder through web which mainly implemented by web-side file manage module.

The kernel computing module includes pre-process sub-module and simulation sub-module. The function of pre-process sub-module is to partition CAD data file to grid data file and initial data file for simulation sub-module. The simulation sub-module read the pre-process sub-module input file, and then start computing.

- Post-process module is designed to display the temperature field file and defect file.
- Network security module authorize user's privilege, log system security, prevent some hostility compute task to submit, and prevent unauthorized user read legal user's file and so on.
- Computing management module manages every user's computing process.
- The function of account module is to compute the rent fee of computing server.

Accurate parameter of material's capability is required to get right and accurate result of simulation. The system provide reference material database, and clients may input their own material database.

The key enabling technology to implement WEB-based casting CAE system

<u>Computing task scheduling algorithms</u>

Because of the resources is limited, the incoming tasks must be scheduled for the availability of computing server. Computing task schedule system maintains three tables: computing server status table, computing process status table, computing task table. Computing server status table maintains the list of computing servers status, computing task table queues computing task with sequence of first in first out (FIFO), computing process status table maintains the list of computing process status.

If the computing task table is not null, the scheduling system will query computing server status table to check whether there is idle server or not, if not, just waiting, otherwise submit computing task to idle computing server. If the client wants to query computing status, scheduling system just feedback the query result of computing process status table.

The following describes some computing task scheduling algorithms implemented in the system:

Round-Robin Scheduling: The round-robin scheduling algorithm sends each incoming computing request to the next compute server in it's list. Thus in a three server cluster (servers A, B and C) request 1 would go to server A, request 2 would go to server B, request 3 would go to server C, and request 4 would go to server A, thus completing the cycling or 'round-robin' of servers. It treats all computing servers equal regardless whether the server is single CPU or multi CPU, the memory is large or small.

Weighted Round-Robin Scheduling: The weighted round-robin scheduling is designed to better handle servers with different processing capacities. Each server is assigned a weight, an integer value that indicates the processing capacity. The default weight is 1. For example, the real servers, A, B and C, have the weights, 4, 3, 2 respectively, a good scheduling sequence will be AABABCABC in a scheduling period. The weighted round-robin scheduling is better than the round-robin scheduling, when the processing capacity of compute servers are different.

XML-based data transferring technology

The eXtensible Markup Language (XML) is the universal format for structured documents and data on the Web. Structured data includes things like spreadsheets, address books, configuration parameters, financial transactions, and technical drawings. One of XML's advantages is the share and reuse of data. XML is designed for describing, not for the presentation of structure data. Different presentation of the data for the same xml file can be achieved on web by using different stylesheet .

The process parameters for simulation, which are input by clients, are transferred and reused between different modules of system. Using xml for data file can implement sharing these parameters between different modules. Figure 4 depicts xml file transferred between modules of the system.

Through HTML/Applet the client side can generate, present, revise, receive, and submit xml file dynamically. The xml file is used for transferring parameters between the servlet of computing task scheduling system and distributed computing object, and for submitting parameters from the servlet of computing task scheduling system to database. Sharing and reusing data between computing management module and distributed computing object also uses xml file.

Internet-based post-processing system

Internet-based post-processing technology is a necessary module of an on-line CAE system. A new concept post-processing system is studied and established for Internet-based casting CAE system. It is designed following B/S architecture. Figure 5 shows all modules of the whole system. There are three modules at server side, and eight at browser side. The server side modules are implemented with Java Serverlet technique and the browser side modules are based on Java Applet technology. Those modules at browser side reside in an applet. The applet is an event driven system. MVC architecture is helpful to event driven system, it can make these systems easier to maintain and reuse. The "M" of MVC means "module", it is the software abstract of real issues. The "V" of MVC means "view", it is the representation of a visualization system. The "C"

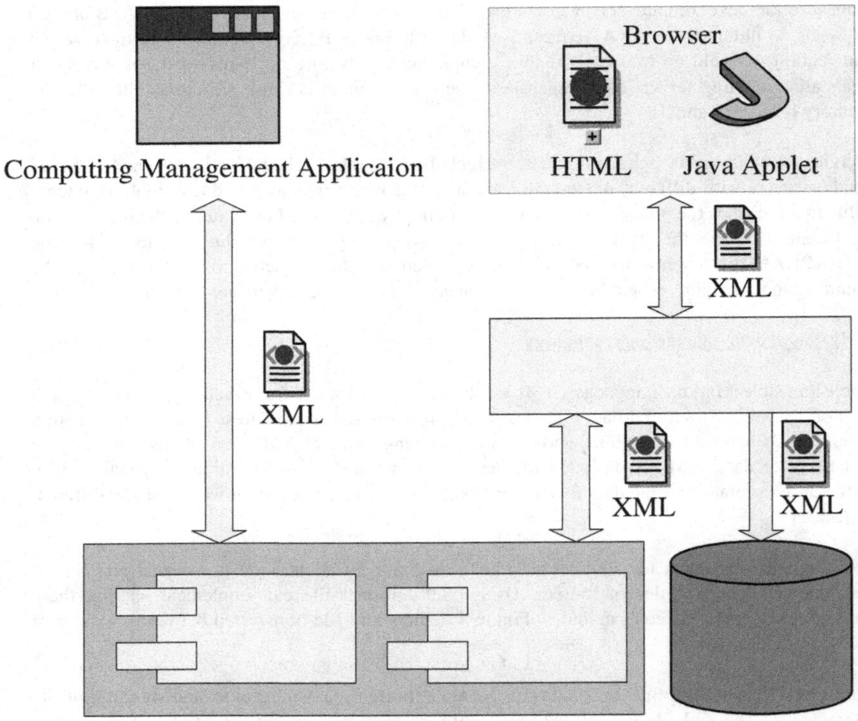

Fig. 4 XML transfer between modules

of MVC means "controller", it handles all events generated by user and system and control the system to run in a right way. The file chooser, uncompress, data loader and visualize module are "M" part of the system. The display, user interface and print module compose "V". The controller module stands for "C". A simplified painter's algorithm is worked out to remove hidden surface. Users of the system can analyze the visualized simulation result in their browser on-line. Rotation, cutting and printing functions are provided.

<u>CORBA-based distributed simulation technology</u>

There are three main specifications: Microsoft's DCOM/COM, Sun's J2EE, and OMG's CORBA (Common Object Request Broker Architecture) in distributed object technology. DCOM/COM can only run on Microsoft Windows platform, so it's usage is limited if there are other platforms in system; Because of lower efficiency and slower execution speed than c/c++ in numerical

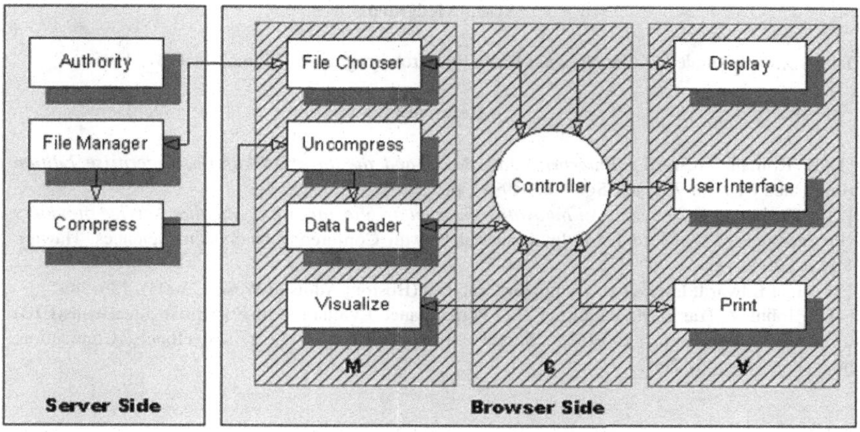

Fig. 5 Modules of internet-based post-processing system

simulation, pure java solution is not a good one for this WEB-CAE system. To obtain the best performance it is an ideal solution to use java in web server side programming and c/c++ in computing server side programming, and to use CORBA mechanism to connect these two sides. ORB (Object Request Broker) is the core technology of CORBA, it provides a mechanism to enable objects to transparently make and receive requests and responses in a distributed environment.

In this WEB-CAE system, the EJB(Enterprise JavaBeans) sits between the client(a servlet deployed in WEB server) and the CORBA server(a computing server implemented the numerical simulation of solidification process). The client communicates to the EJB using RMI/IIOP whereas the EJB communicates to the CORBA server using IIOP. The CORBA object that implemented the algorithm of numerical simulation of solidification process, at instantiation, binds itself to the naming service. After the client servlet that implemented computing schedule system submits computing task, the EJB can invoke methods of the CORBA server object by a reference to the CORBA server object which is obtained by the adapter object through looking it up in the naming service.

Conclusions

Based on the ASP philosophy, multi-tier architecture of the "browser/web server/application server/database" is used for implementing web-based casting CAE system. This new model transforms simulation software from one computer environment to Internet-accessible, its successful application will be a part of e-Commerce in the area of information-based manufacturing.

Acknowledgments

This work was funded by the National Basic Research Program of China 2005CB724105.

References

[1] E. Roman, *Mastering Enterprise JavaBeans and the Java 2 Platform, Enterprise Edition* (Published by John Wiley & Sons Inc, 1999), pp. 29-51.
[2] B. Furht and C. Phoenix, *An Innovative Internet architecture* for *application service providers* (Proceedings of the 33rd IEEE International Hawaii Conference on System Sciences, Hawaii, U.S.A., 2000.1).
[3] Angel Edward, Interactive Computer Graphics (Boston: Addison Wesley, 2003), 180-249.
[4] B. John, V. Trang, M. Vernik, et al., "Performance Evaluation of Enterprise JavaBeans(EJB) CORBA Adapter to CORBA Server Interoperability", Java Developer Connection, http://java.sun.com, 2002.

AUTHOR INDEX
Shape Casting: Third International Symposium

SUBJECT INDEX
Shape Casting: Third International Symposium